Statistics for Social and Behavioral Sciences

Series editor
Stephen E. Fienberg
Carnegie Mellon University
Pittsburgh
Pennsylvania
USA

Statistics for Social and Behavioral Sciences (SSBS) includes monographs and advanced textbooks relating to education, psychology, sociology, political science, public policy, and law.

More information about this series at http://www.springer.com/series/3463

Estela Bee Dagum • Silvia Bianconcini

Seasonal Adjustment Methods and Real Time Trend-Cycle Estimation

 Springer

Estela Bee Dagum
Department of Statistical Sciences
University of Bologna
Bologna, Italy

Silvia Bianconcini
Department of Statistical Sciences
University of Bologna
Bologna, Italy

ISSN 2199-7357   ISSN 2199-7365   (electronic)
Statistics for Social and Behavioral Sciences
ISBN 978-3-319-31820-2   ISBN 978-3-319-31822-6   (eBook)
DOI 10.1007/978-3-319-31822-6

Library of Congress Control Number: 2016938799

Mathematics Subject Classification (2010): 62G08, 62M10, 62P20, 62P25

Printed on acid-free paper

This Springer imprint is published by Springer Nature
The registered company is Springer International Publishing AG Switzerland

To my sons, Alex, Paul, and Leo, with their lovely families for their strong valuable support and lively discussions during the writing of this book

To Paolo who always inspires me to try my best

Preface

In order to assess the current stage of the business cycle at which the economy stands, real time trend-cycle estimates are needed. The basic approach is that of assessing the real time trend-cycle of major socioeconomic indicators (leading, coincident, and lagging) using percentage changes, based on seasonally adjusted data, calculated for months and quarters in chronological sequence. The main goal is to evaluate the behavior of the economic indicators during incomplete phases by comparing current contractions or expansions with corresponding phases in the past. This is done by measuring changes of single time series (mostly seasonally adjusted) from their standing at cyclical turning points with past changes over a series of increasing spans. This differs from business cycle studies where cyclical fluctuations are measured around a long-term trend to estimate complete business cycles. The real time trend corresponds to an incomplete business cycle and is strongly related to what is currently happening on the business cycle stage. Major changes of global character in the financial and economic sector have introduced high levels of variability in time series making difficult to detect the direction of the short-term trend by simply looking at seasonally adjusted data, and the use of trend-cycle data or smoothed seasonally adjusted series has been supported. Failure in providing reliable real time trend-cycle estimates could give rise to dangerous drift of the adopted policies. Therefore, a reliable estimation is of fundamental importance.

This book includes two opening chapters, Chap. 1 which is a general introduction and Chap. 2 on time series components. The remaining nine chapters are divided in two parts, one on seasonal adjustment and the other on real time trend-cycle estimation.

Since the input for trend-cycle estimation is seasonally adjusted data, Part I of this book thoroughly discusses the definitions and concepts involved with three major seasonal adjustment methods as follows:

Chapter 3. Seasonal adjustment, meaning, purpose, and methods.

Chapter 4. Linear smoothing or moving average seasonal adjustment methods.

Chapter 5. Seasonal adjustment based on ARIMA model decomposition: TRAMO-SEATS.

Chapter 6. Seasonal adjustment based on structural time series models.

Two of the seasonal adjustment methods are officially adopted by statistical agencies, namely, X12ARIMA and TRAMO-SEATS, and their respective software default options are illustrated with an application to the US Orders for Durable Goods series. The third method, structural time series models and its software STAMP, is also illustrated with an application to the US Unemployment Males (16 years and over) series.

Part II of the book comprises:

Chapter 7. Trend-cycle estimation.

Chapter 8. Recent developments on nonparametric trend-cycle estimation.

Chapter 9. A unified view of trend-cycle predictors in reproducing kernel Hilbert spaces.

Chapter 10. Real time trend-cycle estimation.

Chapter 11. The effect of seasonal adjustment methods on real time trend-cycle estimation.

Chapter 7 systematically discusses the definitions and concepts of the trend-cycle component of the various seasonal adjustment methods previously introduced.

Chapter 8 concentrates on the last 20 years' developments made to improve the Henderson filter used to estimate the trend-cycle in the software of the US Bureau of Census X11 and its variants, the X11/ X12ARIMA. The emphasis has been on determining the direction of the short-term trend for an early detection of a true turning point. It introduces in detail three major contributions: (1) a nonlinear trend-cycle estimator also known as Nonlinear Dagum Filter (NLDF), (2) a cascade linear filter (CLF) that closely approximates the NLDF, and (3) an approximation to the Henderson filter via the reproducing kernel Hilbert space (RKHS) methodology.

Chapter 9 presents a unified approach for different nonparametric trend-cycle estimators by means of the reproducing kernel Hilbert space (RKHS) methodology. These nonparametric trend-cycle estimators are based on different criteria of fitting and smoothing, and they are (1) density functions, (2) local polynomial fitting, (3) graduation theory, and (4) smoothing spline regression. It is shown how nonparametric estimators can be transformed into kernel functions of order two, which are probability densities and from which corresponding higher-order kernels are derived. This kernel representation enables the comparison of estimators based on different smoothing criteria and has important consequences in the derivation of the asymmetric filters which are applied to the most recent seasonally adjusted data for real time trend-cycle estimation.

Chapter 10 is dedicated to real time trend-cycle estimation. Official statistical agencies generally produce estimates derived from asymmetric moving average techniques which introduce revisions as new observations are incorporated to the series as well as delays in detecting true turning points. This chapter presents a reproducing kernel approach to obtain asymmetric trend-cycle filters that converge fast and monotonically to the corresponding symmetric ones. This is done with time-varying bandwidth parameters because the asymmetric filters are time-varying. It shows that the preferred one is the bandwidth parameter that minimizes the distance between the gain functions of the asymmetric and symmetric filters. The theoretical results are empirically corroborated with a set of leading, coincident, and lagging indicators of the US economy.

Chapter 11 deals with the effects of the seasonal adjustment methods when the real time trend is predicted with nonparametric kernel filters. The seasonal adjustments compared are the two officially adopted by statistical agencies, X12ARIMA and TRAMO-SEATS, applied to a sample of US leading, coincident, and lagging indicators.

The eleven chapters have been written as complete as possible, and each one can be read rather independently. We have also introduced a uniform notation all along the chapters to facilitate the reading of the book as a whole.

This book will prove useful for graduate and final-year undergraduate courses in econometrics and time series analysis and as a reference book for researchers and practitioners in statistical agencies, other government offices, and business. The prerequisites are a good knowledge of linear regression, matrix algebra, and knowledge of ARIMA modeling.

We are indebted to participants and students who during many seminars and presentations raised valuable questions answered in this book. Our most sincere gratitude goes to our colleagues who encouraged us to write this book and all those who provided many valuable and useful suggestions through lively discussions or written comments. Our thanks also go to Veronika Rosteck, associate statistics editor of Springer, for her sustained and valuable support while writing the book.

We are solely responsible for any errors and omissions.

Bologna, Italy   Estela Bee Dagum
February 2016   Silvia Bianconcini

Contents

Acronyms

ACGF	Autocovariance generating function
ADR	Average duration of run
AICC	Akaike information criterion (AIC) with a correction for finite sample sizes
AMB	ARIMA model-based method
AO	Additive outlier
AR	Autoregressive process
ARMA	Autoregressive moving average process
ARIMA	Autoregressive integrated moving average process
BIC	Bayesian information criterion
CLF	Cascade linear filter
ESS	European statistical system
GCV	Generalized cross validation
H13	13-term symmetric Henderson filter
$I(k)$	Integrated of order k process
IGLS	Iterative generalized least squares
i.i.d.	Independent and identically distributed
IO	Innovation outlier
LEI	Index of leading economic indicators
LOESS	Locally weighted regression smoother
LOM	Length of month
LS	Level shift outlier
MA	Moving average process
MCD	The number of months it takes the average absolute change in the trend-cycle to dominate that in the irregular
MMSLE	Minimum mean square linear estimator
MSE	Mean square error
MSAR	Markov switching autoregressive process
NBER	National Bureau of Economic Research
NLDF	Nonlinear Dagum filter
OCV	Ordinary cross validation

OLS	Ordinary least squares
QCD	The number of quarters it takes the average absolute change in the trend cycle to dominate that in the irregular
regARIMA	Autoregressive integrated moving average process with regression components
RKHS	Reproducing kernel Hilbert space
RMSE	Root mean square error
SAR	Seasonal autoregressive process
SMA	Seasonal moving average process
SEATS	Signal extraction in ARIMA time series
SIGEX	Signal extraction
STAMP	Structural time series analyzer, modeler, and predictor
STAR	Smooth transition autoregressive process
STL	Seasonal trend decomposition procedure based on LOESS
TAR	Threshold autoregressive process
TC	Temporary change outlier
TRAMO	Time series regression with ARIMA noise, missing observations, and outliers
UC	Unobserved component
WK	Wiener–Kolmogorov filter
$WN(0, \sigma^2)$	White noise process with zero mean and constant variance σ^2, with uncorrelated components
2×12 m.a.	Centered 12-month moving average
3×3 m.a.	Weighted five-term moving average
3×5 m.a.	Weighted seven-term moving average

Chapter 1
Introduction

Abstract This chapter introduces the main topics of the book that are divided into two parts: (I) seasonal adjustment methods and (II) real time trend-cycle estimation. Most socioeconomic series and mainly indicators (leading, lagging, and coincident) are affected by the impact of seasonal variations. In order to assess what is the stage of the cycle at which the economy stands it is necessary to identify, estimate, and remove the effect of seasonality from the data. The seasonal adjustment methods discussed in Part I are grouped into three main categories: smoothing linear methods, ARIMA model-based and structural time series models. There are three seasonal adjustment softwares, one for each method, that are currently available. Two are officially adopted by statistical agencies, named X12ARIMA and TRAMO-SEATS, and the third major software, called STAMP, is mainly used in econometric studies and for academic research. In Part II, the trend-cycle estimates from these seasonal adjustment methods are analyzed and compared to a new set of trend-cycle symmetric and asymmetric filters based on Reproducing Kernel Hilbert Space (RKHS) methodology.

1.1 Book Structure

This book is divided into two parts: one on seasonal adjustment and the other on real time trend-cycle estimation. It has eleven chapters as follows:

1. Introduction.
2. Time series components.
   PART I: SEASONAL ADJUSTMENT.
3. Seasonal adjustment, meaning, purpose, and methods.
4. Linear smoothing or moving averages seasonal adjustment methods.
5. Seasonal adjustment based on ARIMA model decomposition: TRAMO-SEATS.
6. Seasonal adjustment based on structural time series models.
   PART II: REAL TIME TREND-CYCLE ESTIMATION.
7. Trend-cycle estimation.
8. Recent developments on nonparametric trend-cycle estimation.

© Springer International Publishing Switzerland 2016
E. Bee Dagum, S. Bianconcini, *Seasonal Adjustment Methods and Real Time Trend-Cycle Estimation*, Statistics for Social and Behavioral Sciences,
DOI 10.1007/978-3-319-31822-6_1

1

 9. A unified view of trend-cycle predictors in reproducing kernel Hilbert spaces.
10. Real time trend-cycle estimation.
11. The effect of seasonal adjustment methods on real time trend-cycle estimation.

A summary of the contents of these two parts follows.

1.2   Part I: Seasonal Adjustment Methods

A great deal of data in business, economics, and natural sciences occurs in the form of time series where observations are dependent, and where the nature of this dependence is of interest in itself. The time series is generally compiled for consecutive and equal periods, such as weeks, months, quarters, and years.

From a statistical point of view, a time series is a sample realization of a stochastic process, i.e., a process controlled by probability laws. The observations made as the process continues indicate the way it evolves. In the analysis of economic time series, Persons [48] was the first to distinguish four types of evolution, namely: (a) trend, (b) cycle, (c) seasonal variations, and (d) irregular fluctuations.

The trend corresponds to a variation in some defined sense persisting over a long period of time, that is, a period which is long in relation to the cycle. In some cases, it is a steady growth; while in others, the trend may move downward as well as upward. The cycle, usually referred to as the business cycle, is a quasi-periodic oscillation characterized by alternating periods of expansion and contraction. The seasonal variations represent the composite effect of climatic and institutional events which repeat more or less regularly each year. These three types of fluctuation are assumed to follow systematic patterns (they are the signal of the process).

On the other hand, the irregulars are unforeseeable movements related to events of all kinds. In general, they have a stable random appearance but, in some series, extreme values may be present. These extreme values or outliers have identifiable causes, e.g., floods, unseasonal weather, and strikes; and, therefore, can be distinguished from the much smaller irregular variations.

Among all these components, the influence of the seasonal fluctuations in the human activity has been felt from earlier times. The organization of society, the means of production and communication, the habits of consumption, and other social and religious events have been strongly conditioned by both climatic and conventional seasons. The seasonal variations in agriculture, the high pre-Easter and pre-Christmas retail sales, and the low level of winter construction are all general knowledge.

The main causes of seasonality, such as climatic and institutional factors, are mainly exogenous to the economic system and cannot be controlled or modified by the decision makers in the short run. The impact of seasonality in the economic activity, however, is usually not independent on the stages of the business cycle. It is well-known for example, that the seasonal unemployment among adult males is much larger during periods of recession than during periods of prosperity. Another main feature of seasonality is that the phenomenon repeats with certain regularity every year but it may evolve. The latter is mainly due to systematic intervention of the government and to technological and institutional changes as well; and it is more the general case than the exception for economic time series. Therefore, the assumption of stable seasonality, i.e., of seasonality being represented by a strictly periodic function, can be used as a good approximation for few series only. In effect, even in the extreme cases of those activities where seasonality is mainly caused by climatic conditions, e.g., agriculture, fishing, forestry, the seasonal variations change, for weather itself measured by the temperature, and quantity of precipitation changes.

Once the assumption of stable seasonality is abandoned, new assumptions must be made regarding the nature of its evolution; if seasonality changes, is it slowly or rapidly? Is it gradually or abruptly? Is it in a deterministic or a stochastic manner? Today, the most widely accepted hypothesis is that seasonality moves gradually, slowly, and in a stochastic manner. A third characteristic of seasonality is that the phenomenon can be separated from other forces (trend, cycle, and irregulars) that influence the movement of an economic time series and can be measured.

The seasonal variations are distinguished from trend by their oscillated character, from the cycle by being confined within the limits of an annual period, and from the irregulars, by the fact of being systematic. Most socioeconomic indicators, such as leading, coincident, and lagging, are affected by the impact of seasonal variations. In order to assess what is the stage of the current economic conditions it is necessary to identify, estimate, and remove the effect of seasonality. Seasonal adjustment methods can be grouped into three main categories: smoothing linear methods, ARIMA model-based and structural time series models. There are three seasonal adjustment softwares, one for each method, that are currently available but only two are officially adopted by statistical agencies, named X12ARIMA and TRAMO-SEATS. The third major software, called STAMP, is mainly used in econometric studies and for academic research. These topics are largely discussed in Chaps. 4, 5, and 6, respectively, and only brief summaries are given below.

1.2.1 Smoothing Linear Seasonal Adjustment Methods

The best known and most often applied seasonal adjustment methods are based on smoothing linear filters or moving averages applied sequentially by adding (and subtracting) one observation at a time. These methods assume that the time series components change through time in a stochastic manner. Given a time series $y_t, t = 1, \ldots, n$, for any t far removed from both ends, say $m < t < n - m$, the seasonally adjusted value y_t^a is obtained by applying, in a moving manner, a symmetric average

$W(B)$, that is,

$$y_t^a = W(B)y_t = \sum_{j=-m}^{m} w_j B^j y_t = \sum_{j=-m}^{m} w_j y_{t-j}, \qquad (1.1)$$

where the weights w_j are symmetric, that is, $w_j = w_{-j}$, and the length of the average is $2m + 1$. For current and recent data $(n - m < t \leq n)$, a symmetric linear filter cannot be applied, and therefore truncated asymmetric filters are used. For example, for the last available observation y_n, the seasonally adjusted value is given by

$$y_n^a = W_0(B)y_n = \sum_{j=0}^{m} w_{0,j} y_{n-j}. \qquad (1.2)$$

The asymmetric filters are time-varying in the sense that different filters are applied to the m first and last observations. The end estimates are revised as new observations are added because of: (a) the new innovations and (b) the differences between the symmetric and asymmetric filters. The estimates obtained with symmetric filters are often called "final."

The development of electronic computers contributed to major improvements in seasonal adjustment based on moving averages and facilitated their massive application. In 1954, Julius Shiskin of the US Bureau of Census developed a software called Method I, based mainly on the works of Macauley [44] already being used by the US Federal Reserve Board. Census Method I was followed by Census Method II and 11 more experimental versions (X1, X2, ..., X11). The best known and widely applied was the Census Method II-X11 variant developed by Shiskin et al. [51], but produced poor seasonally adjusted data at the end of the series which is of crucial importance to assess the direction of the short-term trend and the identification of real time turning points in the economy.

Estela Bee Dagum [13] working at Statistics Canada developed the X11ARIMA mainly to correct for this serious limitation and, later, Findley et al. [26] developed X12ARIMA that offers a regARIMA option to estimate deterministic components, such as trading day variations, moving holiday effects and outliers, simultaneously with the ARIMA model for extrapolation. It also includes new diagnostic tests and spectral techniques to assess the goodness of the results. The X12ARIMA method is today the one most often applied by statistical agencies in the world. The US Bureau of Census continued research on the development of seasonal adjustment methods, and recently produced a version called X13ARIMA-SEATS which enables the estimation of the seasonal component either via linear filters as those available in X12ARIMA or based on an ARIMA decomposition model.

All the seasonal adjustment methods based on moving averages discussed in this book have nonlinear elements. Hence, the seasonally adjusted total of an aggregated series is not equal to the algebraic sum of the seasonally adjusted series that enter into the aggregation. The main causes of nonlinearity are generated by:

1. a multiplicative decomposition model for the unobserved components,
2. the identification and replacement of outliers,
3. the ARIMA extrapolations, and
4. automatic selection of the moving average length for the estimation of the trend-cycle and seasonality.

The properties of the combined linear filters applied to estimate the various components were originally calculated by Young [54] for the standard option of Census II-X11 variant. Later, Dagum et al. [18] calculated and analyzed all possible filter combinations of Census II-X11 and X11ARIMA. Given the fact that X11ARIMA and X12ARIMA are methods based on the Census Method II-X11 [51], widely discussed in Chap. 4, a brief summary is given here. The reader is referred to Ladiray and Quenneville [41] for a very detailed description of the X11 method.

The X11 method assumes that the main components of a time series follow a multiplicative or an additive model, that is,

(multiplicative model) $y_t = TC_t \times S_t \times I_t,$
(additive model) $y_t = TC_t + S_t + I_t,$

where y_t stands for the original series, TC_t for the trend-cycle, S_t for the seasonality, and I_t for the irregulars.

There are no mixed models in this program, such as $y_t = TC_t \times S_t + I_t$ or other possible combinations. The estimation of the components is made with different kinds of smoothing linear filters. The filters are applied sequentially and the whole process is repeated twice.

The X11 variant produced seasonal factor forecasts for each month which were given by

$$S_{j,t+1} = S_{j,t} + \frac{1}{2}(S_{j,t} - S_{j,t-1}), \qquad j = 1, \ldots, 12,$$

where j is the month and t the current year. The use of seasonal factor forecasts was very popular in statistical bureaus till the end of the 1970s. In 1975, Dagum [12] proposed the use of concurrent seasonal factors obtained from data that included the most recent values and this was adopted by Statistics Canada. Gradually, other statistical agencies applied concurrent seasonal factors and this is now the standard practice.

The final trend-cycle is obtained by a 9-, 13-, or 23-term Henderson moving average applied to the final monthly seasonally adjusted series (a 5- or 7-term Henderson filter for quarterly data). The selection of the appropriate filter is made on the basis of a preliminary estimate of the I/C ratio (the ratio of the average absolute month-to-month change in the irregular to that in the trend-cycle). A 9-term is applied to less irregular series and a 23-term to highly irregular series. The linear smoothing filters applied by Census Method II-X11 variant to produce seasonally adjusted data can be classified according to the distribution of their set of weights in symmetric (two-sided) and asymmetric (one-sided). The symmetric moving averages are used to estimate the component values that fall in the middle of the span of the average, say $2m + 1$, and the asymmetric moving averages, to the m first and last observations.

The sum of the weights of both kinds of filters is one, and thus, the mean of the original series is unchanged in the filtering process. The sum of the weights of a filter determines the ratio of the mean of the smoothed series to the mean of the unadjusted series assuming that these means are computed over periods long enough to ensure stable results. It is very important in filter design that the filter does not displace in time the components of the output relative to those of the input. In other words, the filter must not introduce phase shifts.

Symmetric moving averages introduce no time displacement for some of the components of the original series and a displacement of 180° for others. A phase shift of 180° is interpreted as a reverse in polarity which means that maxima are turned into minima and vice versa. In other words, peaks (troughs) in the input are changed into troughs (peaks) in the output. For practical purposes, however, symmetric moving averages act as though the time displacement is null. This is so because the sinusoids that will have a phase shift of 180° in the filtering process are cycles of short periodicity (annual or less) and moving averages tend to suppress or significantly reduce their presence in the output. In spectral analysis, the phase is a dimensionless parameter that measures the displacement of the sinusoid relative to the time origin. Because of the periodic repetition of the sinusoid, the phase can be restricted to ±180°. The phase is a function of the frequency of the sinusoid, being the frequency equal to the reciprocal of the length of time or period required for one complete oscillation.

It is inherent to any moving average procedure that the m first and last points of an original series cannot be smoothed with the same set of symmetric $2m + 1$ weights applied to the middle values.

The X11 method uses one-sided filters to smooth these end points. Moving averages with asymmetric weights are bounded to introduce phase shifts for all the components of the original series. This is a very undesirable characteristic since it may cause, for example, a business cycle in the smoothed series to turn down before or after it actually does in the unadjusted data. The asymmetric weights associated with the Henderson filters no longer estimate a cubic within their span, but only a linear trend-cycle within the length of the filters.

The X11ARIMA is a modified version of the Census Method II-X11 variant that was developed by Estela Bee Dagum, at Statistics Canada, mainly to produce a better current seasonal adjustment of series with seasonality that changes rapidly in a stochastic manner. The latter characteristic is often found in main socioeconomic indicators, e.g., retail trade, imports and exports, unemployment, and so on. A detailed description of the X11ARIMA method of seasonal adjustment is given in [13] and [14]. The X11ARIMA method consists of:

(i) modeling the original series with an ARIMA model of the Box and Jenkins type,

(ii) extending the original series 1–3 years with forecasts from the ARIMA model that fits and extrapolates well according to well-defined acceptance criteria, and

(iii) estimating each component with moving averages that are symmetric for middle observations and asymmetric for both end years. The latter are obtained via the convolution of the Census II-X11 variant filters and the ARIMA model extrapolations.

For series that result from the accumulation of daily activities, called flow series, deterministic components such as trading day variations and Easter holiday effects are estimated with dummy variable regression models and removed from the series, so that only the remainder is subject to steps (i)–(iii) above. The X11ARIMA was again modified by Dagum [14]. ARIMA models of the Box and Jenkins type were chosen which were previously shown to be very effective for forecasting a large number of series [46, 50].

The ARIMA models applied to seasonal series belong to the general multiplicative Box and Jenkins type, that is,

$$\phi_p(B)\Phi_P(B^s)(1-B)^d(1-B^s)^D y_t = \theta_q(B)\Theta_Q(B^s)a_t \qquad (1.3)$$

where s denotes the periodicity of the seasonal component (equal to 12 for monthly series and 4 for quarterly data), B is the backshift operator, such that $By_t = y_{t-1}$, $\phi_p(B) = (1 - \phi_1 B - \cdots - \phi_p B^p)$ is the nonseasonal autoregressive (AR) operator of order p, $\Phi_P(B^s) = (1 - \Phi_1 B^s - \cdots - \Phi_P B^{Ps})$ is the seasonal autoregressive (SAR) operator of order P, $\theta_q(B) = (1 - \theta_1 B - \cdots - \theta_q B^q)$ is the nonseasonal moving average (MA) operator of order q, $\Theta_Q(B^s) = (1-\Theta_1 B^s - \cdots -\Theta_Q B^{Qs})$ is the seasonal moving average (SMA) operator of order Q, and the a_t's are i.i.d. with mean zero and variance σ_a^2 (white noise). The $(1 - B)^d(1 - B^s)^D$ term implies nonseasonal differencing of order d and seasonal differencing of order D. If $d = D = 0$ (no differencing), it is common to replace y_t in (1.3) by deviations from its mean. The general multiplicative model (1.3) is said to be of order $(p, d, q)(P, D, Q)_s$.

The X12ARIMA is today the most often applied seasonal adjustment method by statistical agencies. It was developed by Findley et al. [26], and it is an enhanced version of the X11ARIMA method.

The major modifications concern: (1) extending the automatic identification and estimation of ARIMA models for the extrapolation option to many more than the three models available in X11ARIMA and (2) estimating trading day variations, moving holidays, and outliers in what is called regARIMA.

RegARIMA consists of regression models with autoregressive integrated moving average (ARIMA) errors. More precisely, they are models in which the mean function of the time series (or its logs) is described by a linear combination of regressors, and the covariance structure of the series is that of an ARIMA process.

If no regressors are used, indicating that the mean is assumed to be zero, the regARIMA model reduces to an ARIMA model.

Whether or not special problems requiring the use of regressors are present in the series to be adjusted, a fundamentally important use of regARIMA models is to extend the series with forecasts (and backcasts) in order to improve the seasonal adjustments of the most recent (and the earliest) data. Doing this reduces problems inherent in the trend-cycle estimation and asymmetric seasonal averaging processes of the type used by the X11 method near the ends of the series. The provision of this extension was the most important improvement offered by the X11ARIMA program. Its theoretical and empirical benefits have been documented in many publications, such as Dagum [14], Bobbit and Otto [5], and the references therein.

The X12ARIMA method has all the seasonal adjustment capabilities of the X11ARIMA variant. The same seasonal and trend moving averages are available, and the program still offers the X11 calendar and holiday adjustment routines incorporated in X11ARIMA. But several important new options have been included. The modeling module is designed for regARIMA model building with seasonal socioeconomic time series. To this end, several categories of predetermined regression variables are available, including trend constants or overall means, fixed seasonal effects, trading day effects, holiday effects, pulse effects (additive outliers), level shifts, temporary change outliers, and ramp effects. User-specified regression variables can also be included in the models.

The specification of a regARIMA model requires specification of both the regression variables to be included in the model and the type of ARIMA model for the regression errors (i.e., the order $(p, d, q)(P, D, Q)_s$). Specification of the regression variables depends on the user knowledge about the series being modeled. Identification of the ARIMA model for the regression errors follows well-established procedures based on examination of various sample autocorrelation and partial autocorrelation functions produced by the X12ARIMA program. Once a regARIMA model has been specified, X12ARIMA estimates its parameters by maximum likelihood using an Iterative Generalized Least Squares (IGLS) algorithm. Diagnostic checking involves examination of residuals from the fitted model for signs of model inadequacy. X12ARIMA produces several standard residual diagnostics for model checking, as well as provides sophisticated methods for detecting additive outliers and level shifts.

Trading day effects occur when a series is affected by the different day-of-the-week compositions of the same calendar month in different years. Trading day effects can be modeled with seven variables that represent *(no. of Mondays)*, ..., *(no. of Sundays)* in month t. Bell and Hillmer [2] proposed a better parameterization of the same effects using six variables defined as *(no. of Mondays)* − *(no. of Sundays)*, ..., *(no. of Saturdays)* − *(no. of Sundays)*, along with a seventh variable for length of month (LOM) or its deseasonalized version, the leap-year regressor (lpyear). In X12ARIMA the six variables are called the *tdnolpyear* variables. Instead of using a seventh regressor, a simpler and often better way to handle multiplicative leap-year effects is to re-scale the February values of the original time series before transformation to $\bar{m}_{\text{Feb}} y_t / m_t$, where y_t is the original time series before transformation, m_t is the length of month t (28 or 29), and $\bar{m}_{Feb} = 28.25$ is the average length of February. If the regARIMA model includes seasonal effects, these can account for the length-of-month effect except in Februaries, so the trading day model only has to deal with the leap-year effect. When this is done, only the *tdnolpyear* variables need be included in the model. X12ARIMA allows explicit choice of either approach, as well as an option (td) that makes a default choice of how to handle length-of-month effects. When the time series being modeled represents the aggregation of some daily series (typically unobserved) over calendar months they are called *monthly flow series*. If the series instead represents the value of some daily series at the end of the month, called a *monthly stock series*, then different regression variables are appropriate.

Holiday effects in a monthly flow series arise from holidays whose dates vary over time if: (1) the activity measured by the series regularly increases or decreases around the date of the holiday and (2) this affects 2 (or more) months depending on the date the holiday occurs each year. Effects of holidays with a fixed date, such as Christmas, are indistinguishable from fixed seasonal effects. *Easter effects* are the most frequently found holiday effects in American and European economic time series, since the date of Easter Sunday varies between March 22 and April 25. *Labor Day* and *Thanksgiving* also are found in American and Canadian time series.

X12ARIMA provides four other types of regression variables to deal with abrupt changes in the level of a series of temporary or permanent nature: *additive outliers* (AO), *level shifts* (LS), *temporary changes* (TC), and *ramps*. Identifying the location and nature of potential outliers is the object of the outlier detection methodology implemented. This methodology can be used to detect AOs, TCs, and LSs (not ramps). Any outlier detected is automatically added to the model as regression variable. Prespecified AOs, LSs, TCs, and ramps are actually simple forms of interventions as discussed by Box and Tiao [6].

Figure 1.1 shows the original values of the time series US Unemployment Rate for Males (16 years and over) observed from January 1992 to December 2013 together with the final seasonally adjusted series obtained with the X12ARIMA software using the default option. Figure 1.2 exhibits the corresponding seasonal factors.

1.2.2 ARIMA Model-Based Seasonal Adjustment Method

Peter Burman [7] was the first to develop a seasonal adjustment method based on ARIMA model decomposition, named SIGEX (Signal Extraction). Later, working on the same topic Hillmer and Tiao [35] developed what is known as ARIMA model-based seasonal adjustment, largely discussed in Bell and Hillmer [2].

An ARIMA model is identified from the observed data and by imposing certain restrictions, models for each component are derived. Since the components are unknown, to obtain a unique solution Hillmer and Tiao proposed a canonical decomposition which has the property of maximizing the variance of the irregulars

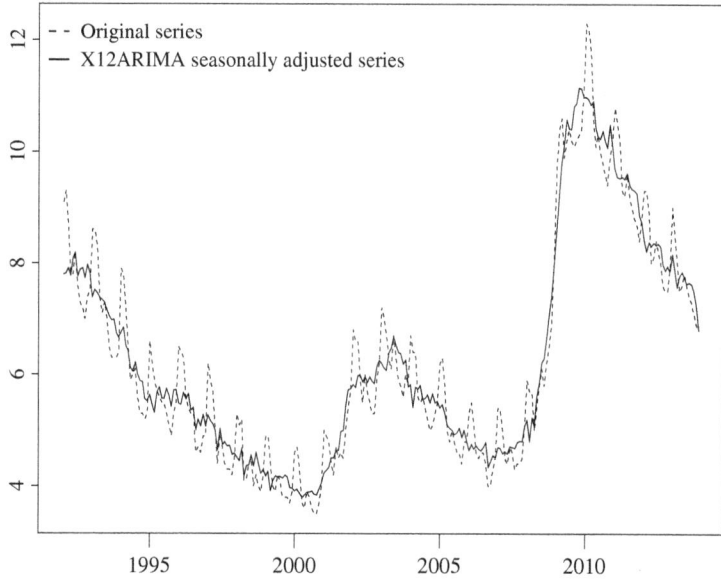

Fig. 1.1 Original and X12ARIMA (default option) seasonally adjusted US Unemployment Rate for Males (16 years and over) observed from January 1992 to December 2013

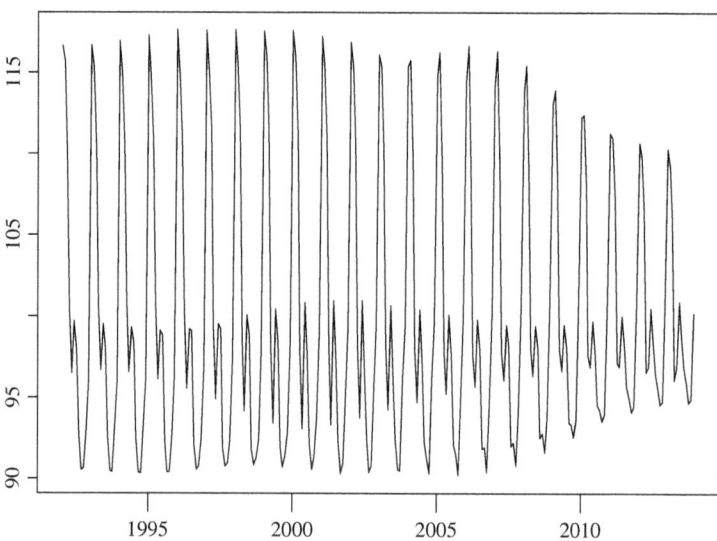

Fig. 1.2 Seasonal factors of the US Unemployment Rate for Males (16 years and over) estimated using the X12ARIMA software (default option)

and minimizing the variance of the estimated components. Because ARIMA model identification and estimation are not robust to outliers, and cannot deal with deterministic components such as trading days and moving holidays, further changes were made by combining dummy variables regression models with ARIMA models. In this regard, Gomez and Maravall [28] developed at the Bank of Spain, a seasonal adjustment software called TRAMO-SEATS which is currently applied mainly by European statistical agencies.

TRAMO stands for "Time series Regression with ARIMA noise, Missing observations and Outliers" and SEATS for "Signal Extraction in ARIMA Time Series." First, TRAMO estimates via regression the deterministic components, which are later removed from the raw data. In the second round, SEATS estimates the seasonal and trend-cycle components from the ARIMA model fitted to the data where the deterministic components are removed. SEATS uses the filters derived from the linearized ARIMA model that describes the stochastic behavior of the time series. It should be mentioned that Eurostat, in collaboration with the National Bank of Belgium, the US Bureau of the Census, the Bank of Spain, and the European Central Bank, has developed an interface of TRAMO-SEATS and X12ARIMA called Demetra+. In the Bank of Spain and Eurostat websites, it is also possible to find a considerable number of papers relevant to TRAMO-SEATS as well as in the European Statistical System (ESS) Guideline. TRAMO is a regression method that performs the estimation, forecasting, and interpolation of missing observations with ARIMA errors. The ARIMA model can be identified automatically or by the user. Given the vector of observations $\mathbf{y} = (y_1, \ldots, y_n)'$, the program fits the regression model

$$y_t = \mathbf{x}_t' \boldsymbol{\beta} + v_t \qquad (1.4)$$

where $\boldsymbol{\beta} = (\beta_1 \cdots \beta_p)'$ is a vector of regression coefficients, $\mathbf{x}_t = (x_{1t} \cdots x_{pt})'$ are p regression variables that define the deterministic part of the model. On the other hand, v_t represents the stochastic part of the model, that is assumed to follow the general ARIMA model

$$\Psi(B)v_t = \pi(B)a_t, \qquad (1.5)$$

where B is the backshift operator, $\Psi(B)$ and $\pi(B)$ are finite polynomials in B, and a_t is assumed to be normally identically distributed, that is, NID$(0, \sigma_a^2)$. The polynomial $\Psi(B)$ contains the unit roots associated with regular and seasonal differencing, as well as the polynomial with stationary autoregressive roots (and complex unit roots, if present). $\pi(B)$ denotes the invertible moving average polynomial. The regression variables $\mathbf{x}_t$ can be given by the user or generated by the program. In the latter case, the variables are for trading day variations, Easter effects, and outliers.

Outliers reflect the effect of some special, non regularly repetitive events, such as implementation of a new regulation, major political or economic changes, modifications in the way the variable is measured, occurrence of a strike or natural disaster, etc. Consequently, discordant observations and various types of abrupt changes are often present in times series data. The location and type of outliers are "a priori" unknown. TRAMO uses an improved Chen and Liu [8] type procedure for outlier detection and correction. The effect of an outlier is modeled by $y_t = \omega\xi(B)I_t(t_0) + v_t$, where $\xi(B)$ is a quotient of polynomials in B that models the type of outlier (its effect over time), and $I_t(t_0)$ is an indicator variable of the occurrence of the outlier, that is,

$$I_t(t_0) = \begin{cases} 1 \text{ if } & t = t_0 \\ 0 \text{ otherwise} \end{cases}$$

ω represents the impact of the outlier at time t_0 and v_t is the outlier free series which follows the model specified in Eq. (1.5). In the automatic detection and correction, by default, three types of outliers can be considered

(1) *Additive Outlier (AO)*: $\xi(B) = 1$;
(2) *Transitory Change (TC)*: $\xi(B) = 1/(1 - \delta B)$, where, by default, $\delta = 0.7$;
(3) *Level Shift (LS)*: $\xi(B) = 1/(1 - B)$.

One can also include a fourth type of outlier, that is, *Innovational Outlier (IO)* for which $\xi(B) = \theta(B)/\phi(B)\delta(B)$ that resembles a shock in the innovations a_t.

The procedure followed by TRAMO consists of:

1. *Pretest for the log/level specification*, that is, a trimmed range mean regression test is performed to select whether the original series will be transformed into log or maintain the level.
2. *Pretest for trading days and Easter effects* made with regressions using the default model for the noise. If the model is subsequently changed, the test is redone. Thus, the output file of TRAMO may say at the beginning "Trading day is not significant," but the final model estimated may contain trading day variables (or vice versa).
3. *Automatic detection and correction of outliers*. The program has a facility for detecting outliers and for removing their effects. The outliers can be entered by the user or they can be automatically identified by the program.
4. *Automatic model selection*. The program further performs an automatic identification of the ARIMA model. This is done in two steps. The first one yields the nonstationary polynomial $(1 - B)^d(1 - B^s)^D$ of model (1.5). This is done by iterating on a sequence of AR and ARIMA models (with constant) which have a multiplicative structure when the data is seasonal. The procedure is based on results of Tiao and Tsay [52] and Tsay [53]. Regular and seasonal differences are obtained up to a maximum order of $(1 - B)^2(1 - B^s)$. The second step identifies an ARMA model for the stationary series (modified for outliers and regression-type effects) following the Hannan–Rissanen procedure [31], with an improvement which consists of using the Kalman filter instead of zeros to calculate the first residuals in the computation of the estimator of the variance of the innovations of model (1.5). Finally, the program combines the facilities for automatic identification and correction of outliers and automatic ARIMA

model identification just described so that it has an option for automatic model identification of a nonstationary series in presence of outliers.

5. *Diagnostic checks.* The main diagnostic tests are on the residuals. The estimated residuals $\hat{a}_t$ are analyzed to test the hypothesis that a_t are independent and identically normally distributed with zero mean and constant variance σ_a^2. Besides inspection of the residual graph, the corresponding sample autocorrelation function is examined. The lack of residual autocorrelation is tested using the Ljung–Box test statistics, and skewness and kurtosis tests are applied to test for normality of the residuals. Specific out-of-sample forecast tests are also performed to evaluate if forecasts behave in agreement with the model.

6. *Optimal forecasts.* If the diagnostics are satisfied, the model is used to compute optimal forecasts for the series, together with their mean square error (MSE). These are obtained using the Kalman filter applied to the original series.

TRAMO is to be used as a preadjustment process to eliminate all the deterministic components such that the residual is a linearized series that can be modeled with an ARIMA process. The latter is decomposed by SEATS in stochastic trend-cycle, seasonality, and irregulars. Both programs can handle routine applications to a large number of series and provide a complete model-based solution to the problems of forecasting, interpolation, and signal extraction for nonstationary time series.

SEATS belongs to the class of procedures based on ARIMA models for the decomposition of time series into unobserved components and consists of the following steps:

1. *ARIMA model estimation.* SEATS starts by fitting an ARIMA model to a series not affected by deterministic components, such as trading day variations, moving holidays, and outliers. Let y_t denote this linearized series, and consider an additive decomposition model (multiplicative if applied to the log transformation of y_t). The complete model can be written in detailed form as

$$\phi_p(B)\Phi_P(B^s)(1-B)^d(1-B^s)^D y_t = \theta_q(B)\Theta_Q(B^s)a_t + c, \tag{1.6}$$

and, in concise form, as

$$\Psi(B)y_t = \pi(B)a_t + c, \tag{1.7}$$

where c is equal to $\Psi(B)\bar{y}$, being $\bar{y}$ the mean of the linearized series y_t. In words, the model that SEATS assumes is that of a linear time series with Gaussian innovations. When used with TRAMO, estimation of the ARIMA model is made by the exact maximum likelihood method described by Gomez and Maravall [27].

2. *Derivation of the ARIMA models for each component.* The program proceeds by decomposing the series that follows the ARIMA model (1.7) into several components. The decomposition can be multiplicative or additive. Next, we shall discuss the additive model, since the multiplicative form is equivalent in the log

transformation of the data. That is,

$$y_t = T_t + C_t + S_t + I_t, \tag{1.8}$$

where T_t denotes the trend component, C_t the cycle, S_t represents the seasonal component, and I_t the irregulars.

The decomposition is done in the frequency domain. The spectrum (or pseudospectrum) is partitioned into additive spectra, associated with the different components which are determined mainly from the AR roots of the model. The trend component represents the long-term evolution of the series and displays a spectral peak at frequency 0, whereas the seasonal component captures the spectral peaks at seasonal frequencies (e.g., for monthly data these are 0.524, 1.047, 1.571, 2.094, 2.618, and 3.142). The cyclical component captures periodic fluctuations with period longer than a year, associated with a spectral peak for a frequency between 0 and $(2\pi/s)$, and short-term variation associated with low order MA components and AR roots with small moduli. Finally, the irregular component captures white noise behavior, and hence has a flat spectrum. The components are determined and fully derived from the structure of the (aggregate) ARIMA model (1.7) for the linearized series directly identified from the data. The program is aimed at monthly or quarterly frequency data and the maximum number of observations that can be processed is 600.

One important assumption is that of orthogonality among the components, and each one will have in turn an ARIMA model. In order to identify the components, the *canonical decomposition* is used which implies that the variance of the irregulars is maximized, whereas the trend, seasonal, and cycle are as stable as possible.

The canonical condition on the trend, seasonal, and cyclical components identifies a unique decomposition, from which the ARIMA models for the components are obtained (including the component innovation variances).

Figure 1.3 shows the original values of the time series US Unemployment Rate for Males (16 years and over) observed from January 1992 to December 2013 together with the final seasonally adjusted series obtained with the TRAMO-SEATS software using the default option. Figure 1.4 exhibits the corresponding seasonal factors.

1.2.3 Structural Time Series Models

The structural time series approach involves decomposing a series into components which have a direct interpretation. A structural model, also called unobserved component (UC) model, consists of a number of stochastic linear processes that stand for the trend, cycle, seasonality, and remaining stationary dynamic features in an observed time series. The trend component typically represents the longer

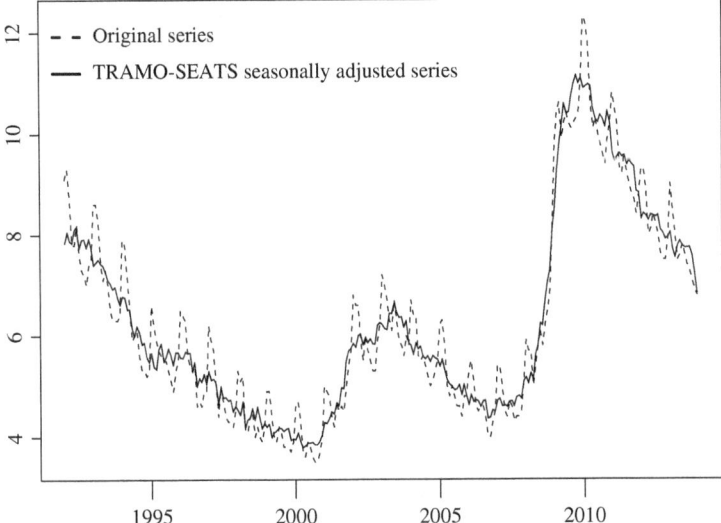

Fig. 1.3 Original and TRAMO-SEATS (default option) seasonally adjusted US Unemployment Rate for Males (16 years and over) observed from January 1992 to December 2013

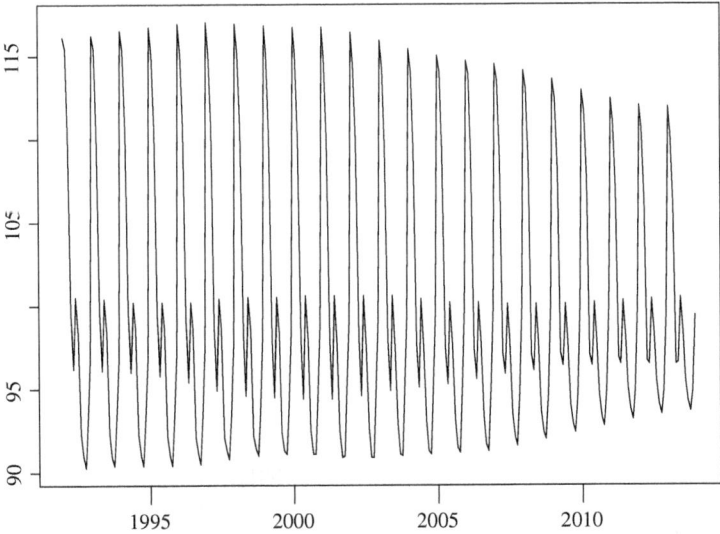

Fig. 1.4 Seasonal factors of the US Unemployment Rate for Males (16 years and over) estimated using the TRAMO-SEATS software (default option)

term movement of the series and is often specified as a smooth function of time. The recurring but persistently changing patterns within the years are captured by the seasonal component. In economic time series, the cycle component can represent the dynamic features associated with the business cycle. Each component is represented as an ARIMA process. Hence, the typical parameters that need to be estimated are the variances of the innovations driving the components, and a selection of other coefficients associated with the ARIMA processes.

The key to handling structural time series models is the *state space form* with the state of the system representing the various unobserved components such as trend, cycle, and seasonality. The estimate of the unobservable state can be updated by means of a *filtering* procedure as new observations become available. Predictions are made by *extrapolating* these estimated components into the future, while *smoothing* algorithms give the best estimate of the state at any point within the sample. The statistical treatment can therefore be based on the Kalman filter and its related methods. A detailed discussion of the methodological and technical concepts underlying structural time series models is given in Chap. 6. The reader is also referred to the monographs by Harvey [32], Kitagawa and Gersh [39], and Kim and Nelson [37]. Durbin and Koopman [24] also provide an exhaustive overview on state space methods for time series, whereas an introduction is given by Commandeur and Koopman [11].

We now introduce the basic structural time series model with explicit specifications for each component. For quarterly or monthly observations $y_t, t = 1, \ldots, n$, it is given by

$$y_t = T_t + S_t + I_t, \tag{1.9}$$

where T_t stands for the trend, S_t for seasonality, and I_t is the irregular component generally assumed to be $\mathrm{NID}(0, \sigma_I^2)$. All these components are stochastic and the disturbances driving them are assumed to be mutually uncorrelated. The definitions of the components are given below, but a full explanation of the underlying rationale can be found in Harvey [32]. The effectiveness of structural time series models compared to ARIMA-type models is discussed in Harvey et al. [33].

The trend component can be specified in many different ways, but the most common specification is an extension of the random walk trend by including a stochastic drift component

$$T_{t+1} = T_t + \beta_t + a_{T_t}, \tag{1.10}$$

$$\beta_{t+1} = \beta_t + \zeta_t \qquad \zeta_t \sim \mathrm{NID}(0, \sigma_\zeta^2), \tag{1.11}$$

where the disturbance series a_{T_t} is normally independently distributed. The initial values T_1 and β_1 are treated as unknown coefficients. When T_t is given by (1.10), Harvey [32] defines $y_t = T_t + I_t$ as the *local linear trend model*. In case $\sigma_\zeta^2 = 0$, the trend (1.10) reduces to an $I(1)$ process given by $T_{t+1} = T_t + \beta_1 + a_{T_t}$, where the drift β_1 is fixed. This specification is referred as a *random walk plus drift* process. If in addition $\sigma_T^2 = 0$, the trend reduces to the deterministic linear trend $T_{t+1} = T_1 + \beta_1 t$. When $\sigma_T^2 = 0$ and $\sigma_\zeta^2 > 0$, the trend T_t in (1.10) remains an $I(2)$ process.

Common specifications for the seasonal component S_t are provided in the following.

Dummy seasonal. Since the seasonal effects should sum to zero over a year, a basic model for this component is given by

$$S_t = -\sum_{j=1}^{s-1} S_{t-j} + \omega_t \qquad t = s, \ldots, n, \tag{1.12}$$

where s denotes the number of "seasons" in a year. In words, the seasonal effects are allowed to change over time by letting their sum over the previous year be equal to a random disturbance term ω_t with mean zero and variance σ_ω^2. Writing out Eq. (1.12) in terms of the lag operator B gives

$$(1 + B + \cdots + B^{s-1})S_t = \gamma(B)S_t = \omega_t \quad t = s, \ldots, n. \tag{1.13}$$

However, since $(1 - B^s) = (1 + B + \cdots + B^{s-1})(1 - B) = \gamma(B)(1 - B)$, the model can also be expressed in terms of the seasonal difference operator as $(1 - B^s)S_t = (1 - B)\omega_t$. The normally distributed disturbance ω_t drives the changes in the seasonal effect over time and is serially and mutually uncorrelated with all other disturbances and for all time periods. In the limiting case where $\sigma_\omega^2 = 0$ for all t, the seasonal effects are fixed over time and are specified as a set of unknown fixed dummy coefficients that sum up to zero.

Trigonometric seasonal. Alternatively, a seasonal pattern can also be modeled by a set of trigonometric terms at the seasonal frequencies, $\lambda_j = 2\pi j/s$, $j = 1, \ldots, [s/2]$, where $[s/2]$ is equal to $s/2$ if s is even, and $(s-1)/2$ if s is odd. The seasonal effect at time t is then described as

$$S_t = \sum_{j=1}^{[s/2]} (S_j \cos \lambda_j t + S_j^* \sin \lambda_j t). \tag{1.14}$$

When s is even, the sine term disappears for $j = s/2$, and so the number of trigonometric parameters, the S_j and S_j^*, is always $(s-1)/2$, which is the same as the number of coefficients in the seasonal dummy formulation. A seasonal pattern (1.14) is the sum of $[s/2]$ cyclical components that can be reformulated as follows:

$$\begin{pmatrix} S_{j,t} \\ S_{j,t}^* \end{pmatrix} = \rho \begin{pmatrix} \cos \lambda_j & \sin \lambda_j \\ -\sin \lambda_j & \cos \lambda_j \end{pmatrix} \begin{pmatrix} S_{j,t-1} \\ S_{j,t-1}^* \end{pmatrix} + \begin{pmatrix} \omega_{j,t} \\ \omega_{j,t}^* \end{pmatrix} \tag{1.15}$$

with $\omega_{j,t}$ and $\omega_{j,t}^*, j = 1, \ldots, [s/2]$, being zero mean white noise processes which are uncorrelated with each other and have a common variance σ_ω^2. Note that, when s is even, the component at $j = s/2$ collapses to

$$S_{j,t} = S_{j,t-1} \cos \lambda_j + \omega_{jt}.$$

The statistical treatment of structural time series models is based on the corresponding state space representation, according to which the observations are assumed to depend linearly on a *state vector* that is unobserved and generated by a

stochastic time-varying process. The observations are further assumed to be subject
to a measurement error that is independent on the state vector. The state vector can
be estimated or identified once a sufficient set of observations becomes available.

The state space form provides a unified representation of a wide range of linear
time series models, as discussed by Harvey [32], Kitagawa and Gersch [39], and
Durbin and Koopman [24]. The general linear state space model for a sequence of
n observations, $y_1, \ldots, y_n$, is specified as follows:

$$y_t = \mathbf{z}_t^T \boldsymbol{\alpha}_t + I_t, \qquad t = 1, \ldots, n \tag{1.16}$$

$$\boldsymbol{\alpha}_{t+1} = \boldsymbol{\Gamma}_t \boldsymbol{\alpha}_t + \mathbf{R}_t \boldsymbol{\varepsilon}_t. \tag{1.17}$$

Equation (1.16) is called the *observation* or *measurement equation* which relates
the observations $y_t, t = 1, \ldots, n$, to the state vector $\boldsymbol{\alpha}_t$ through $\mathbf{z}_t$, that is, an $m \times 1$
vector of fixed coefficients. In particular, $\boldsymbol{\alpha}_t$ is the $m \times 1$ state vector, that contains
the unobserved trend T_t, cycle C_t, and the seasonal component S_t. The irregular
component I_t is generally assumed to follow a white noise process with zero mean
and variance $\sigma_{I_t}^2$.

On the other hand, Eq. (1.17) is called the *state* or *transition equation*, where the
dynamic evolution of the state vector $\boldsymbol{\alpha}_t$ is described through the fixed matrix $\boldsymbol{\Gamma}_t$ of
order $m \times m$. $\boldsymbol{\varepsilon}_t$ is an $r \times 1$ vector of disturbances which are assumed to follow a
multivariate white noise process with zero mean vector and covariance matrix $\boldsymbol{\Sigma}_{\varepsilon_t}$.
It is assumed to be distributed independently on I_t at all time points. The matrix $\mathbf{R}_t$
is an $m \times r$ selection matrix with $r < m$, that in many standard cases is the identity
matrix $\mathbf{I}_m$, being generally $r = m$. Indeed, although matrix $\mathbf{R}_t$ can be specified
freely, it is often composed of a selection from the first r columns of the identity
matrix $\mathbf{I}_m$.

Initial conditions have to be defined for the state vector at the first time point, $\boldsymbol{\alpha}_1$.
It is generally assumed to be generated as $\boldsymbol{\alpha}_1 \sim N(\mathbf{a}_1, \mathbf{P}_1)$, and independently on
the observation and state disturbances, I_t and $\boldsymbol{\varepsilon}_t$. The mean vector $\mathbf{a}_1$ and covariance
matrix $\mathbf{P}_1$ can be treated as given and known in almost all stationary processes for
the state vector. For nonstationary processes and in presence of regression effects in
the state vector, the associated elements in the initial mean vector $\mathbf{a}_1$ can be treated as
unknown and estimated. For an extensive discussion of initialization in state space
analysis, we refer to Durbin and Koopman [24].

By appropriate choices of $\boldsymbol{\alpha}_t, I_t$, and $\boldsymbol{\varepsilon}_t$, of the matrices $\mathbf{z}_t, \boldsymbol{\Gamma}_t, \mathbf{R}_t$, and of the
scalar $\sigma_{I_t}^2$, a wide range of different structural time series models can be introduced.

Consider the basic structural time series model $y_t = T_t + S_t + I_t$ with the trend
component specified through Eqs. (1.10) and (1.11), and S_t as in (1.12) in presence
of quarterly data ($s = 4$). A state vector of five elements and a disturbance vector

of four elements are required, and they are given by

$$\alpha_t = \begin{pmatrix} T_t \\ \beta_t \\ S_{1,t} \\ S_{2,t} \\ S_{3,t} \end{pmatrix}, \quad \varepsilon_t = \begin{pmatrix} \varepsilon_{Tt} \\ \zeta_t \\ \omega_t \end{pmatrix}.$$

The state space formulation of the basic decomposition model is given by Eqs. (1.16) and (1.17) with the system matrices

$$\Gamma_t = \begin{pmatrix} 1 & 0 & 0 & 0 & 0 \\ 0 & 1 & 0 & 0 & 0 \\ 0 & 0 & -1 & -1 & -1 \\ 0 & 0 & 1 & 0 & 0 \\ 0 & 0 & 0 & 1 & 0 \end{pmatrix}, \mathbf{z}_t = \begin{pmatrix} 1 \\ 0 \\ 1 \\ 0 \\ 0 \end{pmatrix}.$$

Formal proofs of the Kalman filter can be found in Anderson and Moore [1], Harvey [32], and Durbin and Koopman [24].

1.2.3.1 Regression Component

The model (1.9) may provide a good description of the time series, although it may sometimes be necessary to include additional components. Seasonal economic time series are often affected by trading day effects and holiday effects which can influence the dynamic behavior of the series. Hence, a set of explanatory variables need to be included in the model to capture specific (dynamic) variations in the time series, as well as outliers and breaks. Therefore, Koopman and Ooms [40] suggest to extend model (1.9) as follows:

$$y_t = T_t + S_t + \mathbf{x}_t'\delta + I_t, \qquad I_t \sim N(0, \sigma_I^2),$$

for $t = 1, \ldots, n$, and where $\mathbf{x}_t$ is a K-dimensional vector of predetermined covariates and δ is a $K \times 1$ vector of regression coefficients, that can be allowed to change over time.

$$\mathbf{a}_{t+1|t} = \Gamma_t \mathbf{a}_{t|t}, \qquad \mathbf{P}_{t+1|t} = \Gamma_t \mathbf{P}_{t|t} \Gamma_t' + \mathbf{R}_t \Sigma_{\varepsilon_t} \mathbf{R}_t^T.$$

Figure 1.5 shows the original values of the time series US Unemployment Rate for Males (16 years and over) observed from January 1992 to December 2013 together with the final seasonally adjusted series obtained with the STAMP software. Figure 1.6 exhibits the corresponding seasonal component estimated using the default option (additive decomposition).

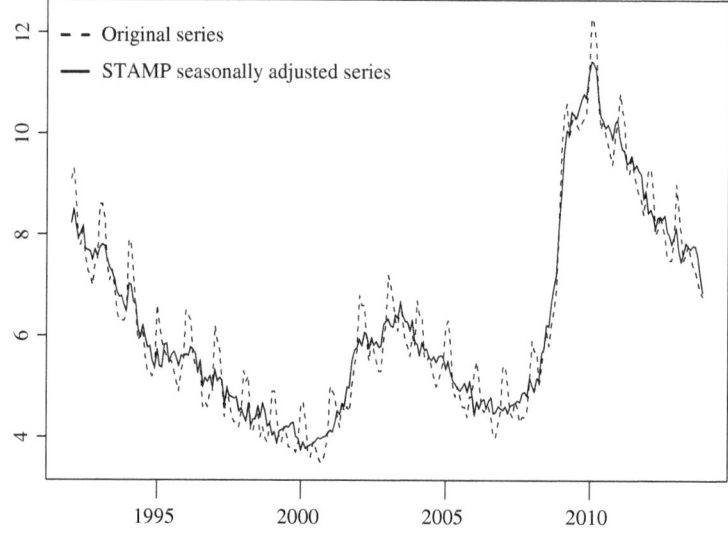

Fig. 1.5 Original and STAMP (default option) seasonally adjusted US Unemployment Rate for Males (16 years and over) observed from January 1992 to December 2013

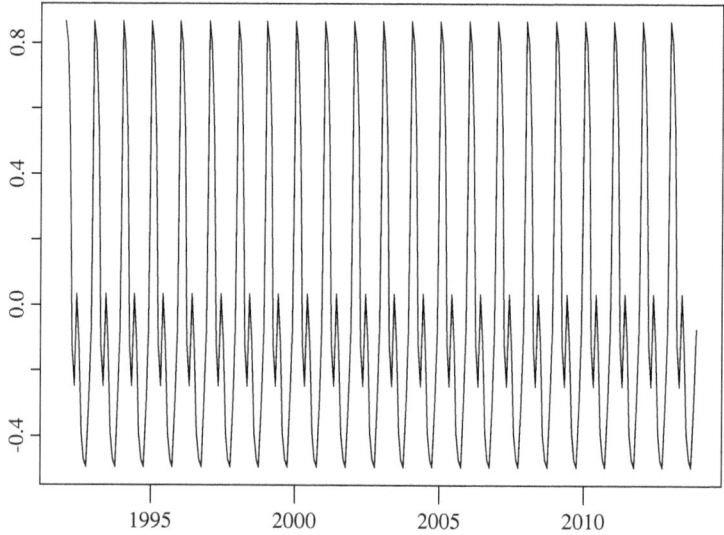

Fig. 1.6 Seasonal effects of the US Unemployment Rate for Males (16 years and over) estimated using the STAMP software (default option)

1.3   Part II: Real Time Trend-Cycle Estimation

The second part of this books deals with recent developments on trend-cycle estimation on real time. It comprises Chaps. 7–11. The basic approach to the analysis of current economic conditions, known as recession and recovery analysis, is that of assessing the real time trend-cycle of major socioeconomic indicators (*leading, coincident, and lagging*) using percentage changes, based on seasonally adjusted units, calculated for months and quarters in chronological sequence. The main goal is to evaluate the behavior of the economic indicators during incomplete phases by comparing current contractions or expansions with corresponding phases in the past. This is done by measuring changes of single time series (mostly seasonally adjusted) from their standing at cyclical turning points with past changes over a series of increasing spans. This differs from business cycle studies where cyclical fluctuations are measured around a long-term trend to estimate complete business cycles. The real time trend corresponds to an incomplete business cycle and is strongly related to what is currently happening on the business cycle stage.

In recent years, statistical agencies have shown an interest in providing trend-cycle or smoothed seasonally adjusted graphs to facilitate recession and recovery analysis. Among other reasons, this interest originated from the recent crisis and major economic and financial changes of global nature which have introduced more variability in the data. The USA entered in recession in December 2007 until June 2009, and this has produced a chain reaction all over the world, with great impact in Europe and China. There are no evidence of a fast recovery as in previous recession: the economic growth is sluggish and with high levels of unemployment. It has become difficult to determine the direction of the short-term trend (or trend-cycle) as traditionally done by looking at month to month (quarter to quarter) changes of seasonally adjusted values, particularly to assess the upcoming of a true turning point. Failure in providing reliable real time trend-cycle estimates could give rise to dangerous drift of the adopted policies. Therefore, a consistent prediction is of fundamental importance. It can be done by means of either univariate parametric models or nonparametric techniques. Since the majority of the statistical agencies use seasonally adjusted software, such as the Census II-X11 method and its variants X11/12ARIMA and X13, this book deals with the nonparametric techniques.

The linear filter developed by Henderson [34] is the most frequently applied and has the property that fitted to exact cubic functions will reproduce their values, and fitted to stochastic cubic polynomials it will give smoother results than those estimated by ordinary least squares. The properties and limitations of the Henderson filters have been extensively discussed by many authors, among them, [10, 14, 15, 20, 21, 25, 29, 36, 41, 43], and Dagum and Bianconcini [16] who were the first to represent the Henderson filter using Reproducing Kernel Hilbert Space (RKHS) methodology.

At the beginning and end of the sample period, the Henderson filter of length, say, $2m + 1$ cannot be applied to the m data points, hence only asymmetric filters can be used. The estimates of the real time trend are then subject to revisions produced by the innovations brought by the new data entering in the estimation and the time-varying nature of the asymmetric filters, in the sense of being different for each of the m data points.

The asymmetric filters applied to the first and last m observations associated with the Henderson filter were developed by Musgrave [45] on the basis of minimizing the mean squared revision between the final estimates, obtained with the symmetric Henderson weights, and preliminary estimates from the asymmetric weights, subject to the constraint that the sum of these weights is equal to one. The assumption made is that at the end of the series, the seasonally adjusted values do not follow a cubic polynomial, but a linear trend-cycle plus a purely random irregular. Several authors have studied the statistical properties and limitations of the Musgrave filters, among others, [4, 20, 21, 23, 30, 42, 49], and Dagum and Bianconcini [16, 17] were the first to introduce an RKHS representation of the asymmetric filters developed by Musgrave [45].

Dagum [15] developed a nonlinear smoother to improve on the classical Henderson filter. The NonLinear Dagum Filter (NLDF) results from applying the 13-term symmetric Henderson filter (H13) to seasonally adjusted series where outliers and extreme observations have been replaced and which have been extended with extrapolations from an ARIMA model. The main purpose of the ARIMA extrapolations is to reduce the size of the revisions of the most recent estimates whereas that of extreme values replacement is to reduce the number of unwanted ripples produced by H13. An unwanted ripple is a 10-month cycle (identified by the presence of high power at $\omega = 0.10$ in the frequency domain) which, due to its periodicity, often leads to the wrong identification of a true turning point. In fact, it falls in the neighborhood between the fundamental seasonal frequency and its first harmonic. On the other hand, a high frequency cycle is generally assumed to be part of the noise pertaining to the frequency band $0.10 \leq \omega < 0.50$. The problem of the unwanted ripples is specific of H13 when applied to seasonally adjusted series. Studies by Dagum, Chhab, and Morry [15] and Chhab et al. [9] showed the superior performance of the NLDF respect to both structural and ARIMA standard parametric trend-cycle models applied to series with different degrees of signal-to-noise ratios. In another study, the good performance of the NLDF is shown relative to nonparametric smoothers, namely: locally weighted regression (LOESS), Gaussian kernel, cubic smoothing spline, and supersmoother [19].

Dagum and Luati developed in 2009 what they called a Cascade Linear Filter (CLF), and distinguished between the Symmetric (SLF) and the Asymmetric Linear Filter (ALF).

A linear filter offers many advantages over a nonlinear one. For one, its application is direct and hence does not require knowledge of ARIMA model identification. Furthermore, linear filtering preserves the crucial additive constraint by which the trend of an aggregated variable should be equal to the algebraic addition of its component trends, thus avoiding the problem of direct versus indirect adjustment. Finally, the properties of a linear filter concerning signal passing and noise suppression can always be compared to those of other linear filters by means of spectral analysis.

These authors studied the properties of the new CLF relative to 13-term Henderson filter (H13) and have shown that the CLF is an optimal trend-cycle estimator among a variety of second and higher order kernels restricted to be of the same length. The theoretical properties were presented by means of spectral analysis, whereas the empirical properties were evaluated on a large sample of real time series pertaining to various socioeconomic areas and with different degrees of variability. It should be noticed that the theoretical properties of CLF cannot be compared with those of NLDF since the latter is data dependent.

From another perspective Dagum and Bianconcini in 2008 and 2013 were the first to introduce an RKHS representation of the symmetric Henderson and the asymmetric filters developed by Musgrave.

An RKHS is a Hilbert space characterized by a kernel that reproduces, via an inner product, every function of the space or, equivalently, a Hilbert space of real valued functions with the property that every point evaluation functional is bounded and linear. Parzen [47] was the first to introduce an RKHS approach in time series analysis applying the famous Loéve theorem by which there is an isometric isomorphism between the closed linear span of a second order stationary stochastic process and the RKHS determined by its covariance function. Parzen demonstrated that the RKHS approach provides a unified framework to three fundamental problems related with: (1) least squares estimation; (2) minimum variance unbiased estimation of regression parameters; and (3) identification of unknown signals perturbed by noise. Parzen's approach is parametric, and basically consists of estimating the unknown signal by generalized least squares in terms of the inner product between the observations and the covariance function. A nonparametric approach of the RKHS methodology was developed by De Boor and Lynch [22] in the context of cubic spline approximation. Later, Kimeldorf and Wahba[38] exploited both developments and treated the general spline smoothing problem from a RKHS stochastic equivalence perspective. These authors proved that minimum norm interpolation and smoothing problems with quadratic constraints imply an equivalent Gaussian stochastic process.

In this book we show how nonparametric estimators can be transformed into kernel functions of order two, that are probability densities, and from which corresponding hierarchies of estimators are derived. The density function provides the "initial weighting shape" from which the higher order kernels inherit their properties. This kernel representation enables the comparison of estimators based on different smoothing criteria, and has important consequences in the derivation of the asymmetric filters which can be applied to the most recent observations. In particular, those obtained by means of RKHS are shown to have superior properties from the view point of signal passing, noise suppression, and revisions relative to the classical ones.

The RKHS approach presented in this book is strictly nonparametric. It makes use of the fundamental theoretical result due to Berlinet [3] according to which a kernel estimator of order p can always be decomposed into the product of a reproducing kernel R_{p-1}, belonging to the space of polynomials of degree at most $p - 1$, and a probability density function f_0 with finite moments up to order $2p$. Hence, the weighted least squares estimation of the nonstationary mean uses weights derived from the density function f_0 from which the reproducing kernel is defined and not from the covariance function.

In the RKHS framework, given the density function, once the length of the symmetric filter is chosen, let us say, $2m + 1$, the statistical properties of the asymmetric filters are strongly affected by the bandwidth parameter of the kernel function from which the weights are derived.

Applied to real data, the kernel acts as a locally weighted average or linear filter that for each target point t gives the estimate

$$\hat{y}_t = \sum_{i=1}^{n} w_{t,i} y_i, \quad t = 1, 2, \ldots, n \tag{1.18}$$

where $w_{t,i}$ denotes the weights to be applied to the observations y_i to get the estimate $\hat{y}_t$ for each point in time $t, t = 1, \ldots, n$.

Once a symmetric span $2m+1$ of the neighborhood has been selected, the weights for the observations corresponding to points falling out of the neighborhood of any target point are null or approximately null. Hence, the estimates of the $n-2m$ central observations are obtained by applying $2m+1$ symmetric weights to the observations neighboring the target point. That is,

$$\hat{y}_t = \sum_{j=-m}^{m} w_j y_{t+j} \quad t = m+1, \ldots, n-m.$$

The weights $w_j, j = -m, \ldots, m$ depend on the shape of the nonparametric estimator K_{p+1} and on the value of a bandwidth parameter b fixed to ensure a neighborhood amplitude equal to $2m + 1$, such that

$$w_j = \frac{K_{p+1}\left(\frac{j}{b}\right)}{\sum_{j=-m}^{m} K_{p+1}\left(\frac{j}{b}\right)}. \tag{1.19}$$

To derive the Henderson kernel hierarchy by means of the RKHS methodology, the density corresponding to the Henderson ideal weighting function, W_j, and its orthonormal polynomials have to be determined. The triweight density function proposed by Loader [43] gives very poor results when the Henderson smoother spans are of short or medium lengths, as in most application cases, ranging from 5 to 23 terms. Hence, Dagum and Bianconcini [16] derived the exact density function corresponding to t W_j. The exact probability density corresponding to the

Henderson's ideal weighting penalty function is given by

$$f_{0H}(t) = \frac{(m+1)}{k} W((m+1)t), \quad t \in [-1, 1]$$

where $k = \int_{-(m+1)}^{(m+1)} W(j)dj$ and $j = (m+1)t$.

These authors also found that the biweight density function gives almost equivalent results to those obtained with the exact density function without the need to be calculated any time that the Henderson smoother length changes. Another important advantage is that the biweight density function belongs to the well-known Beta distribution family, that is,

$$f(t) = \left(\frac{r}{2B(s+1, 1r)} \right) (1 - |t|^r)^s I_{[-1,1]}(t),$$

where $B(a, b) = \int_0^1 t^{a-1}(1-t)^{b-1}dt$ with $a, b > 0$ is the Beta function.

The orthonormal polynomials associated to the biweight function are the Jacobi polynomials, for which explicit expressions for computation are available and their properties have been widely studied in literature.

For the m first and last filters we discuss time-varying bandwidth parameters since the asymmetric filters are time-varying. Three specific criteria of bandwidth selection are chosen based on the minimization of

1. the distance between the transfer functions of asymmetric and symmetric filters,
2. the distance between the gain functions of asymmetric and symmetric filters, and
3. the phase shift function over the domain of the signal.

We deal only with the reduction of revisions due to filter changes that depend on how close the asymmetric filters are respect to the symmetric one [14, 15] and do not consider revisions introduced by the innovations in the new data. Another important aspect discussed deals with the capability of the asymmetric filters to signal the upcoming of a true turning point that depends on the time delay for its identification. This is obtained by calculating the number of months (quarters) it takes for the last trend-cycle estimate to signal a true turning point in the same position of the final trend-cycle data. An optimal asymmetric filter should have a time path that converges fast and monotonically to the final estimate as new observations are added to the series.

It is shown in Chap. 10 that, applied to a set of US leading, coincident, and lagging indicators, the new set of asymmetric kernel filters reduced by one half the size of the revisions and by one third the time delay to detect the June 2009 turning point relative to the Musgrave filters.

References

1. Anderson, B. D. O., & Moore, J. B. (1979). *Optimal filtering*. Englewood Cliffs: Prentice-Hall.
2. Bell, W. H., & Hillmer, S. C. (1984). Issues involved with the seasonal adjustment of economic time series. *Journal of Business and Economic Statistics, 2*, 291–320.
3. Berlinet, A. (1993). Hierarchies of higher order kernels. *Probability Theory and Related Fields, 94*, 489–504.
4. Bianconcini, S., & Quenneville, B. (2010). Real time analysis based on reproducing kernel Henderson filters. *Estudios de Economia Aplicada, 28*(3), 553–574.
5. Bobbit, L., & Otto, M. C. (1990). Effects of forecasts on the revisions of seasonally adjusted values using the X11 seasonal adjustment procedure. In *Proceedings of the American Statistical Association Business and Economic Statistics Session* (pp. 449–453).
6. Box, G. E. P., & Tiao, G. C. (1975). Intervention analysis with applications to economics and environmental problems. *Journal of the American Statistical Association, 70*, 70–79.
7. Burman, J. P. (1980). Seasonal adjustment by signal extraction. *Journal of the Royal Statistical Society Series A, 143*, 321–337.
8. Chen, C., & Liu, L. M. (1993). Joint estimation of model parameters and outlier effects in time series. *Journal of the American Statistical Association, 88*, 284–297.
9. Chhab, N., Morry, M., & Dagum, E. B. (1999). Results on alternative trend-cycle estimators for current economic analysis. *Estadistica, 49–51*, 231–257.
10. Cholette, P. A. (1981). A comparison of various trend-cycle estimators. In O. D. Anderson & M. R. Perryman (Eds.), *Time series analysis* (pp. 77–87). Amsterdam: North Holland.
11. Commandeur, J. J. F., & Koopman, S. J. (2007). *An introduction to state space time series analysis*. Oxford: Oxford University Press.
12. Dagum, E. B. (1975). Seasonal factor forecasts from ARIMA models. In *40th Session International Statistical Institute*, Warsaw (Vol. 3, pp. 203–216).
13. Dagum, E. B. (1978). Modelling, forecasting and seasonally adjusting economic time series with the X-11 ARIMA method. *The Statistician, 27*(3), 203–215.
14. Dagum, E. B. (1988). *The X-11 ARIMA/88 seasonal adjustment method-foundations and user's manual*. Ottawa, ON, Canada: Time Series Research and Analysis Centre, Statistics Canada.
15. Dagum, E. B. (1996). A new method to reduce unwanted ripples and revisions in trend-cycle estimates from X11ARIMA. *Survey Methodology, 22*, 77–83.
16. Dagum, E. B., & Bianconcini, S. (2008). The Henderson smoother in reproducing kernel Hilbert space. *Journal of Business and Economic Statistics, 26*(4), 536–545.
17. Dagum, E. B., & Bianconcini, S. (2013). A unified probabilistic view of nonparametric predictors via reproducing kernel Hilbert spaces. *Econometric Reviews, 32*(7), 848–867.
18. Dagum, E. B., Chhab, N., & Chiu, K. (1996). Derivation and analysis of the X11ARIMA and census X11 linear filters. *Journal of Official Statistics, 12*(4), 329–347.
19. Dagum, E. B., & Luati, A. (2000). Predictive performance of some nonparametric linear and nonlinear smoothers for noisy data. *Statistica, LX*(4), 635–654.
20. Dagum, E. B., & Luati, A. (2009). A cascade linear filter to reduce revisions and turning points for real time trend-cycle estimation. *Econometric Reviews, 28*(1–3), 40–59.
21. Dagum, E. B., & Luati, A. (2012). Asymmetric filters for trend-cycle estimation. In W. R. Bell, S. H. Holan, & T. S. McElroy (Eds.), *Economic time series: Modeling and seasonality* (pp. 213–230). Boca Raton: Chapman & Hall.
22. De Boor, C., & Lynch, R. (1966). On splines and their minimum properties. *Journal of Mathematics and Mechanics, 15*, 953–969.
23. Doherty, M. (2001). The surrogate Henderson filters in X-11. *Australian and New Zealand Journal of Statistics, 43*(4), 385–392.
24. Durbin, J., & Koopman, S. J. (2001). *Time series analysis by state space methods*. Oxford: Oxford University Press.

25. Findley, D. F., & Martin, D. E. K. (2006). Frequency domain analysis of SEATS and X-11/X-12-ARIMA seasonal adjustment filters for short and moderate length time series. *Journal of Official Statistics, 22*, 1–34.
26. Findley, D. F., Monsell, B. C., Bell, W. R., Otto, M. C., & Chen, B. C. (1998). New capabilities and methods of the X12ARIMA seasonal adjustment program. *Journal of Business and Economic Statistics, 16*(2), 127–152.
27. Gomez, V., & Maravall, A. (1994). Estimation, prediction and interpolation for nonstationary series with the Kalman filter. *Journal of the American Statistical Association, 89*, 611–624.
28. Gomez, V., & Maravall, A. (1996). Program TRAMO and SEATS: Instructions for Users. Working paper 9628. Service de Estudios, Banco de Espana.
29. Gray, A., & Thomson, P. (1996). Design of moving-average trend filters using fidelity and smoothness criteria. In P. M. Robinson & M. Rosenblatt (Eds.), *Time series analysis* (in memory of E.J. Hannan) Vol. II (Vol. 115, pp. 205–219). Springer Lecture notes in statistics. New York: Springer.
30. Gray, A. G., & Thomson, P. J. (2002). On a family of finite moving-average trend filters for the ends of series. *Journal of Forecasting, 21*, 125–149.
31. Hannan, E. J., & Rissanen, J. (1982). Recursive estimation of mixed autoregressive moving average order. *Biometrika, 69*, 81–94.
32. Harvey, A. C. (1989). *Forecasting, structural time series models and the Kalman filter.* Cambridge: Cambridge University Press.
33. Harvey, A. C., Koopman, S. J., & Penzer, J. (1998). Messy time series: A unified approach. In T. B. Fomby & R. Carter Hill (Eds.), *Advances in econometrics* (Vol. 13, pp. 103–143). New York, NY, USA: JAI Press.
34. Henderson, R. (1916). Note on graduation by adjusted average. *Transaction of Actuarial Society of America, 17*, 43–48.
35. Hillmer, S. C., & Tiao, G. C. (1982). An ARIMA model-based approach to seasonal adjustment. *Journal of the American Statistical Association, 77*, 63–70.
36. Kenny, P., & Durbin, J. (1982). Local trend estimation and seasonal adjustment of economic and social time series. *Journal of the Royal Statistical Society A, 145*, 1–41.
37. Kim, C. J., & Nelson, C. R. (1999). *State space models with regime switching.* Cambridge, MA: MIT Press.
38. Kimeldorf, G. S., & Wahba, G. (1970). A correspondence between Bayesian estimation on stochastic processes and smoothing by splines. *Annals of Mathematical Statistics, 41*, 495–502.
39. Kitagawa, G., & Gersch, W. (1996). *Smoothness priors analysis of time series.* New York: Springer.
40. Koopman, S. J., & Ooms, M. (2011). Forecasting economic time series using unobserved components time series models. In M. Clements & D. F. Hendry (Eds.), *Oxford handbook of economic forecasting.* Oxford: Oxford University Press.
41. Ladiray, D., & Quenneville, B. (2001). *Seasonal adjustment with the X11 method.* Lecture notes in statistics (Vol. 158). New York: Springer.
42. Laniel, N. (1985). Design Criteria for the 13-term Henderson End Weights. Working paper. Methodology Branch, Statistics Canada, Ottawa.
43. Loader, C. (1999). *Local regression and likelihood.* New York: Springer.
44. Macaulay, F. R. (1931). *The smoothing of time series.* New York: National Bureau of Economic Research.
45. Musgrave, J. (1964). A Set of End Weights to End All End Weights. Working paper. U.S. Bureau of Census, Washington, DC.
46. Newbold, P., & Granger, C. W. J. (1974). Experience with forecasting univariate time series and the combination of forecasts (with discussion). *Journal of the Royal Statistical Society. Series A, 137*, 131–165.
47. Parzen, E. (1959). *Statistical Inference on Time Series by Hilbert Space Methods.* Technical Report No. 53, Statistics Department, Stanford University, Stanford, CA.
48. Persons, W. M. (1919). Indices of business conditions. *Review of Economic Statistics, 1*, 5–107.

49. Quenneville, B., Ladiray, D., & Lefrancois, B. (2003). A note on Musgrave asymmetrical trend-cycle filters. *International Journal of Forecasting, 19*(4), 727–734.
50. Reid, D. J. (1975). A review of short-term projection techniques. In H. A. Gorden (Ed.), *Practical aspects of forecasting* (pp. 8–25). London: Operational Research Society.
51. Shiskin, J., Young, A. H., & Musgrave, J. C. (1967). *The X-11 Variant of the Census Method II Seasonal Adjustment Program*. Technical Paper 15 (revised). US Department of Commerce, Bureau of the Census, Washington, DC.
52. Tiao, G. C., & Tsay, R. S. (1983). Consistency properties of least squares estimates of autoregressive parameters in Arma models. *The Annals of Statistics, 11*, 856–871.
53. Tsay, R. S. (1984). Regression models with time series errors. *Journal of the American Statistical Association, 79*, 118–124.
54. Young, A. H. (1968). A linear approximation to the Census and BLS seasonal adjustment methods. *Journal of the American Statistical Association, 63*, 445–457.

Chapter 2
Time Series Components

Abstract An important objective in time series analysis is the decomposition of a series into a set of unobservable (*latent*) components that can be associated with different types of temporal variations. This chapter introduces the definitions and assumptions made on these unobservable components that are: (1) a long-term tendency or *secular trend*, (2) *cyclical movements* superimposed upon the long-term trend. These cycles appear to reach their peaks during periods of economic prosperity and their troughs during periods of depressions, their rise and fall constituting the business cycle, (3) *seasonal variations* that represent the composite effect of climatic and institutional events which repeat more or less regularly each year, and (4) the *irregular component*. When the series result from the daily accumulation of activities, they can also be affected by other variations associated with the composition of the calendar. The two most important are *trading day variations*, due to the fact that the activity in some days of the week is more important than others, and *moving holidays* the date of which change in consecutive months from year to year, e.g., Easter.

A time series consists of a set of observations ordered in time on a given phenomenon (*target variable*). Usually the measurements are equally spaced, e.g., by year, quarter, month, week, and day. The most important property of a time series is that the ordered observations are dependent through time, and the nature of this dependence is of interest in itself.

Formally, a time series is defined as a sample realization of a set of random variables indexed in time, that is, a stochastic process denoted by $\{Y_1, \ldots, Y_n\}$ or simply $\{Y_t\}$. In this regard, an observed time series is denoted by $\{y_1, \ldots, y_n\}$, where the sub-index indicates the time to which the observation pertains, or in compact form $\{y_t\}$. The first observed value y_1 can be interpreted as the realization of the random variable Y_1, which can also be written as $Y(\omega, t = 1)$ where ω denotes the event belonging to a probabilistic space. Similarly, y_2 is the realization of Y_2, and so on. The random variables $Y_1, Y_2, \ldots, Y_n$ can be characterized by different probability distribution. For socioeconomic time series the probability space is continuous, and the time measurements are discrete. The frequency of measurements is said to be high when it is daily, weekly, or monthly and to be low when the observations are quarterly or yearly.

© Springer International Publishing Switzerland 2016   29
E. Bee Dagum, S. Bianconcini, *Seasonal Adjustment Methods and Real Time
Trend-Cycle Estimation*, Statistics for Social and Behavioral Sciences,
DOI 10.1007/978-3-319-31822-6_2

2.1 Time Series Decomposition Models

An important objective in time series analysis is the decomposition of a series into a set of non-observable (*latent*) components that can be associated with different types of temporal variations. The idea of time series decomposition is very old and was used for the calculation of planetary orbits by seventeenth century astronomers. Persons [37] was the first to state explicitly the assumptions of unobserved components. As Persons saw it, time series was composed of four types of fluctuations:

1. a long-term tendency or *secular trend*;
2. *cyclical movements* superimposed upon the long-term trend. These cycles appear to reach their peaks during periods of industrial prosperity and their troughs during periods of depressions, their rise and fall constituting the business cycle;
3. a *seasonal movement* within each year, the shape of which depends on the nature of the series;
4. *residual variations* due to changes impacting individual variables or other major events, such as wars and national catastrophes affecting a number of variables.

Traditionally, the four variations have been assumed to be mutually independent from one another and specified by means of an additive decomposition model:

$$y_t = T_t + C_t + S_t + I_t, \qquad t = 1, \ldots n \qquad (2.1)$$

where y_t denotes the observed series at time t, T_t the long-term trend, C_t the business cycle, S_t seasonality, and I_t the irregulars.

If there is dependence among the latent components, this relationship is specified through a multiplicative model

$$y_t = T_t \times C_t \times S_t \times I_t, \qquad t = 1, \ldots n \qquad (2.2)$$

where now S_t and I_t are expressed in proportion to the trend-cycle $T_t \times C_t$. In some cases, mixed additive–multiplicative models are used.

Whether a latent component is present or not in a given time series depends on the nature of the phenomenon and on the frequency of measurement. For example, seasonality is due to the fact that some months or quarters of a year are more important in terms of activity or level. Because this component is specified to cancel out over 12 consecutive months or 4 consecutive quarters, or more generally over 365.25 consecutive days, yearly series cannot contain seasonality.

Flow series can be affected by other variations associated with the composition of the calendar. The most important are the *trading day* variations, which are due to the fact that the activity in some days of the week is more important than others. Months with five of the most important days register an excess of activity (*ceteris paribus*) in comparison to months with four such days. Conversely, months with five of the least important days register a short-fall of activity. The length-of-month variation is usually assigned to the seasonal component. The trading day component is usually considered as negligible in quarterly series and even more so in yearly data.

Another important calendar variation is the *moving holiday* or *moving-festival* component. That component is associated with holidays which change date from year to year, e.g., Easter, causing a displacement of activity from 1 month to the previous or the following month. For example, an early date of Easter in March or early April can cause an important excess of activity in March and a corresponding short-fall in April, in variables associated with imports, exports, and tourism.

Under models (2.1) and (2.2), the trading day and moving festival components (if present) are implicitly part of the irregular. In 1965, Young developed a procedure to estimate trading day variations which was incorporated in the X11 seasonal adjustment method [39] and its subsequent versions, the X11ARIMA [17, 18] and X12ARIMA [21] methods. The latter two versions also include models to estimate moving holidays, such as Easter.

If the new components are present, the additive decomposition model becomes

$$y_t = T_t + C_t + S_t + D_t + H_t + I_t \tag{2.3}$$

where D_t and H_t denote the trading day and moving holiday components, respectively. Similarly, the multiplicative decomposition model becomes

$$y_t = T_t \times C_t \times S_t \times D_t \times H_t \times I_t \tag{2.4}$$

where the components S_t, D_t, H_t, and I_t are proportional to the trend-cycle.

Decomposition models (2.3) and (2.4) are traditionally used by seasonal adjustment methods. Other less used decomposition models are the log-additive and the mixed models (additive and multiplicative) where, for example, the systematic relationship among the components is multiplicative, but the irregulars are additive.

Seasonal adjustment actually entails the estimation of all the time series components and the removal of seasonality, trading day, and holiday effects from the observed series. The rationale is that these components which are relatively predictable conceal the current stage of the business cycle which is critical for policy and decision making.

There is another kind of time series decomposition often used for modeling and forecasting univariate ARIMA time series:

$$y_t = \eta_t + e_t \tag{2.5}$$

where η_t and e_t are referred to as the *signal* and the *noise*, according to the electrical engineering terminology. The signal η_t comprises all the systematic components of models (2.1) and (2.4), i.e., T_t, C_t, S_t, D_t, and H_t.

Model (2.5) is classical in signal extraction where the problem is to find the best estimates of the signal η_t given the observations y_t corrupted by noise e_t. The best estimates are usually defined as minimizing the mean square error.

Finally, given its fundamental role in time series modeling, we summarize the well-known decomposition theorem due to Herman Wold [41].

A stochastic process $\{Y_t\}$ is *second order stationary* or *weakly stationary* if the first two moments are not time dependent, that is, the mean and the variance are constant, and the autocovariance function depends only on the time lag and not on the time origin:

$$E(Y_t) = \mu \tag{2.6}$$

$$E(Y_t - \mu)^2 = \sigma_Y^2 < \infty, \tag{2.7}$$

$$E[(Y_t - \mu)(Y_{t+k} - \mu)] = \gamma_k \tag{2.8}$$

where $k = 0, 1, 2, \ldots$, denotes the time lag.

Wold proved that any stochastic process $\{Y_t\}$, stationary to the second order, can be decomposed into two mutually uncorrelated processes $\{Z_t\}$ and $\{V_t\}$, such that

$$Y_t = Z_t + V_t \tag{2.9}$$

where

$$Z_t = \sum_{j=0}^{\infty} \psi_j a_{t-j}, \qquad \psi_0 = 1, \sum_{j=1}^{\infty} \psi_j^2 < \infty \tag{2.10}$$

with $\{a_t\} \sim \mathrm{WN}(0, \sigma_a^2)$.

The component $\{Z_t\}$ is a convergent infinite linear combination of the a_t's, assumed to follow a white noise (WN) process with zero mean, constant variance σ_a^2, and zero autocovariance. Model (2.10) is known as an infinite moving average, $\mathrm{MA}(\infty)$, and the a_t's are the innovations. The component $\{Z_t\}$ is called the non-deterministic or purely linear component since only one realization of the process is not sufficient to determine future values $Z_{t+\ell}, \ell > 0$, without error.

The component $\{V_t\}$ can be represented by

$$V_t = \mu + \sum_{j=1}^{\infty} \left[\alpha_j \sin(\lambda_j t) + \beta_j \cos(\lambda_j t) \right], \qquad -\pi < \lambda < \pi, \tag{2.11}$$

where μ is the constant mean of process $\{Y_t\}$ and $\{\alpha_j\}, \{\beta_j\}$ are mutually uncorrelated white noise processes. The process $\{V_t\}$ is called deterministic because it can be predicted in the future without error from a single realization by means of an infinite linear combination of past values.

Wold theorem demonstrates that the property of stationarity is strongly related to that of linearity. It provides a justification for autoregressive moving average (ARMA) models [5] and some extensions, such as the autoregressive integrated moving average (ARIMA) and regression-ARIMA models (regARIMA).

2.2 The Secular or Long-Term Trend

The concept of trend is used in economics and other sciences to represent long-term smooth variations. The causes of these variations are often associated with structural phenomena such as population growth, technological progress, capital accumulation, and new practices of business and economic organization. For most economic time series, the trends evolve smoothly and gradually, whether in a deterministic or stochastic manner. When there is a sudden change of level and/or slope this is referred to as a structural change. However, it should be noticed that series at a higher levels of aggregation are less susceptible to structural changes. For example, a technological change is more likely to produce a structural change for some firms than for the whole industry.

The identification and estimation of the secular or long-term trend have posed serious challenges to statisticians. The problem is not of statistical or mathematical character but originates from the fact that the trend is a latent (non-observable) component and its definition as a long-term smooth movement is statistically vague. The concept of a long-period is relative, since a trend estimated for a given series may turn out to be just a long business cycle as more years of data become available. To avoid this problem statisticians have used two simple solutions. One is to estimate the trend and the business cycles together, calling it the *trend-cycle*. The other solution is to estimate the trend over the whole series, and to refer to it as the longest non-periodic variation.

It should be kept in mind that many systems of time series are redefined every 15 years or so in order to maintain relevance. Hence, the concept of long-term trend loses importance. For example, the Canadian system of Retail and Wholesale Trade series was redefined in 1989 to adopt the 1980 Standard Industrial Classification (SIC), and in 2003 to conform to the North American Industrial Classification System (NAICS), following the North American Free Trade Agreement. The following examples illustrate the need for such reclassifications. The 1970 Standard Industrial Classification (SIC) considered computers as business machines, e.g., cash registers and desk calculators. The 1980 SIC rectified the situation by creating a class for computers and other goods and services. The last few decades witnessed the birth of new industries involved in photonics (lasers), bio-engineering, nano-technology, and electronic commerce. In the process, new professions emerged, and classification systems had to keep up with these new realities.

There is a large number of deterministic and stochastic models which have been proposed for trend estimation (see Dagum and Dagum [15]). Deterministic models are based on the assumption that the trend can be well approximated by mathematical functions of time, whereas stochastic trends models assume that the trend can be better modeled by applying finite differences of low order together with autoregressive components and moving average errors.

2.2.1   Deterministic Trend Models

The most common representation of a deterministic trend is by means of polynomial functions. The observed time series is assumed to have a deterministic nonstationary mean, i.e., a mean dependent on time. A classical model is the regression error model where the observed data is treated as the sum of the trend and a random component, such that

$$y_t = T_t + e_t, \tag{2.12}$$

where T_t denotes the trend and $\{e_t\}$ is assumed to follow a stationary process, often white noise. The polynomial trend can be written as

$$T_t = \alpha_0 + \alpha_1 t + \cdots + \alpha_p t^p \tag{2.13}$$

where generally $p \leq 3$. The trend is said to be of a deterministic character because the observed series is affected by random shocks which are assumed to be uncorrelated with the systematic part.

Besides polynomial of time, three very well-known growth functions have been widely applied in population and economic studies, namely the modified exponential, the Gompertz, and the logistic function. Historically, the first growth model for time series was proposed by Malthus [34] in the context of population growth. He stated two time path processes, one for the supply of food and the other for population. According to Malthus, the supply of food followed an arithmetic progression and the population a geometric one.

For illustrative purposes, Fig. 2.1 shows a deterministic cubic trend fitted to the annual values of the Real Gross Domestic Product (billions of dollars) for the period starting from 1929 to 2013.

2.2.2   Stochastic Trends

Stochastic models are appropriate when the trend is assumed to follow a nonstationary stochastic process. The nonstationarity is modeled with finite differences of low order (cf. [24] and [35]).

A typical stochastic trend model often used in structural time series modeling is the so-called *random walk with constant drift*. In the classical notation, the model is

$$T_t = T_{t-1} + \beta + a_{T_t}, \qquad t = 1, 2, \ldots, n; a_{T_t} \sim N(0, \sigma_T^2) \tag{2.14}$$

$$\Delta T_t = \beta + a_{T_t}$$

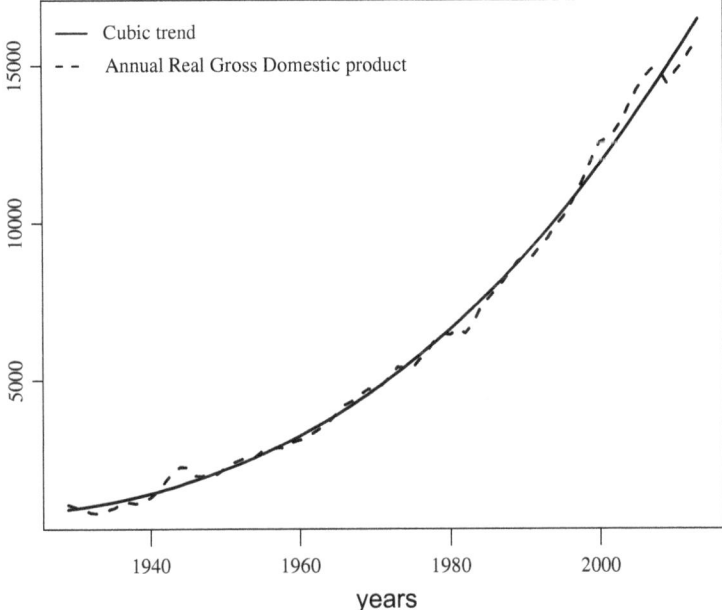

Fig. 2.1 Cubic trend fitted to the annual values of the Real Gross Domestic Product (billions of dollars)

where T_t denotes the trend, β a constant drift, and $\{a_{T_t}\}$ is a normal white noise process. Solving the difference equation (2.14) and assuming $a_{T_0} = 0$, we obtain

$$T_t = \beta t + \Delta^{-1} a_{T_t} = \beta t + \sum_{j=0}^{\infty} a_{T_{t-j}}, \qquad t = 1, \ldots, n, \qquad (2.15)$$

which show that a random walk with constant drift consists of a linear deterministic trend plus a nonstationary infinite moving average.

Another type of stochastic trend belongs to the ARIMA (p, d, q) class, where p is the order of the autoregressive polynomial, q is the order of the moving average polynomial, and d the order of the finite difference operator $(1 - B)$. The backshift operator B is such that $B^n T_t = T_{t-n}$. The ARIMA(p, d, q) model is written as

$$\phi_p(B)(1 - B)^d T_t = \theta_q(B) a_{T_t}, \qquad a_{T_t} \sim N(0, \sigma_T^2), \qquad (2.16)$$

where T_t denotes the trend, $\phi_p(B)$ the autoregressive polynomial in B of order p, θ_q stands for the moving average polynomial in B of order q, and $\{a_t\}$ denotes the innovations assumed to follow a normal white noise process. For example, with $p = 1, d = 2$, and $q = 0$, model (2.16) becomes

$$(1 - \phi_1 B)(1 - B)^2 T_t = a_{T_t} \qquad (2.17)$$

which means that after applying first order differences twice, the transformed series can be modeled by an autoregressive process of order one.

2.3   The Business Cycle

The business cycle is a quasi-periodic oscillation characterized by periods of expansion and contraction of the economy, lasting on average from 3 to 5 years. Because most time series are too short for the identification of a trend, the cycle and the trend are estimated jointly and referred to as the *trend-cycle*. As a result the concept of trend loses importance. The trend-cycle is considered a fundamental component, reflecting the underlying socioeconomic conditions, as opposed to seasonal, trading day, and irregular fluctuations.

The proper identification of cycles in the economy requires a definition of contraction and expansion. The definition used in capitalistic countries to produce the chronology of cycles is based on fluctuations found in the aggregate economic activity. A cycle consists of an expansion phase simultaneously present in many economic activities, followed by a recession phase and by a recovery which develops into the next expansion phase. This sequence is recurrent but not strictly periodic. Business cycles vary in intensity and duration. For example, in the USA, the 1981 recession was very acute but of short duration, whereas the 2008 recession was severe and of long duration. The latter crisis was precipitated by the sub-prime loan problem in the financial sector and this has caused an increased interest in the linkage between financial and real economic activities.

The analysis of business and economic cycles has preoccupied economists and statisticians for a long time. In 1862, the French economist Clement Juglar identified the presence of economic cycles 8–11 years long, although he was cautious not to claim any strict rigid regularity. The period going from 1950/51 till 1973 was characterized by high worldwide growth, and at least the problem of depression was declared dead; first in the late 1960s, when the Phillip curve was seen as being able to steer the economy which was followed by stagflation in the 1970s, which discredited the theory. Secondly in the early 2000s, following the stability and growth in the 1980 and 1990 in what came to be known as The Great Moderation. This phrase was sometimes used to describe the perceived end to economic volatility created by twentieth century banking systems. The term was coined by Harvard economists: James Stock and Mark Watson in their article written in 2002, "*Has the Business Cycle Changed and Why.*" The validity of this concept as a permanent shift has been questioned by the economic and financial crisis that started at the end of 2007. In the mid-1980s major economic variables such as GDP, industrial production, monthly payroll employment, and the unemployment rate began a decline in volatility (see [4]). Stock and Watson [40] viewed the causes of the moderation to be "improved policy, identifiable good luck in the form of productivity and commodity price shocks, and other unknown forms of good luck." The greater predictability in economic and financial performance had caused firms to hold less capital and to be less concerned about liquidity positions. This, in turn, is thought to have been a factor in encouraging increased debt levels and a reduction in risk premium required by investors.

The period of the Great Moderation ranges between 1987 and 2007, and it is characterized by predictable policy, low inflation, and modest business cycles. However, at the same time various regions in the world have experienced prolonged depressions, most dramatically the economic crisis in former Eastern Bloc countries following the end of the Soviet Union in 1991. For several of these countries the period 1989–2010 has been an ongoing depression, with real income still lower than in 1989. In economics a depression is a more severe downturn than a recession, which is seen by economists as part of a normal business cycle.

Considered a rare and extreme form of recession, a depression is characterized by its length, and by abnormally large increases in unemployment, falls in the availability of credit, shrinking output and investment, numerous bankruptcies, significantly reduced amounts of trade and commerce, especially international, as well as highly volatile relative currency value fluctuations most often due to devaluations.

In 1946, economists Arthur F. Burns and Wesley C. Mitchell [9] provided the new standard definition of business cycles in their book "Measuring Business Cycles": "Business cycles are a type of fluctuation found in the aggregate economic activity of nations that organize their work mainly in business enterprises: a cycle consists of expansions occurring at about the same time in many economic activities, followed by similarly general recessions, contractions, and revival which merge into the expansion phase of the next cycle; in duration business cycles vary from more than one year to ten to twelve years; they are not divisible into shorter cycles of similar characteristics with amplitudes approximating their own."

In 1954 Schumpeter [38] stated that an economic cycle has four stages : (1) expansion (increases in production and prices, low interests rates); (2) crisis (stock exchanges crash and multiple bankruptcies of firms occur); (3) recession (drops in prices and in output, high interests rates); and (4) recovery (stocks recover because of the fall in prices and incomes). In this model, recovery and prosperity are associated with increases in productivity, consumer confidence, aggregate demand, and prices.

In a broad sense, there have been two ways by which economic and business cycles have been studied, one analyzing complete cycles and the other, studying the behavior of the economic indicators during incomplete phases by comparing current contractions or expansions with corresponding phases in the past in order to assess current economic conditions.

Because completed economic and business cycles are not directly observable, for their identification and estimation, it is needed to remove from seasonally adjusted series a long-term trend. This was originally done by means of a long moving average using the Bry and Boschan method [7] adopted by the National Bureau of Economic Research (NBER). Later, other methods were developed, such as those of Hodrick and Prescott [28], Baxter and King [2], and Buttherworth [11] filters.

The index of Leading Economic Indicators (LEI) is intended to predict future economic activity. Typically, three consecutive monthly changes in the same direction suggest a turning point in the economy. For example, consecutive negative readings would indicate a possible recession. In the USA, LEI is a composite of the following 11 leading indicators: average workweek (manufacturing), initial unemployment claims, new orders for consumer goods, vendor performance, plant and equipment orders, building permits, change in unfilled durable orders, sensitive material prices, stock prices (S&P 500), real M2, and index of consumer expectations.

On the other hand, the Index of Coincident Indicators includes nonagricultural employment, index of industrial production, personal income, and manufacturing and trade sales.

Until recently the statistical analysis of macroeconomic fluctuations was dominated by linear time series methods. Over the past 15 years, however, economists have increasingly applied tractable parametric nonlinear time series models to business cycle data; most prominent in this set of models are the classes of threshold autoregressive (TAR) models, Markov switching autoregressive (MSAR), and smooth transition autoregressive (STAR) models.

The basic approach to the analysis of current economic conditions (known as recession and recovery analysis, see [36]) is that of assessing the short-term trend of major economic indicators (leading, coincident, and lagging) using percentage changes, based on seasonally adjusted units and calculated for months and quarters in chronological sequence. The main goal is to evaluate the behavior of the economic indicators during incomplete phases by comparing current contractions or expansions with corresponding phases in the past. This is done by measuring changes of single time series (mostly seasonally adjusted) from their standing at cyclical turning points with past changes over a series of increasing spans. In recent years, statistical agencies have shown an interest in providing further smoothed seasonally adjusted data (where most of the noise is suppressed) and trend-cycle estimates, to facilitate recession and recovery analysis. Among other reasons, this interest originated from major economic and financial changes of global nature which have introduced more variability in the data and, consequently, in the seasonally adjusted values, making very difficult to determine the direction of the short-term trend which includes the impact of the trend jointly with that of the business cycle, for an early detection of a turning point.

There are two approaches for current economic analysis modeling, the parametric one, that makes use of filters based on models, such as ARIMA models (see, among several others, Maravall [35], Kaiser and Maravall [29] and [30]) or state space models (see, e.g., Harvey [24] and Harvey and Trimbur [25]).

The other approach is nonparametric and based on digital filtering techniques. For example, the estimation of the trend-cycle with the Census Method II-X11, and its variants X11ARIMA and X12ARIMA, is done by the application of linear filters due to Henderson [26]. The Henderson filters are applied to seasonally adjusted data where the irregulars have been replaced taking into account the extreme values.

Stochastic linear approaches, parametric and nonparametric, are discussed in detail in Chap. 7.

2.3.1 Deterministic and Stochastic Models for the Business Cycle

Similarly to the trend, the models for cyclical variations can be deterministic or stochastic. Deterministic models often consist of sine and cosine functions of different amplitude and periodicities. For example, denoting the cycle by c_t, a deterministic model is

$$C_t = \sum_{j=1}^{2} \left[\alpha_j \cos(\lambda_j t) + \beta_j \sin(\lambda_j t) \right], \tag{2.18}$$

where $\lambda_1 = 2\pi/60$ and $\lambda_2 = 2\pi/40$. Model (2.18) takes into consideration two dominant cycles found in the European and American economies, those of 60 and 40 months, respectively.

Stochastic models of the ARIMA type, involving autoregressive models of order 2 with complex roots, have also been used to model the trend-cycle. For example,

$$C_t = \phi_1 C_{t-1} + \phi_2 C_{t-2} + a_{C_t}, \qquad a_{C_t} \sim N(0, \sigma_C^2) \tag{2.19}$$

where C_t denotes the cycle, $\{a_{C_t}\}$ is assumed Gaussian white noise, and the following conditions apply to the parameters: $\phi_1 + \phi_2 < 1$, $\phi_2 - \phi_1 < 1$, and $-1 < \phi_1 < 0$ (see [5]).

2.4 The Seasonal Variations

Seasonality originates from climate seasons and conventional events of religious, social, and civic nature, which repeat regularly from year to year.

The climatic seasons influence trade, agriculture, the consumption patterns of energy, fishing, mining, and related activities. For example, the consumption of heating oil increases in winter, and the consumption of electricity increases in the summer months because of air conditioning.

Institutional seasons such as Christmas, Easter, civic holidays, the school and academic year have a large impact on retail trade and on the consumption of certain goods and services, namely travel by plane, hotel occupancy, and consumption of gasoline.

The four main causes of seasonality are attributed to the weather, composition of the calendar, major institutional deadlines, and expectations. Seasonality is largely exogenous to the economic system but can be partially offset by human intervention. For example, seasonality in money supply can be controlled by central bank decisions on interest rates. In other cases, the seasonal effects can be offset by international and inter-regional trade. To some extent seasonality can evolve through technological and institutional changes. For example, the developments of appropriate construction materials and techniques made it possible to continue building in winter. The development of new crops, which better resist cold and dry weather, has influenced the seasonal pattern. The partial or total replacement of some crops by chemical substitutes, e.g., substitute of sugar, vanilla, and other flavors, reduces seasonality in the economy.

As for institutional change, the extension of the academic year to include some summer months affected the seasonal pattern of unemployment for the population of 15–25 years of age. Similarly, the practice of spreading holidays over the whole year impacted on seasonality.

The changing industrial mix of an economy also transforms the seasonal pattern, because some industries are more seasonal than others. In particular, economies which diversify and depend less on "primary" industries (e.g., fishing and agriculture) typically become less seasonal.

Another important characteristic of seasonality is the feasibility of its identification even if it is a latent (not directly observable) variable. The identification and estimation of seasonality, however, is not done independently on the other components affecting the time series under study, that is, trend, business cycle, trading day variations, moving holidays, and irregulars. The seasonal variations can be distinguished from the trend by their oscillatory character, from the business cycle by having annual periodicity, and from the irregulars by being systematic.

Seasonality entails large costs to society and businesses. One cost is the necessity to build warehouses to store inventories of goods to be sold as consumers require them, for example, grain elevators. Another cost is the under-use and over-use of the factors of production: capital and labor. Capital in the form of unused equipment, buildings, and land during part of the year has to be financed regardless. For example, this is the case in farming, food processing, tourism, and electrical generation. The cold climate increases the cost of buildings and infrastructure, e.g., roads, transportation systems, water and sewage systems, schools, and hospitals; not to mention the damage to the same caused by the action of ice. The labor force is over-used during the peak seasons of agriculture and construction for example; and, under-used in trough seasons sometimes leading to social problems.

In order to determine whether a series contains seasonality, it is sufficient to identify at least 1 month (or quarter) which tends to be systematically higher or lower than other month.

For illustrative purposes, Fig. 2.2 exhibits the seasonal variations affecting the US New Orders for Durable Goods (NODG) from February 1992 till December 2013 that are characterized by three main peaks and two main troughs. The peaks are observed in March, June and, in recent years, also in December when orders are highest, whereas the troughs correspond to the months of January and July.

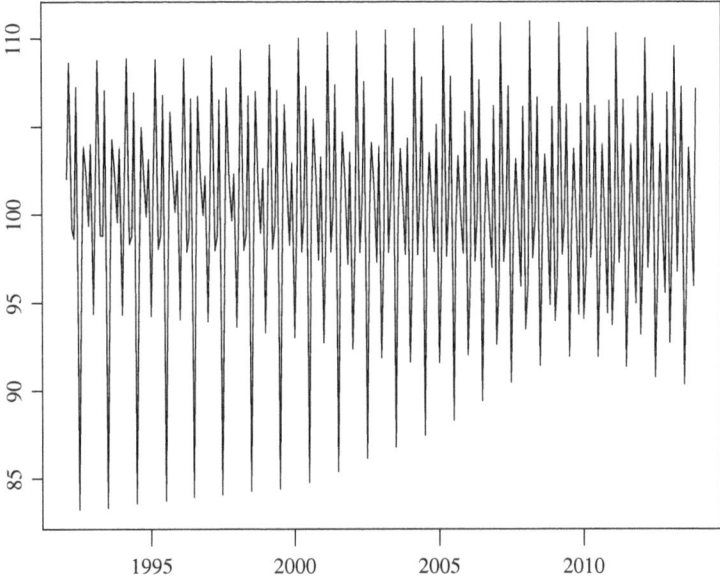

Fig. 2.2 Seasonal pattern of US New Orders for Durable Goods (January 1992–December 2013)

The seasonal pattern measures the relative importance of the months of the year. The constant 100 % represents an average month or a nonseasonal month. The peak month in March 1993 is 108.6 or 8.6 % larger than on an average month; the trough month is July with 83.2 in 1992, almost 17 % lower than on an average month. The seasonal amplitude, the difference between the peak and trough months of the seasonal pattern, has changed from 25.4 % in 1992 to 19.3 % in 2013. It is apparent that the seasonal amplitude is decreasing since 2005.

2.4.1 Seasonal Adjustment Methods

The seasonal adjustment methods developed so far are based on univariate time series decomposition models with no causal explanation. It is difficult to classify existing methods into mutually exclusive categories. However, it is possible to group the majority of seasonal adjustment methods into two main classes: one based on moving averages or linear filters and the other on explicit models with a small number of parameters for each component. These methods will be extensively discussed in the following chapters and only a brief summary is provided in this section.

2.4.1.1 Moving Average Methods

The best known and most often applied seasonal adjustment methods are based on moving averages or smoothing linear filters applied sequentially by adding (and subtracting) one observation at a time. These methods assume that the time series components change through time in a stochastic manner. Given a time series, $y_t, t = 1, \ldots, n$, for any t far removed from both ends, say $m+1 \leq t \leq n-m$, the seasonally adjusted value y_t^a is obtained by application of a symmetric moving average $W(B)$, as follows:

$$y_t^a = W(B)y_t = \sum_{j=-m}^{m} w_j y_{t-j} \qquad (2.20)$$

where the weights w_j are symmetric, that is, $w_j = w_{-j}$, and the length of the average is $2m + 1$.

For current and recent data $(n - m < t \leq n)$ a symmetric linear filter cannot be applied, and therefore truncated asymmetric filters are used. For example, for the last available observation y_n, the seasonally adjusted value is given by

$$y_n^a = W_0(B)y_n = \sum_{j=0}^{m} w_{0j} y_{n-j}. \qquad (2.21)$$

The asymmetric filters are time-varying in the sense that different filters are applied for the $m + 1$ first and last observations. The end estimates are revised as new observations are added because of: (1) the new innovations and (2) the differences between the symmetric and asymmetric filters. The estimates obtained with symmetric filters are often called "final."

The development of electronic computers contributed to major improvements in seasonal adjustment based on moving averages and facilitated their massive application. In 1954, Julius Shiskin of the US Bureau of Census developed a software called Method I, based mainly on the works of Macauley [33] already being used by the US Federal Reserve Board. Census Method I was followed by Census Method II and eleven more experimental versions (X1 , X2,..., X11). The best known and widely applied was the Census Method II-X11 variant developed by Shiskin et al. [39], but produced poor seasonally adjusted data at the end of the series which is of crucial importance to assess the direction of the short-term trend and the identification of turning points in the economy. Estela Bee Dagum developed in 1978 [16] a variant she called the X11ARIMA to correct for this serious limitation.

The X11ARIMA method consists of:

(i) modeling the original series with an ARIMA model of the Box and Jenkins type,
(ii) extending the original series 1–3 years with forecasts from the ARIMA model that fits and extrapolates well according to well-defined acceptance criteria, and
(iii) estimating each component with moving averages that are symmetric for middle observations and asymmetric for both end years. The latter are obtained via the convolution of Census II-X11 variant and the ARIMA model extrapolations.

For flow series, deterministic components such as trading day variations and Easter holiday effects are estimated with dummy variable regression models and removed from the series, so that only the remainder is subject to steps (i)–(iii) above. The X11ARIMA was extended again by Dagum in 1988 [18] and, in 1998, David Findley et al. [21] developed X12ARIMA that offers a regARIMA option to estimate deterministic components, such as trading day variations and moving holiday effects, simultaneously with the ARIMA model for extrapolation. It also includes new diagnostic tests and spectral techniques to assess the goodness of the results. The X12ARIMA method is today the one most often applied by statistical agencies in the world. The US Bureau of Census continued research on the development of seasonal adjustment methods and recently produced a beta version called X13ARIMA-SEATS which enables the estimation of the seasonal component either via linear filters as those available in X12ARIMA or based on an ARIMA decomposition model.

All the seasonal adjustment methods based on moving averages discussed in this book are nonlinear. Hence, the seasonally adjusted total of aggregated series is not equal to the algebraic sum of the seasonally adjusted series that enter into the aggregation. The main causes of nonlinearity are

1. a multiplicative decomposition model for the unobserved components,
2. the identification and replacement of extreme values,
3. the ARIMA extrapolations, and
4. automatic selection of moving average length for the estimation of the trend-cycle and seasonality.

The properties of the combined linear filters applied to estimate the various components were originally calculated by Young [43] for the standard option of Census II-X11 variant. Later, Dagum et al. [19] calculated and analyzed all possible filter combination of Census II-X11 and X11ARIMA. Cleveland and Tiao [14] and Burridge and Wallis [10] found ARIMA models that approximated well some of the linear filters used for the trend-cycle and seasonal component of Census II-X11.

2.4.1.2 Model-Based Seasonal Adjustment Methods

The best known model-based methods are: (1) the regression methods with global or locally deterministic models for each component, and (2) stochastic model-based methods that use ARIMA models.

2.4.1.2.1 Regression Methods

Seasonal adjustment by regression methods is based on the assumption that the systematic components of time series can be closely approximated by simple function of time over the entire span of the raw series. In general, two types of mathematical functions are considered. One is a polynomial of fairly low degree to represent the trend component; the other, linear combinations of sine and cosine functions, with different periodicity and fixed amplitude and phase, to represent business cycles and seasonality. To overcome the limitation of using global deterministic representations for the trend, cycle, and seasonality, regression methods were extended to incorporate stochastic representations by means of local polynomials (spline functions) for successive short segments of the series and introducing changing seasonal amplitudes. A major breakthrough in this direction was made by Akaike [1] who introduced prior constraints to the degree of smoothness of the various components and solved the problem with the introduction of a Bayesian model. Another important contribution is the regression method with locally deterministic models estimated by the LOESS (locally weighted regression) smoother developed by Cleveland et al. [13].

The simplest and often studied seasonal adjustment regression method assumes that the generating process of seasonality can be represented by strictly periodic functions of time of annual periodicity. The problem is to estimate s seasonal coefficients (s being the seasonal periodicity, e.g., 12 for monthly series and 4 for quarterly series) subject to the seasonal constraint that they sum to zero. This regression seasonal model can be written as

$$y_t = S_t + e_t, \qquad t = 1, \ldots, n$$

$$S_t = \sum_{j=1}^{s} \gamma_j d_{jt} \quad \text{subject to} \quad \sum_{j=1}^{s} \gamma_j = 0 \qquad (2.22)$$

here the d_j's are dummy variables taking a value of unity in season j and a value of zero otherwise; $\{e_t\} \sim \mathrm{WN}(0, \sigma_e^2)$ is a white noise process with zero mean, constant variance, and non-autocorrelated. The seasonal parameters γ_j's can be interpreted as the expectation of Y_t in each season and represent the seasonal effects of the series under question. Since the parameters are constant, the seasonal pattern repeats exactly over the years, that is,

$$S_t = S_{t-s}. \qquad (2.23)$$

Equation (2.22), often included in econometric models, assumes that the effects of the seasonal variations are deterministic because they can be predicted with no error. However, for most economic and social time series, seasonality is not deterministic but changes gradually in a stochastic manner. One representation of stochastic seasonality is to assume in regression (2.22) that the γ_j's are random variables instead of constant. Hence, the relationship given in Eq. (2.23) becomes

$$S_t = S_{t-s} + \omega_t, \qquad \forall t > s \qquad (2.24)$$

where $\{\omega_t\} \sim WN(0, \sigma_\omega^2)$ and $E(\omega_t e_t) = 0, \forall t$. The stochastic seasonal balance constraint is here given by $\sum_{j=0}^{s-1} S_{t-j} = \omega_t$ with expected value equal to zero.

Equation (2.24) assumes seasonality to be generated by a nonstationary stochastic process. In fact, it is a random walk process and can be made stationary by seasonal differencing $(1 - B^s)$, where B denotes the backward shift operator. In reduced form, the corresponding regression model is a linear combination of white noise processes of maximum lag s. Since $(1 - B^s) = (1 - B)(1 + B + \cdots + B^{s-1})$, model-based seasonal adjustment methods attribute only $\gamma(B) = \sum_{j=0}^{s-1} B^j$ to the seasonal component leaving $(1 - B)$ as part of a nonstationary trend. Thus, the corresponding seasonal stochastic model can be written as

$$\gamma(B)S_t = \omega_t. \qquad (2.25)$$

Equation (2.25) represents a seasonality that evolves in an erratic manner and it was included in seasonal adjustment methods based on structural model decomposition (see, e.g., [23] and [31]). This seasonal behavior is seldom found in observed time series where seasonality evolves smoothly through time. To represent this latter, [27] introduced a major modification

$$\gamma(B)S_t = \eta_s(B)b_t, \qquad (2.26)$$

where the left-hand side of Eq. (2.26) follows an invertible moving average process of maximum order $s - 1$, denoted by MA($s - 1$). Equation (2.26) is discussed extensively in [3] and can be generalized easily by replacing the MA part with stationary invertible autoregressive moving average (ARMA) processes of the [5] type. Depending on the order of the ARMA process a large variety of evolving stochastic seasonal variations can be modeled.

Other seasonal models used in regression methods, where the components are assumed to be local polynomials of time, are based on a smoothness criterion. For example, given $y_t = S_t + e_t, t = 1, \ldots, n$, restrictions are imposed on S_t, such that $\sum_{t=1}^n (S_t - S_{t-s})^2$ and $\sum_t [\gamma(B)S_t]^2$ be small. The solution is given by minimizing a weighted linear combination of both criteria as follows:

$$\sum_{t=1}^n (y_t - S_t)^2 + \alpha \sum_{t=1}^n (S_t - S_{t-s})^2 + \beta \sum_{t=1}^n [S(B)S_t]^2 \qquad (2.27)$$

where the ill-posed problem of choosing the α and β parameters was solved by Akaike [1] using a Bayesian model.

2.4.1.2.2 Stochastic Model-Based Methods

Stochastic model-based methods were mainly developed during the 1980s following two different approaches. One, originally known as seasonal adjustment by SIGEX (Signal Extraction), was developed by Burman [8] and as ARIMA model-based seasonal adjustment by Hillmer and Tiao [27], largely discussed by Bell and Hillmer [3]. The other is referred to as structural model decomposition method (see, e.g., [23] and [31]). The main difference between these two approaches is that in the latter simple ARIMA models are directly specified for each unobserved component, whereas in the former an overall ARIMA model is obtained from observable data and, by imposing certain conditions, models for each component are derived. Since the components are unknown, to obtain a unique decomposition Hillmer and Tiao proposed a canonical decomposition which has the properties of maximizing the variance of the irregulars and minimizing the variance of the other components.

ARIMA models identification and estimation are sensitive to outliers or extreme values and cannot deal with deterministic components, such as trading days and moving holidays. Therefore, further developments were made by combining dummy variables regression models with ARIMA models to deal with these cases. In this regard, Gomez and Maravall [22] developed at the Bank of Spain a seasonal adjustment software called TRAMO-SEATS which is currently applied mainly by European statistical agencies. TRAMO stands for "Time Series Regression with ARIMA noise, Missing observations and Outliers" and SEATS for "Signal Extraction in ARIMA Time Series." First, TRAMO estimates via regression dummy variables and direct ARIMA modeling, the deterministic components which are after removed from the input data. In a second round, SEATS estimates the seasonality and trend-cycle components from an ARIMA model fitted to the modified data where the deterministic components are removed. SEATS uses the filters derived from the linearized ARIMA model that describes the behavior of the time series. It should be mentioned that Eurostat, with the collaboration of the Bank of Belgium, the US Bureau of the Census, the Bank of Spain, and the European Central Bank, is developing an interface of TRAMO-SEATS and X12ARIMA called Demetra+, and more recently JDemetra+. At the Bank of Spain and Eurostat web site it is also possible to find a considerable number of papers relevant to TRAMO-SEATS as well as in the European Statistical System (ESS) Guidelines.

On the other hand, the structural model decomposition method starts directly with an observation equation (sometimes called *measurement equation*) that relates the observed time series to the unobserved components. Simple ARIMA or stochastic trigonometric models are assumed for each component. The structural model is cast in a state space form and estimated with Kalman filtering and smoothing. Koopman et al. [32] developed STAMP, that stands for Structural Time series Analyzer, Modeler and Predictor, and includes several types of models for each component. STAMP is set up in an easy-to-use form which enables the user to concentrate on model selection and interpretation. STAMP 8 is an integrated part of the OxMetrics modular software system for time series data analysis, forecasting, financial econometric modeling, and the statistical analysis of cross-section and panel data. STAMP is not used officially by statistical agencies, but mainly by econometricians and academics.

2.5 Calendar Variations

The main calendar variations are moving holidays and trading days.

2.5.1 The Moving Holiday Component

The *moving holiday* or *moving festival* component is attributed to calendar variations, namely due to the fact that some holidays change date in successive years. For example, Easter can fall between March 23 and April 25. The Chinese New Year date depends on the lunar calendar. Ramadan falls 11 days earlier from year to year. In the Moslem world, Israel and in the Far East, there are many such festivals. For example, Malaysia contends with as many as 11 moving festivals, due to its religious and ethnic diversity. These festivals affect the time series and may cause a displacement of activity from 1 month to the previous or the following month. For example, an early date of Easter in March or early April can cause an important excess of activity in March and a corresponding short-fall in April, in variables associated with imports, exports, and tourism. When Easter falls late in April (e.g., beyond the 10-th), the effect is captured by the seasonal factor of April. In the long run, Easter falls in April 11 times out of 14. Some of these festivals have a positive impact on certain variables, for example, air traffic, sales of gasoline, hotel occupancy, restaurant activity, sales of flowers and chocolate (in the case of Easter). The impact may be negative on other industries or sectors which close or reduce their activity during these festivals.

The festival effect may affect only the day of the festival itself, or a number of days preceding and/or following the festival. In the case of Easter, travelers tend to leave a few days before and return after Easter, which affects air traffic and hotel occupancy, etc., for a number of days. Purchases of flowers and other highly perishable goods, on the other hand, are tightly clustered immediately before the Easter date.

The effect of moving festivals can be seen as a seasonal effect dependent on the date(s) of the festival. For illustrative purposes, Fig. 2.3 displays the Easter effect on US Imports of Goods from Canada observed from January 1985 to December 2013. In this particular case, the Easter effect is rather mild. In some of the years,

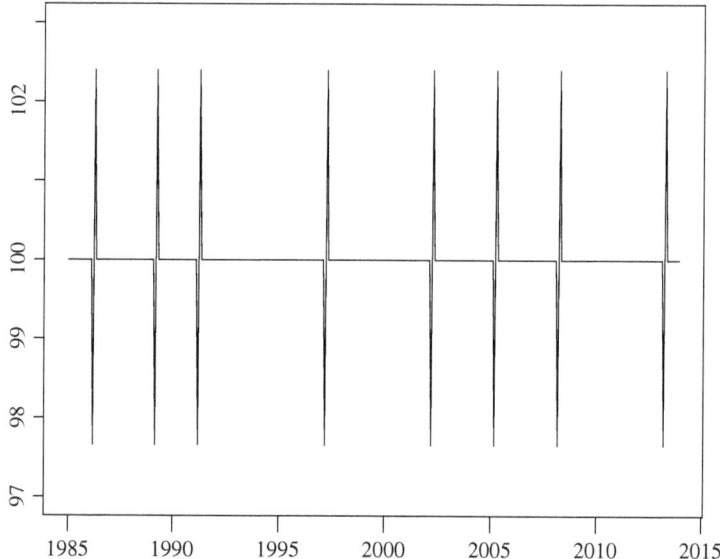

Fig. 2.3 Moving Holiday component of US Imports of Goods from Canada

Table 2.1 Dates of Easter and presence of effect in March

April 7 1985	April 23 2000, no effect
March 30 1986	April 15 2001, no effect
April 19 1987 no effect	March 31 2002
April 4 1988	April 20 2003, no effect
March 26 1989	April 11 2004, no effect
April 15, 1990 no effect	March 27 2005
March 31 1991	April 16 2006, no effect
April 19 1992, no effect	April 8 2007, no effect
April 11 1993, no effect	April12 2009, no effect
April 3 1994	April 4 2010
April 16 1995, no effect	April 24 2011, no effect
April 7 1996	April 8 2012, no effect
March 30 1997	March 31 2013
April 12 1998, no effect	April 20 2014, no effect
April 4 1999	

the effect is absent because Easter fell too late in April. The dates of Easter appear in Table 2.1.

In the case illustrated, the effect is felt 1 day before Easter and on Easter Sunday but not after Easter. This is evidenced by years 1985, 1986, 1989, 1991, 1997, 2002, 2005, and 2013, where Easters falls early in April or in March. Imports are substantially affected by Easter, because customs do not operate from Good Friday to Easter Monday. Easter can also significantly affect quarterly series, by displacing activity from the second to the first quarter.

Generally, festival effects are difficult to estimate, because the nature and the shape of the effect are often not well-known. Furthermore, there are few observations, i.e., one occurrence per year.

2.5.2   The Trading Day Component

Time series may be affected by other variations associated with the composition of the calendar. The most important calendar variations are the trading day variations, which are due to the fact that the activities of some days of the week are more important than others. Trading day variations imply the existence of a daily pattern analogous to the seasonal pattern. However, these daily factors are usually referred to as *daily coefficients*.

Depending on the socioeconomic variable considered, the activity of some days may be 60 % more important than an average day and other days, 80 % less important.

If the most important days of the week appear five times in a month (instead of four), the month registers an excess of activity *ceteris paribus*. If the least important days appear five times, the month records a short-fall. As a result, the monthly *trading day component* can cause variations of +8 % or −8 % (say) between neighboring months and also between same months of neighboring years. The trading day component is usually considered as negligible and very difficult to estimate in quarterly series.

Figure 2.4 displays the monthly trading day component obtained from the following daily pattern: 101.55, 102.26, 100.75, 99.99, 97.85, 97.73, and 99.95 for Monday to Sunday (in percentage), respectively. The daily pattern indicates that Tuesday, Friday, and Saturday are approximately 2.2 % more important than an average day (100 %); and that Wednesday, Thursday, and Sunday are less important.

For the multiplicative, the log-additive and the additive time series decomposition models, the monthly trading day component is respectively obtained in the following manner:

$$D_t = \sum_{\tau \in t} d_\tau / n_t = (2800 + \sum_{\tau \in t_5} d_\tau) / n_t \qquad (2.28)$$

$$D_t = \exp\left(\sum_{\tau \in t} d_\tau / n_t\right) = \exp\left[\left(\sum_{\tau \in t_5} d_\tau\right) / n_t\right] \qquad (2.29)$$

$$D_t = \sum_{\tau \in t} d_\tau = \sum_{\tau \in t_5} d_\tau \qquad (2.30)$$

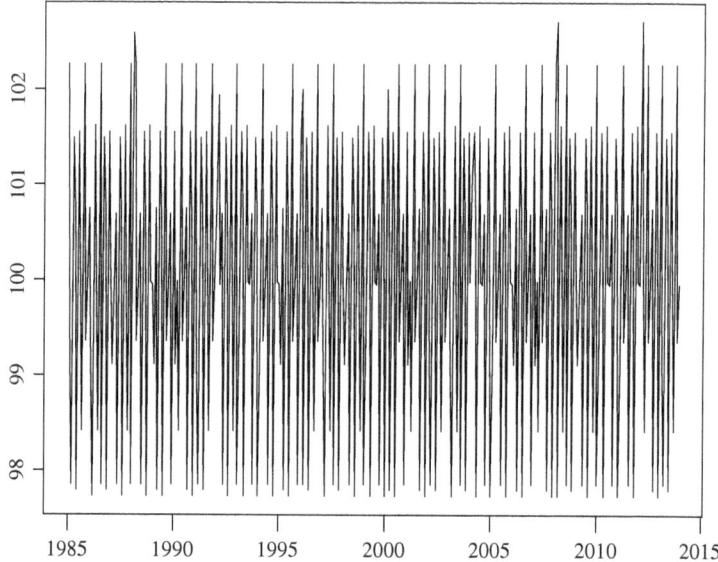

Fig. 2.4 Trading day estimates of US Imports of Goods from Canada

where n_t is the number of days in month t, t_5 is the set of days of the week that appears five times in month t, and d_τ are the daily coefficients in the month. Setting n_t equal to the number of days in month t implies that the length-of-month effect is captured by the multiplicative seasonal factors, except for Februaries. To adjust Februaries for the length-of-month, the seasonal factors of that month are multiplied by 29/28.25 and 28/28.25 for the leap and non-leap years, respectively.

The other option is to set n_t equal to 30.4375, so that the multiplicative trading day component also accounts for the length-of-month effect. The number 2800 in Eq. (2.28) is the sum of the first 28 days of the months expressed in percentage.

The monthly trading day estimates of the US Imports of Goods from Canada shown in Fig. 2.4 were obtained with the log-additive model (2.29). They display a drop of 5.45 % between February and March 1996 and an increase of 6.52 % between January and February 2000.

One can identify several instances where the change between same-months is significant. Indeed, same-month year-ago comparisons are never valid in the presence of trading day variations, not even as a rule of thumb. Furthermore, it is apparent that the monthly trading day factors in the figure are identical for quite a few months. Indeed for a given set of daily coefficients, there are only 22 different monthly values for the trading day component, for a given set of daily coefficients: seven values for 31-day months (depending on which day the month starts), seven for 30-day months, seven for 29-day months, and one for 28-day month. In other words, there are at most 22 possible arrangements of days in monthly data.

Many goods and services are affected by daily patterns of activity, which entail higher costs for producers, namely through the need of higher inventories, equipment, and staff on certain days of the week. For example, there is evidence that consumers buy more gasoline on certain days of the week, namely on Thursdays, Fridays, Saturdays, and holidays, which results in line-ups and shortages at the pumps. In order to cope with the problem, gasoline retailers raise their price on those days to promote sales on other days. Furthermore, the elasticity of demand for gasoline is low. In other words, to reduce consumption by a small percentage, prices must be raised by a disproportionate percentage, which upsets some consumers. On the other hand, consumers can buy their gasoline on other days. The alternative for retailers is to acquire larger inventories, larger tanks, more pumps, and larger fleets of tanker trucks, all of which imply higher costs and translate into much higher prices. In other words, there are savings associated with more uniform daily patterns; and costs, with scattered daily patterns. A similar consumer behavior prevails for the purchases of food, which probably results in more expensive prices, namely through higher inventories, larger refrigerators, more numerous cash registers, and more staff, than otherwise necessary. Deaths also occur more often on certain days of the week. Car accidents, drowning, skiing, and other sporting accidents tend to occur on weekend days and on holidays. In principle, stock series pertaining to 1 day display a particular kind of trading day variations. Among other things, inventories must anticipate the activity (flow) of the following day(s). For such stock series, the monthly trading day factor coincides with the daily weight of the day.

2.5.2.1   A Classical Model for Trading Day Variation

A classical deterministic model for trading day variations was developed by Young [42],

$$y_t = D_t + e_t, \qquad t = 1, \ldots, n \tag{2.31}$$

$$D_t = \sum_{j=1}^{7} \alpha_j N_{jt} \tag{2.32}$$

where $e_t \sim \text{WN}(0, \sigma_e^2)$, $\sum_{j=1}^{7} \alpha_j = 0, \alpha_j, j = 1, \ldots, 7$, denote the effects of the 7 days of the week, Monday to Sunday, and N_{jt} is the number of times day j is present in month t. Hence, the length of the month is $N_t = \sum_{j=1}^{7} N_{jt}$, and the cumulative monthly effect is given by (2.32). Adding and subtracting $\bar{\alpha} = (\sum_{j=1}^{7} \alpha_j)/7$ to Eq. (2.32) yields

$$D_t = \bar{\alpha} N_t + \sum_{j=1}^{7} (\alpha_j - \bar{\alpha}) N_{jt}. \tag{2.33}$$

Hence, the cumulative effect is given by the length of the month plus the net effect due to the days of the week. Since $\sum_{j=1}^{7} (\alpha_j - \bar{\alpha}) = 0$, model (2.31) takes into account the effect of the days present five times in the month. Model (2.33) can then be written as

$$D_t = \bar{\alpha} N_t + \sum_{j=1}^{6} (\alpha_j - \bar{\alpha})(N_{jt} - N_{7t}), \qquad (2.34)$$

with the effect of Sunday being $\alpha_7 = -\sum_{j=1}^{6} \alpha_j$.

Deterministic models for trading day variations assume that the daily activity coefficients are constant over the whole range of the series. Stochastic model for trading day variations has been rarely proposed. Dagum et al. [20] developed a model where the daily coefficients change over time according to a stochastic difference equation.

2.6 The Irregular Component

The irregular component in any decomposition model represents variations related to unpredictable events of all kinds. Most irregular values have a stable pattern, but some extreme values or outliers may be present. Outliers can often be traced to identifiable causes, for example, strikes, droughts, floods, and data processing errors. Some outliers are the result of displacement of activity from one month to the other.

Figure 2.5 shows the irregular component of US New Orders for Durable Goods (January 1992–December 2013). It includes extreme values and also outliers, namely in January 2001, October 2008, and January 2009. The former is identified as an additive outlier, whereas the latter are identified as level shift outliers.

As illustrated in Fig. 2.5, the values of the irregular component may be very informative, as they quantify the effect of events known to have happened. Note that it is much easier to locate outliers in the irregular component than in the raw series because the presence of seasonality hides the irregular fluctuations.

2.6.1 Redistribution of Outliers and Strikes

Some events can cause displacements of activity from one month to the next months, or vice versa. This phenomenon is referred to as redistribution of outliers. We also deal with the strike effect under this headline. The outliers must be modeled and temporally removed from the series in order to reliably estimate the systematic components, namely the seasonal and trading day components. Events such as

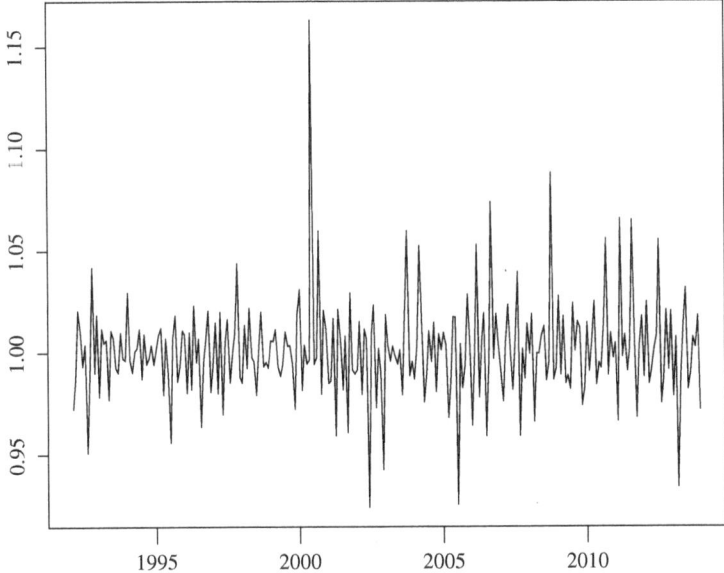

Fig. 2.5 Irregular component of US New Orders for Durable Goods (January 1992–December 2013)

major snow storms and power blackouts usually postpone activity to the next month, without much longer term effect.

2.6.2   Models for the Irregulars and Outliers

The irregulars are most commonly assumed to follow a white noise process $\{I_t\}$ defined by

$$E(I_t) = 0, E(I_t^2) = \sigma_I^2 < \infty, E(I_tI_{t-k}) = 0 \quad if \quad k \neq 0$$

If σ_I^2 is assumed constant (homoscedastic condition), $\{I_t\}$ is referred to as white noise in the strict sense.

If σ_I^2 is finite but not constant (heteroscedastic condition), $\{I_t\}$ is called white noise in the weak sense.

For inferential purposes, the irregular component is often assumed to be normally identically distributed and not correlated, which implies independence. Hence, $I_t \sim NID(0, \sigma_I^2)$.

There are different models proposed for the presence of outliers depending on how they impact the series under question. If the effect is transitory, the outlier is said to be *additive*, and if permanent, to be *multiplicative*.

Box and Tiao [6] introduced the following intervention model to deal with different types of outliers:

$$y_t = \sum_{j=0}^{\infty} h_j x_{t-j} + \eta_t = \sum_{j=0}^{\infty} h_j B^j x_t + \eta_t = H(B)x_t + \eta_t,$$

where the observed series $\{y_t\}$ consists of an input series $\{x_t\}$ considered a deterministic function of time and a stationary process $\{\eta_t\}$ of zero mean and uncorrelated with $\{x_t\}$. In such a case the mean of $\{y_t\}$ is given by the deterministic function $\sum_{j=0}^{\infty} h_j x_{t-j}$. The type of function assumed for $\{x_t\}$ and weights $\{h_j\}$ depend on the characteristic of the outlier or unusual event and its impact on the series.

If the outlier at time t_0 is additive, in the sense that it will not permanently modify the level of $\{y_t\}$, that is, $E(y_t) = 0, \forall t \le t_0$, and also $E(y_t) \to 0$ for $t \to t_0$, then an appropriate x_t is the impulse function defined by

$$x_t = P_t(t_0) = \begin{cases} 1, & t = t_0, \\ 0, & t \ne t_0. \end{cases}$$

If, instead, the outlier at time t_0 is multiplicative in the sense that it will modify the level of $\{y_t\}$, that is, $E(y_t) = 0, \forall t \le t_0$, and also $E(y_t) \to c \ne 0$ for $t \to t_0$, then an appropriate x_t is the step function defined by

$$x_t = S_t(t_0) = \begin{cases} 1, & t \ne t_0, \\ 0, & t < t_0. \end{cases}$$

In fact, $S_t(t_0) - S_{t-1}(t_0) = (1 - B)S_t(t_0) = P_t(t_0)$.

Once the deterministic function is chosen, the weights $\{h_t\}$ can follow different patterns dependent on the impact of the outlier. For example, if the outlier is additive and present at time $t = t_0$, an appropriate model is

$$H(B)x_t = \omega P_t(t_0) = \begin{cases} \omega, & t = t_0 \\ 0, & t \ne t_0 \end{cases}$$

where ω represents the effect of the outlier.

If the outlier is seasonal and additive starting at $t = t_0$, an appropriate model is

$$H(B)x_t = \omega P_t(t_0 + ks) = \begin{cases} \omega, & t = t_0, t_0 + s, t_0 + 2s, \ldots \\ 0, & \text{otherwise} \end{cases}$$

For additive redistribution of outliers (displacement of activity), during period $t = t_0, t = t_0 + 1, \ldots, t = t_0 + k$, an appropriate model is

$$H(B)x_t = \omega_0 P_t(t_0) + \sum_{i=1}^{k} \omega_i P_t(t_0 + i) = \begin{cases} \omega_0 = -\sum_{i=1}^{k} \omega_i, & t = t_0 \\ \omega_1, & t = t_0 + 1 \\ \cdots \\ \omega_k, & t = t_0 + k \\ 0, & \text{otherwise} \end{cases}$$

where the weights $\omega_0, \ldots, \omega_k$ measure the outlier effects during the $k + 1$ periods $t_0, t_0 + 1, \ldots, t_0 + k$. The global effect is such that the sum of all the outlier effects cancel out.

For a sudden permanent change of level at time $t = t_0$, a step function can be used

$$H(B)x_t = \omega_0 S_t(t_0) = \begin{cases} \omega, & t \geq t_0 \\ 0, & \text{otherwise} \end{cases}$$

where ω represent the level difference of the series before and after the outlier. Since $(1 - B)S_t(n) = P_t(n)$, we can write

$$H(B)x_t = \frac{\omega}{(1 - B)} P_t(t_0) = \begin{cases} \omega, & t = t_0 \\ 0, & \text{otherwise} \end{cases}$$

where $\omega(1-B)^{-1} = \omega(1+B+B^2+\cdots)P_t(t_0) = \sum_{j=0}^{\infty} B^j P_t(t_0) = \sum_{j=0}^{\infty} \omega P_t(t_0+j)$. For a sudden transitory change at time $t = t_0$, an appropriate model is

$$H(B)x_t = \frac{\omega}{(1 - \delta B)} P_t(t_0) = \begin{cases} 0, & t < t_0 \\ \omega, & t = t_0 \\ \delta^2 \omega, & t = t_0 + 1 \\ \delta^2 \omega, & t = t_0 + 2 \end{cases}$$

where ω denotes the initial effect, and $0 < \delta < 1$ is the rate of decrease of the initial effect.

For a gradual permanent level change at time $t = t_0$, an appropriate model is

$$H(B)x_t = \frac{\omega}{(1 - \delta B)} S_t(t_0) = \begin{cases} 0, & t < t_0 \\ \omega, & t = t_0 \\ (1 + \delta)\omega, & t = t_0 + 1 \\ (1 + \delta + \delta^2)\omega, & t = t_0 + 2 \\ \cdots \\ \frac{\omega}{(1-\delta)}, & t \to \infty \end{cases}$$

The intervention model proposed by Box and Tiao [6] requires the location and the dynamic pattern of an event to be known. Chen and Liu [12] consider the estimation problem when both the location and the dynamic pattern are not known a priori. The proposed iterative procedure is less vulnerable to the spurious and masking effects during outlier detection and allows to jointly estimate the model parameters and multiple outlier effects. This procedure is the one actually implemented in X12ARIMA and TRAMO-SEATS software.

References

1. Akaike, H. (1980). Seasonal adjustment by a Bayesian modelling. *Journal of Time Series Analysis, 1*, 1–13.
2. Baxter, M., & King, R. G. (1995). Measuring business cycles approximate band-pass filters for economic time series. NBER Working Papers 5022, National Bureau of Economic Research, Inc.
3. Bell, W. H., & Hillmer, S. C. (1984). Issues involved with the seasonal adjustment of economic time series. *Journal of Business and Economic Statistics, 2*, 291–320.
4. Bernanke, B. S. (2004). *Gradualism*. Speech 540, Board of Governors of the Federal Reserve System (U.S.).
5. Box, G. E. P., & Jenkins, G. M. (1970). *Time series analysis: Forecasting and control*. San Francisco, CA, USA: Holden-Day (Second Edition 1976).
6. Box, G. E. P., & Tiao, G. C. (1975). Intervention analysis with applications to economics and environmental problems. *Journal of the American Statistical Association, 70*, 70–79.
7. Bry, G., & Boschan, C. (1971). Cyclical analysis of time series: Selected procedures and computer programs. NBER Technical Paper n. 20.
8. Burman, J. P. (1980). Seasonal adjustment by signal extraction. *Journal of the Royal Statistical Society Series A, 143*, 321–337.
9. Burns, A. F., & Wesley, M. C. (1946). *Measuring business cycles*. New York: National Bureau of Economic Research.
10. Burridge, P., & Wallis, K. F. (1984). Unobserved components models for seasonal adjustment filters. *Journal of Business and Economic Statistics, 2*(4), 350–359.
11. Butterworth, S. (1930). On the theory of filter amplifiers. *Experimental Wireless and the Radio Engineer, 7*, 536–541.
12. Chen, C., & Liu, L. M. (1993). Joint estimation of model parameters and outlier effects in time series. *Journal of the American Statistical Association, 88*(241), 284–297.
13. Cleveland, R. B., Cleveland, W. S., McRae, J. E., & Terpenning, I. (1990). STL: A seasonal trend decomposition procedure based on LOESS. *Journal of Official Statistics, 6*(1), 3–33.
14. Cleveland, W. P., & Tiao, G. C. (1976). Decomposition of seasonal time series: A model for the census X-11 program. *Journal of the American Statistical Association, 71*, 581–587.
15. Dagum, C., & Dagum, E. B. (1988). Trend. In S. Kotz & N. L. Johnson (Eds.), *Encyclopedia of statistical sciences* (Vol. 9, pp. 321–324). New York: Wiley.
16. Dagum, E. B. (1978). Modelling, forecasting and seasonally adjusting economic time series with the X11ARIMA method. *The Statistician, 27*(3), 203–216.
17. Dagum, E. B. (1980). The X11ARIMA seasonal adjustment method. Statistics Canada, Ottawa, Canada. Catalogue No. 12-564.
18. Dagum, E. B. (1988). *The X11ARIMA/88 seasonal adjustment method-foundations and user's manual*. Ottawa, Canada: Time Series Research and Analysis Centre, Statistics Canada.
19. Dagum, E. B., Chhab, N., & Chiu, K. (1990). Derivation and properties of the X11ARIMA and census X11 linear filters. *Journal of Official Statistics, 12*(4), 329–347.

20. Dagum, E.B., Quenneville, B., & Sutradhar, B. (1992). Trading-day variations multiple regression models with random parameters. *International Statistical Review, 60*(1), 57–73.
21. Findley, D. F., Monsell, B. C., Bell, W. R., Otto, M. C., & Chen, B. C. (1998). New capabilities and methods of the X12ARIMA seasonal adjustment program. *Journal of Business and Economic Statistics, 16*(2), 127–152.
22. Gomez, V., & Maravall, A. (1996). *Program TRAMO and SEATS*. Madrid: Bank of Spain.
23. Harvey, A. C. (1981). *Time series models*. Oxford, UK: Phillip Allan Publishers Limited.
24. Harvey, A. C. (1985). Trends and cycles in macroeconomic time series. *Journal of the Business and Economic Statistics, 3*, 216–227.
25. Harvey, A. C., & Trimbur, T. M. (2003). General model-based filters for extracting cycles and trends in economic time series. *The Review of Economics and Statistics, 85*(2), 244–255.
26. Henderson, R. (1916). Note on graduation by adjusted average. *Transaction of Actuarial Society of America, 17*, 43–48.
27. Hillmer, S. C., & Tiao, G. C. (1982). An ARIMA-model-based approach to seasonal adjustment. *Journal of the American Statistical Association, 77*, 63–70.
28. Hodrick, R. J., & Prescott, E. C. (1997). Postwar U.S. business cycles: An empirical investigation. *Journal of Money, Credit and Banking, 29*(1), 1–16.
29. Kaiser, R., & Maravall, A. (2002). A complete model-based interpretation of the Hodrick-Prescott filter: Spuriousness reconsidered. Banco de Espana Working Papers 0208, Banco de Espana.
30. Kaiser, R., & Maravall, A. (2005). Combining filter design with model-based filtering (with an application to business-cycle estimation). *International Journal of Forecasting, 21*(4), 691–710.
31. Kitagawa, G., & Gersch, W. (1984). A smoothness priors state-space modelling of time series with trend and seasonality. *Journal of the American Statistical Association, 79*, 378–389.
32. Koopman, S. J., Harvey, A. C., Doornik, J. A., & Shephard, N. (1998). *Structural time series analysis STAMP (5.0)*. London: Thomson Business Press.
33. Macaulay, F. R. (1931). *The smoothing of time series*. New York: National Bureau of Economic Research.
34. Malthus, T. R. (1798). *An essay on the principle of population*. Oxford World's Classics reprint: xxix Chronology.
35. Maravall, A. (1993). Stochastic and linear trend models and estimators. *Journal of Econometrics, 56*, 5–37.
36. Moore, G. H. (1961). *Business cycle indicators*. Princeton, NJ: Princeton University Press.
37. Persons, W. M. (1919). Indices of business conditions. *Review of Economic Statistics, 1*, 5–107.
38. Schumpeter, J. A. (1954). *History of economic analysis*. New York: Oxford University Press Inc.
39. Shiskin, J., Young, A. H., & Musgrave, J. C. (1967). The X-11 variant of the Census Method II seasonal adjustment program. Technical Paper 15 (revised). US Department of Commerce, Bureau of the Census, Washington, DC.
40. Stock, J. H., & Watson, M. W. (2003). *Has the business cycle changed and why?*. NBER macroeconomics annual 2002 (Vol. 17, pp. 159–230). Cambridge: MIT Press.
41. Wold, H. O. (1938). *A study in the analysis of stationary time series*. Uppsala, Sweden: Almquist and Wiksell (Second Edition 1954).
42. Young, A. H. (1965). Estimating trading-day variations in monthly economic time series. US Bureau of the Census, Technical Report No. 12.
43. Young, A. H. (1968). A linear approximation to the Census and BLS seasonal adjustment methods. *Journal of the American Statistical Association, 63*, 445–457.

Part I
Seasonal Adjustment Methods

Chapter 3
Seasonal Adjustment: Meaning, Purpose, and Methods

Abstract This chapter deals with the causes and characteristics of seasonality. For gradual changes in seasonality whether of a stochastic or a deterministic type, various models have been proposed that can be grouped into two broad categories: (1) models that assume that the generating process of seasonality varies only in amplitude and (2) models that assume that the generating process varies in both amplitude and phase. The basic assumptions of both groups of models are studied with particular reference to the two seasonal adjustment methods officially adopted by statistical agencies, the X12ARIMA and TRAMO-SEATS. The economic significance of seasonality and the need for seasonal adjustment are also discussed. Since seasonality ultimately results mainly from noneconomic forces (climatic and institutional factors), external to the economic system, its impact on the economy as a whole cannot be modified in a short period of time. Therefore, it is of interest for policy making and decision taking to have the seasonal variations removed from the original series. The main reason for seasonal adjustment is the need of standardizing socioeconomic series because seasonality affects them with different timing and intensity.

3.1 Seasonality, Its Causes and Characteristics

A great deal of data in business, economics, and natural sciences occurs in the form of time series where observations are dependent and where the nature of this dependence is of interest in itself. The time series is generally compiled for consecutive and equal periods, such as weeks, months, quarters, and years. From a statistical point of view, a time series is a sample realization of a stochastic process, i.e., a process controlled by probability laws. The observations made as the process continues indicate the way it evolves.

In the analysis of economic time series, Persons [44] was the first to distinguish four types of evolution, namely: (a) the trend, (b) the cycle, (c) the seasonal variations, and (d) the irregular fluctuations.

Among all these components, the influence of the seasonal fluctuations in the human activity has been felt from earlier times. The organization of society, the means of production and communication, the habits of consumption, and other

© Springer International Publishing Switzerland 2016

E. Bee Dagum, S. Bianconcini, *Seasonal Adjustment Methods and Real Time Trend-Cycle Estimation*, Statistics for Social and Behavioral Sciences, DOI 10.1007/978-3-319-31822-6_3 social and religious events have been strongly conditioned by both climatic and conventional seasons. The seasonal variations in agriculture, the high pre-Easter and pre-Christmas retail sales, and the low level of winter construction are all general knowledge. The main causes of seasonality, the climatic and institutional factors are mainly exogenous to the economic system and cannot be controlled or modified by the decision makers in the short run. The impact of seasonality in the economic activity, however, is usually not independent on the stages of the business cycle. It is well-known, for example, that the seasonal unemployment among adult males is much larger during periods of recession than during periods of prosperity.

Another main feature of seasonality is that the phenomenon repeats with certain regularity every year but it may evolve. The latter is mainly due to systematic intervention of the government and to technological and institutional changes as well; and it is more the general case than the exception for economic time series. Therefore, the assumption of stable seasonality, i.e., of seasonality being representable by a strictly periodic function, can be used as a good approximation for few series only. In effect, even in the extreme cases of those activities where seasonality is mainly caused by climatic conditions, e.g., agriculture, fishing, forestry, the seasonal variations change, for weather itself measured by the temperature, and quantity of precipitation changes. There are many other reasons that can produce temporal shifts in the seasonal effects. A decline in the participation of the primary sector in the Gross National Product will modify seasonal patterns in the economy as a whole, as will a change in the geographical distribution of industry in a country extending over several climatic zones. Changes in technology alter the importance of climatic factors. Customs and habits change with a different distribution of income and, thus, modify the demand for certain goods and its corresponding seasonal pattern.

Once the assumption of stable seasonality is abandoned, new assumptions must be made regarding the nature of its evolution; if seasonality changes, is it slowly or rapidly? Is it gradually or abruptly? Is it in a deterministic or a stochastic manner? Today, the most widely accepted hypothesis is that seasonality moves gradually, slowly, and in a stochastic manner. For gradual changes in seasonality whether of a stochastic or a deterministic type, various models have been studied. These models can be grouped into two broad categories:

1. models that assume the generating process of seasonality varies in amplitude only and
2. models that assume the generating process varies in both amplitude and phase.

Kuznets [38] and Wald [49] were the first to propose an estimation procedure that assumed seasonality changes in amplitude while the pattern (i.e., the proportionality relationship between the seasonal effects of each month) remained constant over time.

On the other hand, the majority of the methods officially adopted by statistical agencies belongs to the second group. Seasonality is assumed to change either in a stochastic or in a deterministic manner. The stochastic approach has been analyzed among many others by Hannan [23], Hannan et al. [24], Brewer et al. [5],

Cleveland [9], and Cleveland and Tiao [10]. Pierce [45] proposed a mixed model where seasonality can be in part deterministic and in part stochastic.

A third characteristic of seasonality is that the phenomenon can be distinguished from other movements (trend, cycle, and irregulars) that influence an economic time series and can be measured. The seasonal variations are distinguished from trend by their oscillating character, from the cycle by being confined within the limits of an annual period, and from the irregulars, by the fact of being systematic. For example, the causes of seasonality in labor force series are found in the main factors that influence their demand and supply. The seasonal variations in employment and unemployment of adult males are strongly conditioned by the stages of the business cycle and the weather. There is practically no seasonality in the supply of adult males labor but there is a large amount of seasonal variation in the demand side. This group is mainly absorbed by the primary sector (construction, agriculture, mining, fishing, and forestry) where seasonality is mostly climatic with very large oscillation, and by the industrial sector, where seasonality is mostly induced by the seasonality in the primary sector but it is also strongly affected by the stages of the cycle. On the other hand, seasonality for females and young males originates from the side of the demand as well as of the supply. In fact, females and young males are mainly employed by the tertiary sector where seasonality tends to be originated by institutional and religious events (Christmas, Easter, Federal Taxes deadlines). Similarly, from the viewpoint of the supply, seasonal variations occur because this group tends to move in and out of the labor force in accordance with the school year.

3.2   The Economic Significance of Seasonality and the Need for Seasonally Adjusted Series

The seasonal variations affect the economic sectors in different degrees. Nor is it only the intensity of the primary effects of the seasons that differs for the various economic activities, producing seasonal swings of varying size or amplitude, the timing of these effects is also different, creating different patterns as the season begins in 1 month or another. Thus, for example, in winter occur both the seasonal low in construction and the seasonal peak in retail trade. The seasonal amplitudes for both sectors also differ significantly, being much larger for construction than for retail trade.

Similar to the business cycle, the presence of seasonality in the economy introduces a disequilibrium in the allocation of the resources, implying an extra cost. Because of the seasonal variations, the economy has a supply of labor, equipment, and raw materials in excess to what it would be required if the activity proceeds at an even rate throughout the year. Since these surpluses may arise also from the different impacts of the other components (trend, cycle, and irregulars) on the economic processes, it becomes very difficult to estimate the actual cost imposed by seasonality. Some attempts of measurement have been made by Kuznets [38], Baron

[3], Dawson et al. [16], and Judek [33] among many others. For instance, Kuznets [38] suggested two measures of the degree of overcapacity in industry attributable to seasonal variations: one gives the total difference in activity between each of the months and the months of peak output; the other, the amount by which the months below 100 depart from 100. The basic idea is that the seasonal peak months measure the normal plus the latent capacity of the industry, therefore, an output below the seasonal peak indicates an inadequate utilization of the maximum possible capacity. On the other hand, an output below a seasonal index of 100 indicates an inadequate utilization of the normal capacity of the industry. The same idea is applicable to the surplus of labor. If the seasonal peak in employment indicates the number of workers attached to a given industry, then the average monthly departure from the peak measures the average monthly seasonal unemployment and hence the surplus of labor that is in the industry because of seasonal swings. If, however, only the annual average is considered the correct estimate of labor attached to the industry, then only departures below 100 should be considered as measuring seasonal unemployment and excess of labor supply.

Because seasonality ultimately results mainly from noneconomic forces (climatic and institutional factors), external to the economic system, its impact on the economy as a whole cannot be modified in a short period of time. Therefore, it is of interest to decision makers to have the seasonal variations removed from the original series to obtain a seasonally adjusted series. Seasonal adjustment means the removal of seasonal variations in the original series jointly with trading day variations and moving holiday effects. The main reason for seasonal adjustment is the need of standardizing socioeconomic series because seasonality affects them with different timing and intensity. In this manner, the variations of a seasonally adjusted series are due to variations only in the trend, the cycle, and the irregulars. It should be noticed, however, that the removal of seasonal variations from a time series does not indicate how the series would have evolved as there had been no seasonal variations, rather it shows more clearly the trend-cycle abstracting from seasonality.

The information given by seasonally adjusted series plays a very important role in the analysis of current economic conditions, particularly, in determining the stage of the cycle at which the economy stands. Such knowledge is useful in forecasting subsequent cyclical movements and provides the basis for decision making to control the level of the economic activities. It is particularly important around turning points since, e.g., failure to recognize the downturn in the business cycle may lead to the adoption of policies to curb expansion when, in fact, a recession is under way. Analysts who wish to get a picture of the economic situation not distorted by exogenous variables related to the seasonal variations, may make comparisons with the same month of the year before. However, apart from other considerations, the basic problem with same-month year-ago comparisons is that they show only what has happened a year after, not what was happening during the year not what is happening currently. To evaluate the current state of the economy, the analyst should be able to measure cyclical changes for each month over less than 1 year span, e.g., to compare May with April (1-month span) or May with February (3-month span).

Decision making based on the raw data can lead to wrong policies, particularly if the series is strongly affected by seasonal variations. The average absolute month percentage change in the seasonal variation can be much greater than the corresponding changes in the irregular or trend-cycle. Results of several studies of selected economic US indicators show that the average absolute month-to-month percentage changes in the seasonal component run between 3 and 7 times the average absolute percentage changes in the trend-cycle or in the irregulars over the same time spans (see Shiskin [46]). The measurement of seasonality is also very useful for short-term decision making. The knowledge of the seasonal pattern of the economic activities facilitates a better planning of the economic resources during the periods of peak loads and of inactivity. This knowledge can also be applied for a better integration of economic activities characterized by opposite seasonal patterns. Resources which formerly served only one purpose can serve additional purposes and thus reduce the burden imposed by seasonality to the whole economy.

3.3   Basic Assumptions of Main Seasonal Adjustment Methods

The majority of the seasonal adjustment methods developed until now is based on univariate time series models, where the estimation of seasonality is made in a simple and mechanical manner and not based on a causal explanation of the phenomenon under study. Very few attempts have been made to follow the latter approach and none of them reached further than the experimental stage. Among others, Mendershausen [43] tried to regress the seasonal effects for each month on a set of exogenous variables (meteorological and social variables) in order to build an explanatory model for seasonality but his empirical results were not very fruitful.

On the other hand, univariate time series methods of seasonal adjustment try to estimate the generating mechanism of the observations under the simple assumption that the series is composed of a systematic part which is a well-determined function of time and, a random part which obeys a probability law. The feasibility of this decomposition was proved in a well-known theorem due to Herman Wold [52]. Wold showed that any second order stationary process $\{Y_t\}$ can be uniquely represented as the sum of two mutually uncorrelated processes

$$Y_t = Z_t + V_t,$$

where $\{Z_t\}$ is a convergent infinite moving average process, and $\{V_t\}$ a deterministic process. The decomposition is linear and is determined entirely by the second moments of the process.

In a broad sense, seasonal methods can be classified as *deterministic* or *stochastic* depending on the assumptions made concerning how seasonality evolves through time. Deterministic methods assume that the seasonal pattern can be predicted with no error or with variance of the prediction error null. On the contrary, stochastic methods assume that seasonality can be represented by a stochastic process, a process governed by a probability law and, consequently, the variance of the prediction error is not null. The best known seasonal adjustment methods belong to the following types:

(a) regression methods which assume global or local simple functions of time,
(b) stochastic-model-based methods which assume simple autoregressive integrated moving average (ARIMA) models, and
(c) moving average methods which are based on linear filtering and hence do not have explicit parametric models.

Only methods (a), which assume global simple functions for each component, are deterministic; the others are considered stochastic. Moving average or linear filtering methods are those adopted mostly by statistical agencies to produce officially seasonally adjusted series.

3.3.1 Regression Methods

The use of regression methods for the decomposition of a time series is old. In the late 1930s, Fisher [20] and Mendershausen [43] proposed to fit polynomials by the least squares method to estimate the seasonal effects. To deal with moving seasonality, Cowden [11] suggested to fit polynomials to each month seasonal obtained from the ratios (differences) of the original series and a centered twelve term moving average. Jones [30] and Wald [49] considered the problem of fitting simultaneously a polynomial of a relatively low degree to estimate the trend plus twelve constants (one for each month) representing stable seasonality.

In the 1960s, the use of multiple regression techniques for the seasonal adjustment of economic time series was strongly advocated. The main reasons for this were the widespread interest in the use of monthly and quarterly series for econometric model building and the fast development of electronic computers. In effect, the former posed the problem of whether to use as input seasonally adjusted data directly, or to use first, raw data, and then seasonally adjusted. These two alternatives do not give the same results unless the seasonal adjustment method is strictly linear (conditioned fulfilled by the regression methods). The electronic computers reduced significantly the heavy burden imposed by the computations in the least squares method despite the existence of several simplifying procedures such as the orthogonal polynomials and the Buys-Ballot table. Important contributions were made by Hannan [22], Lovell [40], Ladd [39], Jorgenson [31], Henshaw [28], Stephenson and Farr [48], and Wallis [50]. Their contributions facilitated the application of least squares methods for the seasonal adjustment of economic time series.

The regression methods assume that the systematic part of a time series can be approximated closely by simple functions of time over the entire span of the series. In general, two types of functions of time are considered. One is a polynomial of fairly low degree that fulfills the assumption that the economic phenomenon moves slowly, smoothly, and progressively through time (the trend). The other type of function is a linear combination of sines and cosines of different frequencies representing oscillations, strictly periodic or not, that affect also the total variation of the series (the cycle and seasonality).

For certain regression functions such as fixed periodic functions, polynomial time trends, and interaction of the two types, ordinary least squares estimates are asymptotically efficient if the random part is second order stationary, i.e., the mean and variance are constant and the covariance is function only of the time lag. Therefore, by taking differences to ensure a stationary process, ordinary least squares are generally adequate. If the relationship among the components is multiplicative, the standard procedure is to take logarithms and then, differences, in order to transform the generating mechanism of the series into an additive form with a stationary random term.

A large number of model specifications have been considered for monthly series. Young [53] analyzed and compared the seasonal adjustment obtained by regression methods with the results given by the Census II-X11 variant which belongs to the category of linear smoothing techniques. The models analyzed have different degrees of flexibility concerning the pattern of the seasonals and trend-cycle and the relationship between the components is assumed to be multiplicative.

Model I

$$\log y_t = \log S_t + \log C_t + \log I_t \tag{3.1}$$

$$\log S_t = \sum_{k=1}^{6} [\alpha_{1,k} \sin \lambda_k + \alpha_{2,k} \cos \lambda_k t]$$

$$+ \sum_{k=1}^{2} [\beta_{1,k} t^k \sin \lambda_k t + \beta_{2,k} t^k \cos \lambda_k t], \tag{3.2}$$

where $\lambda_k = \frac{2\pi k}{12}$,

$$\log C_t = \gamma_0 + \sum_{k=1}^{5} \gamma_k t^k + \delta_1 \sin \omega t + \delta_2 \cos \omega t, \tag{3.3}$$

where $\omega = \frac{2\pi}{72}$.

This model contains a deterministic stable seasonality represented by the sines and cosines of periodicity 12, 6, 4, 3, 2.4, and 2 plus a deterministic moving seasonality that is assumed to follow a second degree polynomial through time. The trend-cycle expression contains a fifth degree polynomial and sines and cosines with a periodicity of 72 months that take into account a 5 year cycle.

Model II

$$\log y_t = \log S_t + \log C_t + \log I_t \tag{3.4}$$

$$\log S_t = \sum_{k=1}^{12} \alpha_k d_k \tag{3.5}$$

$$\log C_t = \sum_{k=1}^{10} \gamma_k t^k. \tag{3.6}$$

Instead of using sines and cosines, this model contains a deterministic stable seasonality represented by 12 dummy variables d_k where $d_k = 1$ if t is the k-th month of the year and equals 0 otherwise. The trend-cycle curve is represented by a 10-th degree polynomial.

Model III

$$\log y_t = \log S_t + \log C_t + \log I_t \tag{3.7}$$

$$\log S_t = \sum_{k=1}^{12} \alpha_k d_k + \sum_{k=1}^{12} \beta_{0,k} d_k + \sum_{k=1}^{12} \beta_{1,k} d_k t \tag{3.8}$$

$$\log C_t = \sum_{k=1}^{5} \gamma_k t^k. \tag{3.9}$$

The seasonal model corresponds to a deterministic stable seasonality plus a deterministic linearly moving seasonality. The trend-cycle is represented by a fifth degree polynomial.

Young [53] concluded that for central observations some of the model specifications for the trend-cycle and the seasonals approached closely the weights applied by the Census II - X11 program, but not for the observations at both ends. Moreover, he found that the amount of revision between the current and the historical ("final") seasonal factors was significantly larger with the regression models than with Census II - X11, although the historical factors were obtained from the regression models. In fact, the results showed that the current factors of the Census II - X11 were better predictors of the historical regression factors than the current regression factors.

In 1972, Stephenson and Farr [48] developed a flexible regression model for changing trend and seasonality, but still yielded results that were not, in overall, superior to those of the Census II - X11 program.

Lovell [41] suggested a simple model of the following form:

$$y_t = \sum_{k=1}^{12} \alpha_{1,k} d_k + \sum_{k=1}^{12} \alpha_{2,k} [d_k C^*] + I_t \tag{3.10}$$

subject to

$$\sum_{k=1}^{12} \alpha_{2,k} = 0$$

where C^* is a 12 month lagging average of the series (a trend level estimate) and the d_k's are monthly dummies. This model assumes a seasonal pattern that is stable plus a slowly moving trend conditioned to seasonality for the whole span of the series. The estimate of the trend C^* is the same for all the series. The seasonally adjusted data are calculated by adding the mean of the original series to the residuals I_t.

This model which is extremely simple as compared to other models specifications for regression methods did not give better results than those obtained by moving averages procedures. As suggested by Early [17], the revisions in the seasonally adjusted values were high and not necessarily monotonic for earlier years.

To overcome the limitation of using global deterministic representations for the trend, cycle, and seasonality, regression methods were extended to incorporate stochastic representations by means of local polynomials (spline functions) for successive short segments of the series and introducing changing seasonal amplitudes. A major breakthrough in this direction was made by Akaike [1] who introduced prior constraints to the degree of smoothness of the various components and solved the problem with the introduction of a Bayesian model. Another important contribution is the regression method with locally deterministic models estimated by the LOESS (locally weighted regression) smoother developed by Cleveland et al. [8].

The simplest and often studied seasonal adjustment regression method assumes that the generating process of seasonality can be represented by strictly periodic functions of time of annual periodicity. The problem is to estimate s seasonal coefficients (s being the seasonal periodicity, e.g., 12 for monthly series, 4 for quarterly series) subject to the seasonal constraint that they sum to zero. This regression seasonal model can be written as

$$y_t = S_t + I_t, \qquad t = 1, \ldots, n$$

$$S_t = \sum_{j=1}^{s} \gamma_j d_{jt} \quad \text{subject to} \quad \sum_{j=1}^{s} \gamma_j = 0, \qquad (3.11)$$

where the d_j's are dummy variables taking a value of unity in season j and a value of zero otherwise; $\{I_t\} \sim WN(0, \sigma_e^2)$ is a white noise process with zero mean, constant variance, and non-autocorrelated. The seasonal parameters γ_j's can be interpreted as the expectation of Y_t in each season, and represent the seasonal effects of the series under question. Since the parameters are constant, the seasonal pattern repeats exactly over the years, that is,

$$S_t = S_{t-s}. \qquad (3.12)$$

Equation (3.11), often included in econometric models, assumes that the effects of the seasonal variations are deterministic because they can be predicted with no error. However, for most economic and social time series, seasonality is not deterministic but changes gradually in a stochastic manner. One representation of stochastic seasonality is to assume in Eq. (3.11) when the γ_j's are random variables instead of constant. Hence, the relationship given in Eq. (3.12) becomes

$$S_t = S_{t-s} + \omega_t, \qquad \forall t > s, \tag{3.13}$$

where $\{\omega_t\} \sim \mathrm{WN}(0, \sigma_\omega^2)$ and $E(\omega_t I_t) = 0, \forall t$. The stochastic seasonal balance constraint is here given by $\sum_{j=0}^{s-1} S_{t-j} = \omega_t$, with expected value equal to zero.

Equation (3.13) assumes seasonality to be generated by a nonstationary stochastic process. In fact, it is a random walk process and can be made stationary by applying seasonal differences $(1 - B^s)$, where B denotes the backward shift operator. In reduced form, the corresponding regression model is a linear combination of white noise processes of maximum lag s. Since $(1 - B^s) = (1 - B)(1 + B + \cdots + B^{s-1})$, model-based seasonal adjustment methods attribute only $(1 + B + \cdots + B^{s-1}) = \sum_{j=0}^{s-1} B^j$ to the seasonal component, leaving $(1 - B)$ as part of a nonstationary trend. Thus, the corresponding seasonal stochastic model can be written as

$$(1 + B + \cdots + B^{s-1})S_t = \omega_t. \tag{3.14}$$

Equation (3.14) represents a seasonality that evolves in an erratic manner and it was included in seasonal adjustment methods based on structural model decomposition (see, e.g., Harvey [25] and Kitagawa and Gersch [36]). This seasonal behavior is seldom found in observed time series where seasonality evolves smoothly through time. To represent this latter, Hillmer and Tiao [29] introduced a major modification

$$(1 + B + \cdots + B^{s-1})S_t = \eta_s(B)b_t, \qquad b_t \sim \mathrm{WN}(0, \sigma_b^2) \tag{3.15}$$

where the left-hand side of Eq. (3.15) follows an invertible moving average process of maximum order $s - 1$, denoted by $\mathrm{MA}(s - 1)$. Equation (3.15) is discussed extensively in Bell and Hillmer [4], and can be generalized easily by replacing the MA part with stationary invertible autoregressive moving average (ARMA) processes of the Box and Jenkins type. Depending on the order of the ARMA process, a large variety of evolving stochastic seasonal variations can be modeled.

Other seasonal models used in regression methods, where the components are assumed to be local polynomials of time, are based on a smoothness criterion. For example, given $Y_t = S_t + I_t, t = 1, \ldots, n$, restrictions are imposed on S_t, such that $\sum_{t=1}^{n} (S_t - S_{t-s})^2$ and $\sum_t [(1 + B + \cdots + B^{s-1})S_t]^2$ have to be small. The solution is given by minimizing a weighted linear combination of both criteria as follows:

$$\sum_{t=1}^{n} (y_t - S_t)^2 + \alpha \sum_{t=1}^{n} (S_t - S_{t-s})^2 + \beta \sum_{t=1}^{n} [(1 + B + \cdots + B^{s-1})S_t]^2 \tag{3.16}$$

where the ill-posed problem of choosing α and β parameters was solved by Akaike [1] using a Bayesian model. Mixed seasonal models with deterministic and stochastic effects are given by Pierce [45].

Although the performance of regression methods might be improved by more closely tailoring the model specifications to the particular series, it is difficult to derive regression estimates which, from the standpoint of revisions, are better than moving averages estimates. The specifications made concerning the trend-cycle and seasonality in the regression models apply to the whole span of the series. Because of this, the procedure lacks of flexibility and the most recent estimated values are always influenced by distant observations. Regression methods assume a deterministic behavior of the components, in contrast to moving averages techniques that assume a stochastic pattern. The latter, therefore, tend to follow better cyclical movements and changing seasonality.

3.3.2 Stochastic Model-Based Methods

Stochastic model-based methods were mainly developed during the 1980s following two different approaches. One, originally known as seasonal adjustment by SIGEX (Signal Extraction), was developed by Burman [6] and as ARIMA model-based seasonal adjustment by Hillmer and Tiao [29], the other is referred to as structural model decomposition method (see, e.g., Harvey [25] and Kitagawa and Gersch [36]). The main difference between these two approaches is that in the latter simple ARIMA models are directly specified for each unobserved component, whereas in the former an overall ARIMA model is obtained from observable data and, by imposing certain restrictions, models for each component are derived. Since the components are unknown, to obtain a unique decomposition Hillmer and Tiao [29] proposed a canonical decomposition which has the properties of maximizing the variance of the irregulars and minimizing the variance of the stochastic seasonal models.

ARIMA models identification and estimation are sensitive to outliers or extreme values and cannot deal with deterministic components, such as trading days and moving holidays. Therefore, further developments were made by combining dummy variables regression models with ARIMA models to deal with these cases. In this regard, Gomez and Maravall [21] developed at the Bank of Spain a seasonal adjustment software called TRAMO-SEATS which is currently applied mainly by European statistical agencies. TRAMO stands for "Time Series Regression with ARIMA noise, Missing observations and Outliers" and SEATS for "Signal Extraction in ARIMA Time Series." Firstly, TRAMO estimates via regression dummy variables and direct ARIMA modeling, the deterministic components which are after removed from the input data. In a second round, SEATS estimates the seasonal and trend-cycle components from an ARIMA model fitted to the data where the deterministic components are removed. SEATS uses the filters derived from the ARIMA model that describes the behavior of the linearized time series.

It should be mentioned that Eurostat, with the collaboration of the Bank of Belgium, the US Bureau of the Census, the Bank of Spain, and the European Central Bank, has developed an interface of TRAMO-SEATS and X12ARIMA called Demetra+, and more recently an update version called JDemetra+ . At the Bank of Spain and Eurostat web site it is also possible to find a considerable number of papers relevant to TRAMO-SEATS as well as in the European Statistical System (ESS) Guidelines.

The structural model decomposition method starts directly with an observation equation, also called *measurement equation*, that relates the observed time series to the unobserved components. Simple ARIMA or stochastic trigonometric models are assumed for each component. The structural model is cast in a state space form and estimated with Kalman filtering and smoothing. Koopman et al. [37] developed STAMP, that stands for Structural Time series Analyzer, Modeler and Predictor, and includes several types of models for each component. STAMP is set up in an easy-to-use form which enables the user to concentrate on model selection and interpretation. STAMP 8 is an integrated part of the OxMetrics modular software system for time series data analysis, forecasting, financial econometric modeling, and the statistical analysis of cross-section and panel data. STAMP is not used officially by statistical agencies, but mainly by econometricians and academics.

3.3.3 Linear Smoothing Methods

Linear smoothing filters or moving averages for seasonal adjustment were already known in the early 1920s but seldom used [18, 35]. One of the main reasons for this was the fact that the best known seasonal annihilator, the centered 12 months moving average, was found to be a poor trend-cycle estimator. It cannot follow closely peaks and troughs of short-term business cycles (of 5 years of periodicity or less). Moreover, unless the erratic fluctuations are small, it has not enough terms to smooth the data successfully. Another reason was that moving averages are sensitive to outliers and, therefore, an "a priori" treatment of extreme values is required. This sort of limitation lead King [35] and other researchers to use moving medians instead of moving arithmetic means for the estimation of seasonality.

However, by the end of the 1920s, the development of new smoothing filters and different techniques of application gave a big push to the use of this procedure for the seasonal adjustment of economic time series. Joy and Thomas [32] described a method based on moving averages and applied to one hundred series of the Federal Reserve Board. According to the authors, the method was able to handle "long-time changes in seasonal fluctuations" which appear to have occurred in a number of important industries .

By the same time, Frederick Macaulay [42] developed another method also based on linear smoothing filters. His book, *The Smoothing of Time Series*, published in 1931, became a classic on the subject. Gradually, government agencies and statistical bureaus started to apply smoothing procedures for the seasonal adjustment of their series. The procedure, however, was costly, time consuming, and mainly subjective, because the adjustments were mostly made by hand.

The basic properties of linear smoothing filters or moving averages were extensively discussed in the following three classical books: (1) Whittaker and Robinson [51], (2) Macauley [42], and (3) Kendall and Stuart [34].

Only a summary of the basic principles of smoothing techniques will be presented here, and given the complexity of the subject, we shall make it as simple as possible.

The majority of the seasonal adjustment methods officially adopted by statistical bureaus makes the assumption that although the signal of a time series is a smooth function of time, it cannot be approximated well by simple mathematical functions over the entire range. Therefore, they use the statistical technique of smoothing. A smooth curve is one which does not change its slope in a sudden or erratic manner. The most commonly used measure of smoothness is based on the smallness of the sum of squares of the third differences of successive points on the curve [42].

Given a time series, $y_t, t = 1, \ldots, n$, for any t far removed from both ends, say $m < t < n - m$, the seasonally adjusted value y_t^a is obtained by application of a symmetric moving average $W(B)$

$$y_t^a = W(B)y_t = \sum_{j=-m}^{m} w_j y_{t-j}, \tag{3.17}$$

where the weights w_j are symmetric, that is, $w_{m,j} = w_{m,-j}$, and the length of the average is $2m + 1$.

For current and recent data $(n - m < t \leq n)$, a symmetric linear filter cannot be applied, and therefore truncated asymmetric filters are used. For example, for the last available observation y_n, the seasonally adjusted value is given by

$$y_n^a = W_0(B)y_n = \sum_{j=0}^{m} w_{0,j} y_{n-j}. \tag{3.18}$$

The asymmetric filters are time-varying in the sense that different filters are applied for the m first and last observations. The end estimates are revised as new observations are added because of the new innovations and the differences between the symmetric and asymmetric filters. The estimates obtained with symmetric filters are often called *final*.

The development of electronic computers contributed to major improvements in seasonal adjustment based on moving averages and facilitated their massive application. In 1954, Julius Shiskin [47] of the US Bureau of Census developed a software called Method I, based mainly on the works of Macauley [42], already being used by the US Federal Reserve Board. Census Method I was followed by Census Method II and eleven more experimental versions (X1 , X2,. . ., X11). The best known and widely applied was the Census II - X11 variant developed by Shiskin et al. [47], but produced poor seasonally adjusted data at the end of the series which is of crucial importance to assess the direction of the short-term trend and the identification of turning points in the economy. Estela Bee Dagum [12] developed the X11ARIMA to correct for this serious limitation.

The X11ARIMA method consists of:

(i) modeling the original series with an ARIMA model of the Box and Jenkins type,
(ii) extending the original series 1–3 years with forecasts from the ARIMA model that fits and extrapolate well according to well-defined acceptance criteria, and
(iii) estimating each component with moving averages that are symmetric for middle observations and asymmetric for both end years. The latter are obtained via de convolution of Census II - X11 variant and the ARIMA model extrapolations.

For flow series, deterministic components such as trading day variations and moving holiday effects are estimated with dummy variable regression models, and removed from the series, so that only the remainder is subject to steps (i)–(iii) above. The X11ARIMA software was developed by Dagum in 1980 [13] and was extended again in 1988 [14]. In 1998, Findley et al. [19] developed X12ARIMA that offers a regARIMA option to estimate deterministic components such as trading day variations and moving holiday effects, simultaneously with the ARIMA model for extrapolation. It also includes new diagnostic tests and spectral techniques to assess the goodness of the results. The X12ARIMA method is today one of the most often applied by statistical agencies in the world. The US Bureau of Census continued research on the development of seasonal adjustment methods and recently produced a beta version called X13ARIMA-SEATS which enables the estimation of the seasonal component either via linear filters as those available in X12ARIMA or based on an ARIMA model decomposition.

All the moving average methods here discussed are nonlinear. Hence, the seasonally adjusted total of aggregated series is not equal to the algebraic sum of the seasonally adjusted series that enters into the aggregation. The main causes of nonlinearity are:

(i) a multiplicative decomposition model for the unobserved components,
(ii) the identification and replacement of extreme values,
(iii) the ARIMA extrapolations, and
(iv) automatic selection of moving average length for the estimation of the trend-cycle and seasonality.

The properties of the combined linear filters applied to estimate the various components were originally calculated by Young [53] for a standard option. Later, Dagum et al. [15] calculated and analyzed all possible filter combinations of Census II - X11 and X11ARIMA. Cleveland and Tiao [10] and Burridge and Wallis [7] found ARIMA models that approximated well some of the linear filters used for the trend-cycle and seasonal component of Census II - X11.

The set of weights of a smoothing linear filter can be obtained following two different techniques: (a) by fitting a polynomial and (b) by summation formulae.

The process of fitting a polynomial to generate the set of weights of a moving average requires that the span or length of the average as well as the degree of the polynomial be chosen in advance. For a given span, say $2m + 1$ and a polynomial of degree p not greater than $2m$, the coefficients are estimated by the method of least squares, and then the polynomial is used to determine the smoothed value in the middle of the span. The weights are function only of the span of the average $2m + 1$ and the degree of the polynomial, p, and do not depend on the values of the observations. The smoothed value is a weighted average of the observations, the weights being independent on which part of the series is taken. For a given p, the variance of the smoothed series decreases with increasing $2m + 1$ and, for a given $2m + 1$, the variance goes up with increasing p (see, e.g., [2]).

The set of weights for linear smoothing filters based on summation formulae was mainly developed by actuaries. The basic principle for the summation formulae is the combination of operations of differencing and summation in such a manner that when differences above a certain order is neglected, they will reproduce the functions operated on. The merit of this procedure is that the smoothed values thus obtained are functions of a large number of observed values whose errors to a considerable extent cancel out.

The smoothing linear filters developed by actuaries using summation formulae, have the good properties that when fitted to second or third degree parabolas, will fall exactly on those parabolas. If fitted to stocastic, non mathematical data, it will give smoother results than can be obtained from the weights which give the middle point of a second degree parabola fitted by least squares method. Recognition of the fact that smoothness of the resulting filtering depends directly on the smoothness of the weight diagram, led Robert Henderson [26] to develop a formula which makes the sum of squares of the third differences of the smoothed series a minimum for any number of terms. To fulfill this requirement is equivalent to minimize the sum of the squares of the third differences of the set of weights of the filter.

There is a relationship between smoothing by the mid ordinate of third degree parabolas fitted by least squares and smoothing by the mid ordinate of third degree parabolas from summation formulae. In the least squares, the assumption made is that all deviations between observed and fitted values are equally weighted, and thus the sum of squares of the deviations is made a minimum. In the summation formulae, the deviations between observed and fitted values are not equally weighted and if different weights are applied, the sum of squares of the deviations is made a minimum.

The Henderson's trend-cycle moving averages give the same results as if weighted least squares had been applied where the weights assigned to the deviations are those that give the smoothest possible diagram. The latter in the sense that the sum of squares of the third differences is made a minimum. In smoothing real data there is always a compromise between: (1) how good the fit should be (in the sense of minimizing the sum of squares of the deviations between observed and fitted values); and (2) how smooth the fitted curve should be. The lack of fit is measured by the sum of squares of the deviations between observed and smoothed values and the lack of smoothness by the sum of squares of the third differences of the smoothed curve.

This problem was suggested by Henderson and Sheppard [27], and first analyzed by Whittaker and Robinson [51], who thus put the theory of smoothing within the framework of the theory of probability. The degree of fitness was weighted by a constant k to be determined and the problem was to minimize k times the sum of squares of the deviations between the observed and fitted curve plus the sum of squares of the third differences of the fitted (smoothed) curve. That is to minimize:

$$k \sum_{t=1}^{n}(y_t - \hat{y}_t)^2 + \sum_{t=3}^{n}((1 - B)^3 \hat{y}_t)^2 \qquad (3.19)$$

The smoothed curve obtained with the Whittaker procedure is such that each of its values is equal to the corresponding unadjusted values plus $1/k$ times a sixth difference of the smoothed curve. The weights were calculated for the whole range of the data and not in a moving manner. The set of weights eliminates stable seasonal fluctuations only if k is very small. For $k = 0.0009$, more than 95 % of the seasonal effects corresponding to a 12 month sine curve are eliminated, but cycles of less than 36 month periodicity are very poorly fitted as shown by Macauley [42]. This problem can be expressed in a more adequate way by minimizing a loss function given by a convex combination between the lack of smoothness and the lack of fit. The principle should be applied in a moving manner in order to make it more flexible. Hence, we want to minimize

$$k \sum_{t=1}^{n}(y_t - \hat{y}_t)^2 + (1 - k) \sum_{t=3}^{n}((1 - B)^3 y_t)^2, \qquad 0 \le k \le 1. \qquad (3.20)$$

The larger is k, the more importance is given to closeness of fit as compared to smoothness; if $k = 1$, the problem reduces to fitting only and, if $k = 0$, the problem reduces to smoothing only.

References

1. Akaike, H. (1980). Seasonal adjustment by a Bayesian modelling. *Journal of Time Series Analysis, 1*, 1–13.
2. Anderson, T. W. (1971). *The statistical analysis of time series*. New York: Wiley.
3. Baron, R.V. (1973). Analysis of seasonality and trends. In *Statistical series, W.I. Methodology. Causes and effects of seasonality*. Technical Publication, No. 39. Israel: Central Bureau of Statistics.
4. Bell, W. H., & Hillmer, S. C. (1984). Issues involved with the seasonal adjustment of economic time series. *Journal of Business and Economic Statistics, 2*, 291–320.

5. Brewer, K. R.W., Hagan, P. J., & Perazzelli, P. (1975). Seasonal adjustment using Box-Jenkins models. In *Proceedings of International Institute of Statistics*, Warsaw. 40th Session (pp. 133–139). Contributed Papers.

6. Burman, J. P. (1980). Seasonal adjustment by signal extraction. *Journal of the Royal Statistical Society Series A, 143*, 321–337.

7. Burridge, P., & Wallis, K. F. (1984). Unobserved components models for seasonal adjustment filters. *Journal of Business and Economic Statistics, 2*(4), 350–359.

8. Cleveland, R. B., Cleveland, W. S., McRae, J. E., & Terpenning, I. (1990). STL: A seasonal trend decomposition procedure based on LOESS. *Journal of Official Statistics, 6*(1), 3–33.

9. Cleveland, W. P. (1972). *Analysis and Forecasting of Seasonal Time Series* (215 pp.). Unpublished doctoral thesis, Department of Statistics, University of Wisconsin.

10. Cleveland, W. P., & Tiao, G. C. (1976). Decomposition of seasonal time series: A model for the census X-11 program. *Journal of the American Statistical Association, 71*, 581–587.

11. Cowden, D. J. (1942). Moving seasonal indexes. *Journal of the American Statistical Association, 37*, 523–524.

12. Dagum, E. B. (1978). Modelling, forecasting and seasonally adjusting economic time series with the X11ARIMA method. *The Statistician, 27*(3), 203–216.

13. Dagum, E. B. (1980). *The X11ARIMA seasonal adjustment method*. Ottawa, ON, Canada: Time Series Research and Analysis Center, Statistics Canada.

14. Dagum, E. B. (1988). *The X11ARIMA/88 seasonal adjustment method-foundations and user's manual*. Ottawa, ON, Canada: Time Series Research and Analysis Centre, Statistics Canada.

15. Dagum, E. B., Chhab, N., & Chiu, K. (1990). Derivation and properties of the X11ARIMA and census X11 linear filters. *Journal of Official Statistics, 12*(4), 329–347.

16. Dawson, D. A., Denton, F. T., Feaver, C., & Robb, A. (1975). *Seasonal patterns in the Canadian Labour force* (108 pp.). Economic Council of Canada, Discussion Paper. No. 38.

17. Early, J. (1976). Comment. Brooking Papers on Economic Activity, 1, pp. 239–243.

18. Falkner, H. D. (1924). The measurement of seasonal variation. *Journal of the American Statistical Association, 19*, 167–179.

19. Findley, D. F., Monsell, B. C., Bell, W. R., Otto, M. C., & Chen, B. C. (1998). New capabilities and methods of the X12ARIMA seasonal adjustment program. *Journal of Business and Economic Statistics, 16*(2), 127–152.

20. Fisher, A. (1937). A brief note on seasonal variation. *The Journal of Accountancy, 64*, 174–199.

21. Gomez, V., & Maravall, A. (1996). *Program TRAMO and SEATS*. Madrid: Bank of Spain.

22. Hannan, E. J. (1960). The estimation of seasonal variations. *Australian Journal of Statistics, 2*, 1–15.

23. Hannan, E. J. (1964). The estimation of changing seasonal pattern. *Journal of the American Statistical Association, 59*, 1063–1077.

24. Hannan, E. J., Terrell, R. D., & Tuckwell, N. E. (1970). The seasonal adjustment of economic time series. *International Economic Review, 11*, 24–52.

25. Harvey, A. C. (1981). *Time series models*. Oxford, UK: Phillip Allan Publishers Limited.

26. Henderson, R. (1916). Note on graduation by adjusted average. *Transaction of Actuarial Society of America, 17*, 43–48.

27. Henderson, R., & Sheppard, H. N. (1919). *Graduation of mortality and other tables*. Actuarial Studies, No. 4. New York: Actuarial Society of America.

28. Henshaw, R. C. (1966). Application of the general linear model to seasonal adjustment of economic time series. *Econometrica, 34*, 381–395.

29. Hillmer, S. C., & Tiao, G. C. (1982). An ARIMA model-based approach to seasonal adjustment. *Journal of the American Statistical Association, 77*, 63–70.

30. Jones, H. L. (1943). Fitting polynomial trends to seasonal data by the method of least squares. *Journal of the American Statistical Association, 38*, 453–465.

31. Jorgenson, D. N. (1964). Minimum variance, linear unbiased, seasonal adjustment of economic time series. *Journal of the American Statistical Association, 59*, 402–421.

32. Joy, A., & Thomas, N. (1928). The use of moving averages in the measurement of seasonal variations. *Journal of the American Statistical Association, 23*, 242–252.

33. Judek, S. (1975). *Canada's seasonal variations in employment and unemployment, 1953–73* (118 pp.). Research Paper No. 75. Canada: University of Ottawa.
34. Kendall, M. G., & Stuart, A. (1966). *The advanced theory of statistics* (Vol. 3, 552 pp.). New York: Hafner Publishing Co.
35. King, W. F. (1924). An improved method for measuring the seasonal factor. *Journal of the American Statistical Association, 19*, 301–313.
36. Kitagawa, G., & Gersch, W. (1984). A smoothness priors state-space modelling of time series with trend and seasonality. *Journal of the American Statistical Association, 79*, 378–389.
37. Koopman, S. J., Harvey, A. C., Doornik, J. A., & Shephard, N. (1998). *Structural time series analysis STAMP (5.0)*. London: Thomson Business Press.
38. Kuznets, S. (1933). *Seasonal variations in industry and trade* (455 pp.). New York: National Bureau of Economic Research.
39. Ladd, G. W. (1964). Regression analysis of seasonal data. *Journal of the American Statistical Association, 59*, 402–421.
40. Lovell, M. C. (1963). Seasonal adjustment of economic time series and multiple regression analysis. *Journal of the American Statistical Association, 58*, 993–1010.
41. Lovell, M. C. (1976). Least squares seasonally adjusted unemployment data. Brooking Papers on Economic Activity, l, pp. 225–237.
42. Macaulay, F. R. (1931). *The smoothing of time series*. New York: National Bureau of Economic Research.
43. Mendershausen, H. (1939). Eliminating changing seasonals by multiple regression analysis. *The Review of Economic Studies, 21*, 71–77.
44. Persons, W. M. (1919). Indices of business conditions. *Review of Economic Statistics, 1*, 5–107.
45. Pierce, D. (1978). Seasonal adjustment when both deterministic and stochastic seasonality are present. In A. Zellner (Ed.), *Proceedings of the NBER/Bureau of the Census Conference on Seasonal Analysis of Economic Time Series*, Washington, DC (pp. 242–269).
46. Shiskin, J. (1973). Measuring current economic fluctuations. In *Annals of economic and social measurements* (Vol. 2, pp. 1–16.). Cambridge: National Bureau of Economic Research.
47. Shiskin, J., Young, A. H., & Musgrave, J. C. (1967). *The X-11 variant of the Census Method II seasonal adjustment program*. Technical Paper 15 (revised). Washington, DC: US Department of Commerce, Bureau of the Census.
48. Stephenson, J. A., & Farr, H. T. (1972). Seasonal adjustment of economic data by application of the general linear statistical model. *Journal of the American Statistical Association, 67*, 37–45.
49. Wald, A. (1936). *Berechnung und Ausschaltung von Saisonschwankungen*. Beitrage Zur Konjunkturforschung. 9, Vienna: Verlag Julius Springer.
50. Wallis, K. E. (1978). Seasonal adjustment and multiplicative time series analysis. In A. Zellner (Ed.), *Proceedings of the NBER/Bureau of the Census Conference on Seasonal Analysis of Economic Time Series*, Washington, DC (pp. 347–357).
51. Whittaker, E., & Robinson, G. (1924). *The calculus of observation: A treatise on numerical mathematics* (397 pp.). London: Blackie and Sons Ltd. (also published by Dover Publications, New York, 1967).
52. Wold, H. O. (1938). *A study in the analysis of stationary time series*. Uppsala, Sweden: Almquist and Wiksell (Second Edition 1954).
53. Young, A. H. (1968). A linear approximation to the Census and BLS seasonal adjustment methods. *Journal of the American Statistical Association, 63*, 445–457.

Chapter 4
Linear Filters Seasonal Adjustment Methods: Census Method II and Its Variants

Abstract The best known and most often applied seasonal adjustment methods are based on smoothing linear filters or moving averages applied sequentially by adding (and subtracting) one observation at a time. This chapter discusses with details the basic properties of the symmetric and asymmetric filters of the Census Method II-X11 method which belong to this class. It also discusses the basic assumptions of its two more recent variants, X11ARIMA and X12ARIMA. The latter consists of two linked parts: the regARIMA model for estimation of the deterministic components (mainly calendar effects), and the decomposition part of the linearized series for the stochastic components (trend-cycle, seasonality, and irregulars) performed using the X11 filters combined with those of the ARIMA model extrapolations. An illustrative example of the seasonal adjustment with the X12ARIMA software default option is shown with the US New Orders for Durable Goods series. The illustrative example concentrates on the most important tables of this software that enable to assess the quality of the seasonal adjustment.

4.1 Introduction

The best known and most often applied seasonal adjustment methods are based on smoothing linear filters or moving averages applied sequentially by adding (and subtracting) one observation at a time. These methods assume that the time series components change through time in a stochastic manner. Given a time series, $y_t, t = 1, \ldots, n$, for any t far removed from both ends, say, $m < t < n - m$, the seasonally adjusted value y_t^a is obtained by application of a symmetric moving average $W(B)$ as follows:

$$y_t^a = W(B)y_t = \sum_{j=-m}^{m} w_j B^j y_t = \sum_{j=-m}^{m} w_j y_{t-j}, \tag{4.1}$$

where the weights w_j are symmetric, that is, $w_j = w_{-j}$, and the length of the average is $2m + 1$.

For current and recent data ($n - m < t \le n$) a symmetric linear filter cannot be applied, and therefore truncated asymmetric filters are used. For example, for the

© Springer International Publishing Switzerland 2016
E. Bee Dagum, S. Bianconcini, *Seasonal Adjustment Methods and Real Time Trend-Cycle Estimation*, Statistics for Social and Behavioral Sciences,
DOI 10.1007/978-3-319-31822-6_4

last available observation y_n, the seasonally adjusted value is given by

$$y_n^a = W_0(B)y_n = \sum_{j=0}^{m} w_{0,j}y_{n-j}. \qquad (4.2)$$

The asymmetric filters are time-varying in the sense that different filters are applied to the m first and last observations. The end estimates are revised as new observations are added because of: (1) the new innovations, and (2) the differences between the symmetric and asymmetric filters. The estimates obtained with symmetric filters are often called "final."

The development of electronic computers contributed to major improvements in seasonal adjustment based on moving averages and facilitated their massive application.

In 1954, Julius Shiskin of the US Bureau of Census developed software called Method I, based mainly on the works of Macauley [22] already being used by the US Federal Reserve Board. Census Method I was followed by Census Method II and eleven more experimental versions (X1 , X2, ..., X11). The best known and widely applied was the Census Method II-X11 variant developed by Shiskin et al. [26], but produced poor seasonally adjusted data at the end of the series which is of crucial importance to assess the direction of the short-term trend and the identification of turning points in the economy. Dagum [6] developed the X11ARIMA mainly to correct for this serious limitation and, later Findley et al. [9] developed X12ARIMA that offers a regARIMA option to estimate deterministic components, such as trading day variations, moving holiday effects and outliers, simultaneously with the ARIMA model for extrapolation. It also includes new diagnostic tests and spectral techniques to assess the goodness of the results. The X12ARIMA method is today the one most often applied by statistical agencies in the world. The US Bureau of Census continued research on the development of seasonal adjustment methods, and recently produced a beta version called X13ARIMA-SEATS which enables the estimation of the seasonal component either via linear filters as those available in X12ARIMA or based on an ARIMA model decomposition.

All the seasonal adjustment methods based on moving averages discussed in this book have nonlinear elements. Hence, the seasonally adjusted total of an aggregated series is not equal to the algebraic sum of the seasonally adjusted series that enter into the aggregation. The main causes of nonlinearity are generated by:

1. a multiplicative decomposition model for the unobserved components,
2. the identification and replacement of extreme values,
3. the ARIMA extrapolations, and
4. automatic selection of the moving average length for the estimation of the trend-cycle and seasonality.

The properties of the combined linear filters applied to estimate the various components were originally calculated by Young [28] for the standard option of Census II-X11 variant. Later, Dagum et al. [8] calculated and analyzed all possible filter combinations of Census II-X11 and X11ARIMA.

4.1.1 Main Steps to Produce a Seasonally Adjusted Series

Given the fact that X11ARIMA and X12ARIMA are methods that are based on the Bureau of the Census Method II-X11 [26], we shall first briefly discuss the Census II-X11 variant and refer the reader to Ladiray and Quenneville [18] for a very detailed description.

The Census II-X11 method assumes that the main components of a time series follow a multiplicative or an additive model, that is,

(multiplicative model) $y_t = TC_t \times S_t \times I_t$
(additive model) $y_t = TC_t + S_t + I_t$

where y_t stands for the original series, TC_t for the trend-cycle, S_t for the seasonal, and I_t for the irregular.

There are no mixed models in this program, such as $y_t = TC_t \times S_t + I_t$ or other possible combinations. The estimation of the components is made with different kinds of smoothing linear filters.

The main steps followed to produce monthly seasonally adjusted data are:

1. Compute the ratios (differences) between the original series and a centered 12 months moving average (2×12 m.a.), that is, a 2 months average of 12 months (average) as a first estimate of the seasonal and irregular components.
2. Apply a weighted 5-term moving average (3×3 m.a.) to the seasonal–irregular ratios (differences) for each month separately, to obtain an estimate of the seasonal factors (effects). Compute a 2×12 m.a. of the preliminary factors for the entire series. To obtain the six missing values at either end of this average, repeat the first (last) available value six times.
3. Adjust these seasonal factors (differences) to sum to 12 (0.000 approximately) over any 12-month period by dividing (subtracting) the centered 12 months average into (from) the factors (differences).
4. Divide (subtract) the seasonal factors (differences) into (from) the seasonal–irregular ratios (differences) to obtain an estimate of the irregular component.
5. Compute a moving 5-year standard deviation (σ) of the estimates of the irregular component and test the irregulars in the central year of the 5-year period against 2.5σ. Remove values beyond 2.5σ as extreme and recompute the moving 5-year σ.
6. Assign a zero weight to irregulars beyond 2.5σ and a weight of one (full weight) to irregulars within 1.5σ. Assign a linearly graduated weight—between 0 and 1 to irregulars between 2.5σ and 1.5σ.
7. For each separate month apply a weighted 5-term moving average (3×3 m.a.) to the seasonal–irregular ratios (differences) with extreme values replaced by the corresponding values in step 6, to estimate preliminary seasonal factors (effects).
8. Repeat step 3.
9. To obtain a preliminary seasonally adjusted series divide (subtract) the series obtained at step 8 into (from) the original series.

10. Apply a 13-term Henderson moving average to the seasonally adjusted series and divide (subtract) the resulting trend-cycle into (from) the original series to obtain a second estimate of the seasonal–irregular ratios (differences).
11. Apply a weighted 7-term moving average (3 × 5 m.a.) to the seasonal–irregular ratios (differences) for each month separately to obtain a second estimate of the seasonal component.
12. Repeat step 3.
13. Divide (subtract) the series obtained at step 11 into (from) the original series to obtain a seasonally adjusted series.

The filters are applied sequentially and the thirteen steps are repeated twice.

The Census II-X11 variant produces seasonal factor forecasts for each month which are given by

$$S_{j,t+1} = S_{j,t} + \frac{1}{2}(S_{j,t} - S_{j,t-1}), \qquad j = 1, \ldots, 12,$$

where j is the month and t the current year. The use of seasonal factor forecasts was very popular in statistical Bureaus till the end of 1970s. In 1975, Dagum proposed the use of concurrent seasonal factors obtained from data that included the most recent value and this was adopted by Statistics Canada. Gradually, other statistical agencies applied concurrent seasonal factors and this is now the standard practice.

The final trend-cycle is obtained by 9-, 13-, or 23-term Henderson moving averages applied to the final monthly seasonally adjusted series (5-term and 7-term Henderson filters are used for quarterly data). The selection of the appropriate filter is made on the basis of a preliminary estimate of the I/C ratio (the ratio of the average absolute month-to-month change in the irregular to that in the trend-cycle). A 9-term is applied to less irregular series and a 23-term to highly irregular series.

Using a linear approximation of the Census Method II, Young [28] arrived at the conclusion that a 145-term moving average was needed to estimate one seasonal factor with central weights if the trend-cycle component is adjusted with a 13-term Henderson moving average. The first and last 72 seasonal factors (6 years) are estimated using sets of asymmetrical end weights. It is important to point out, however, that the weights given to the most distant observations are very small and, therefore, the moving average can be very well approximated by taking one half of the total number of terms plus one. So, if a 145-term moving average is used to estimate the seasonal factor of the central observation, a good approximation is obtained with only 73 terms, i.e., 6 years of observations. This means that the seasonal factor estimates from unadjusted series whose observations end at least 3 years later can be considered "final", in the sense that they will not change significantly when new observations are added to the raw data.

4.2 Basic Properties of the Two-Sided Linear Smoothing Filters of Census Method II-X11 Variant

The linear smoothing filters applied by Census Method II-X11 variant to produce seasonally adjusted data can be classified according to the distribution of their set of weights in symmetric (two-sided) and asymmetric (one-sided). The symmetric moving averages are used to estimate the component values that fall in the middle of the span of the average, say, $2m + 1$, and the asymmetric moving averages, to the m first and last observations.

The sum of the weights of both kinds of filters is one and thus, the mean of the original series is unchanged in the filtering process. The sum of the weights of a filter determines the ratio of the mean of the smoothed series to the mean of the unadjusted series assuming that these means are computed over periods long enough to ensure stable results. It is very important in filter design that the filter does not displace in time the components of the output relative to those of the input. In other words, the filter must not introduce phase shifts.

Symmetric moving averages introduce no time displacement for some of the components of the original series and a displacement of 180° for others. A phase shift of 180° is interpreted as a reverse in polarity which means that maxima are turned into minima and vice versa. In other words, peaks (troughs) in the input are changed into troughs (peaks) in the output. For practical purposes, however, symmetric moving averages act as though the time displacement is null. This is so because the sinusoids that will have a phase shift of 180° in the filtering process are cycles of short periodicity (annual or less) and moving averages tend to suppress or significantly reduce their presence in the output. In spectral analysis, the phase is a dimensionless parameter that measures the displacement of the sinusoid relative to the time origin. Because of the periodic repetition of the sinusoid, the phase can be restricted to ±180°. The phase is a function of the frequency of the sinusoid, being the frequency equal to the reciprocal of the length of time or period required for one complete oscillation.

The next two sections discuss the basic properties of the two-sided and one-sided filters applied by the Census II-X11 method. The study is based on a spectral analysis of the corresponding filters.

4.2.1 The Centered 12 Months Moving Average

The centered 12 months moving average used for a preliminary estimate of the trend-cycle (step 1) reproduces exactly a linear trend and annihilates a stable seasonal in an additive model. However, if the relationship among the components is multiplicative, this filter will reproduce only a constant trend and eliminate a stable seasonality. Unless the cycles are long (5 years or more) and the irregulars quite small, this filter will not follow closely the peaks and troughs and will not smooth the data successfully. If a sine curve of 3-year periodicity and amplitude 100 is input to this filter, the output is a sine curve of equal periodicity but with amplitude reduced to 82.50. For a 5-year cycle the amplitude is reduced to 93.48 and for a 10-year cycle is 98.33. In general, any simple moving average of m equal weights fitted to a parabola falls $\left(\frac{m^2-1}{12}\right) c$ inside the parabola, see Macauley [22]. Hence, this filter fitted to a second degree parabola $y_t = a + bt + ct^2$ will give a smoothed curve that falls 12.17 units inside the parabola, i.e., $\hat{y}_t = y_t + 12.17c$. The bias introduced by this filter will not affect significantly the preliminary estimate of the seasonal component if the trend is nearly linear or if the direction of the curvature changes frequently. In the latter case, the bias will be sometimes positive and sometimes negative, so it may average out in applying the seasonal filters to the Seasonal–Irregular (SI) ratios for each month separately over several years.

4.2.2   The 9-, 13-, and 23-Term Henderson Moving Averages

The Henderson [14] moving averages are applied to obtain an improved estimate of the trend-cycle (step 11). These linear smoothing filters were developed by summation formulae. Their weights minimize the sum of squares of the third differences of the smoothed curve. The latter is equivalent to say that these filters reproduce the middle value of a function that is a third degree parabola within the span of the filter. The Henderson's moving averages give the same results as would be obtained by smoothing the middle value of a third degree parabola fitted by weighted least squares, where the weights given to the deviations are as smooth as possible.

The fact that the trend-cycle is assumed to follow a cubic over an interval of short duration (between 1 and 2 years approximately) makes these filters very adequate for economic time series.

None of the Henderson's filters used by the Census II-X11 program eliminates the seasonal component, but since they are applied to already seasonally adjusted data, this limitation becomes irrelevant. On the other hand, they are extremely good to pass sines of any periodicity longer than a year. Thus, the 13 months Henderson, which is the most frequently applied, will not reduce the amplitude of sines of 20 months periodicity or more that stand for trend-cycle variations. Moreover, it eliminates almost all the irregular variations that can be represented by sines of very short periodicity, 6 months or less.

4.2.3   The Weighted 5-Term (3 × 3) and 7-Term (3 × 5) Moving Averages

The weighted 5-term moving average is a three-term moving average of a three-term moving average (3×3 m.a.). Similarly, the weighted 7-term moving average is a three-term moving average of a five-term moving average (3×5 m.a.). These two filters are applied to the seasonal–irregular ratios (or differences) for each month separately over several years. Their weights are all positive and, consequently, they reproduce exactly the middle value of a straight line within their spans. This property enables the Census II-X11 method to estimate a linearly moving seasonality within 5 and 7 years spans. Therefore, these filters can approximate quite adequately gradual seasonal changes that follow nonlinear patterns over the whole range of the series (more than 7 years).

The weighted 5-term moving average (3×3 m.a.) is a very flexible filter that allows for fairly rapid changes in direction. But, since the span of the filter is short, the irregulars must be small for the SI to be smoothed successfully. In effect, assuming that sinusoids of 5 years of periodicity or less represent the irregular variations, given the fact that by definition seasonality is expected to move slowly, this filter still passes about 20 % of the amplitude of a 5-year sine wave. Sinusoids of 3 years periodicity or less are completely eliminated.

The weighted 7-term moving average (3×5 m.a.) is less flexible and it is applied for the final estimate of the seasonal factors. Sinusoids of periodicity shorter than 5 years are completely eliminated. For series the seasonal factors of which are nearly stable, this program also provides other optional sets of weights which are applied to longer spans.

4.3 Basic Properties of the One-Sided Linear Smoothing Filters of Census Method II-X11 Variant

It is inherent to any moving average procedure that the first and last m points of an original series cannot be smoothed with the same set of symmetric $2m + 1$ weights applied to the middle values. The Census II-X11 method uses one-sided filters to smooth these end points. Moving averages with asymmetric weights are bounded to introduce phase shifts for all the components of the original series. This is a very undesirable characteristic since it may cause, for example, a business cycle in the smoothed series to turn down before or after it actually does in the unadjusted data.

The asymmetric weights applied in conjunction with the Henderson symmetric filters, developed by Musgrave [23], no longer estimate a cubic within their span. In fact, the weight associated with the 13-term Henderson filter, applied to estimate the last data point, amplifies sinusoids of 3 to 1 year periodicities up to 10 %, passes between 37 and 17 % of the amplitude of sine waves representing irregular variations, and it has a linearly decreasing phase shift for sinusoids of long periodicity. The latter means that the smoothed series lags the original series a number of units of time equal to the slope of the line. For an 18-month cycle, the phase shift is about 4 %, i.e., a lag of more than half of a month.

The set of end weights used in conjunction with the Henderson filters can estimate well only a linear trend-cycle within the length of the filters [23].

Concerning the other two seasonal one-sided filters, they also introduce phase shifts and can only estimate stable or nearly stable seasonality. There are other reasons for the poor quality of both the current and seasonal factor forecasts generated by the Census II-X11 method. As mentioned above, the program repeats the first and last six preliminary seasonal factors obtained with the 3×3 m.a. In other words, for a series ending, say, in December 1977, the seasonal factors from July to December of 1977 are equal to the corresponding seasonal factors of the previous year. This operation is repeated any time that the centered 12-term moving average is applied to estimate the trend-cycle. In the past, when seasonal factor forecasts were used to produce concurrent seasonal adjustment, the last 6 months were subject to larger errors than the corresponding factors for the first 6 months because they were based on observations lagging 2 years with respect to the forecasting year. It was then preferable to forecast only 6 months instead of 12. By concurrent seasonal adjustment is meant the adjustment of the original series made with seasonal factors obtained from a series that includes the last observed point.

4.4   The X11ARIMA Method

4.4.1   General Outline and Basic Assumptions

A detailed description of the X11ARIMA method of seasonal adjustment is given in [6, 7]. The X11ARIMA is a modified version of the Census Method II-X11 variant that was developed mainly to produce a better current seasonal adjustment of series with seasonality that changes rapidly in a stochastic manner. The latter characteristic is often found in main socioeconomic indicators, e.g., retail trade, imports and exports, unemployment, and so on.

The X11ARIMA method consists of:

(i) modeling the original series with an ARIMA model of the Box and Jenkins type,

(ii) extending the original series 1–3 years with forecasts from the ARIMA model that fits and extrapolates well according to well-defined acceptance criteria, and

(iii) estimating each component with moving averages that are symmetric for middle observations and asymmetric for both end years. The latter are obtained via the convolution of Census II-X11 variant and the ARIMA model extrapolations.

For series that result from the accumulation of daily activities, called flow series, deterministic components such as trading day variations and moving holiday effects are estimated with dummy variable regression models and removed from the series, so that only the remainder is subject to steps (i)–(iii) above. The X11ARIMA was extended again by Dagum [7].

An alternative way to cope with the problem of estimating rapidly changing stochastic seasonality is to introduce more adequate one-sided filters but this alternative presents the old problem of end weights to be chosen. In effect, it is well known the trade-off between the phase shift introduced by one-sided filters, and the power to pass low frequencies without modification. Therefore, it was preferable to extend the original series with forecasts. The next step in the development of the X11ARIMA was to decide what kind of forecasting model should be used for the original series. The selection was made in accordance with the following requirements:

1. the forecasting model must belong to the "simplest" class in terms of its description of the real world. No explanatory variables must be involved, the series should be described simply by its past values and lagged random disturbances,
2. it should be efficient for short-term forecasting, one or two years at most,
3. it must generate forecasts that are robust to small changes in parameter values,
4. it must be parsimonious in the number of parameters,
5. it must generate optimum forecasts, in the sense of minimum mean square error forecasts, and
6. it must generate forecasts that follow well the intra-year movement although they could miss the level.

All these requirements were fulfilled by the ARIMA models except for point 5, that is satisfied for only the one-step ahead forecasts. ARIMA models of the Box and Jenkins type were chosen which were previously very effective for forecasting a large number of series [24, 25]. In the Box and Jenkins notation, the general multiplicative ARIMA model, for a series with seasonality of period s, is defined to be of order $(p, d, q)(P, D, Q)_s$, where d and D are the orders of the regular and seasonal difference operator, respectively, p and P are the orders of the autoregressive polynomials in the regular and seasonal lag operator B and B^s, respectively, whereas q and Q are the orders of the moving average polynomials in the regular and seasonal lag operator B and B^s, respectively.

The main advantages of X11ARIMA over the Census II-X11 variant are:

1. The availability of a statistical model that describes the stochastic behavior of the original series. The expected value and the variance of the original series can be calculated and thus, confidence intervals can be constructed for the observations. This is very important to detect outliers at the end of the series.
2. The one-step forecast from these models is a minimum mean square error forecast and can be used as a benchmark for preliminary figures.
3. The total error in the seasonal factor forecasts, and in the concurrent seasonal factors, is significantly reduced for all months. Generally, a reduction of about 30 % in the bias and of 20 % in the absolute values of the total error has been found for Canadian and American series.
4. If concurrent seasonal factors are applied to obtain current seasonally adjusted data, there is no need to revise the series more than twice. For many series, just

one revision will give seasonal factors that are "final" in a statistical sense. There are several reasons for the significant reduction of the error in the seasonal factor forecasts and concurrent seasonal factors. The X11ARIMA produces seasonal factor forecasts from the combination of two filters:

a. the filters of the autoregressive integrated moving averages (ARIMA) models used to forecast the raw data; and
b. the filters that Census Method II-X11 variant applies for concurrent seasonal adjustment.

In this manner, the seasonal factor forecasts are obtained from the forecasted raw values with filters that even though are still one-sided, they are closer to the two-sided filters as compared to the forecasting function of the Census II-X11 variant.

5. The trend-cycle estimate for the last observation is made with the symmetric weights of the Henderson moving average which can reproduce a cubic within their time span. This is relevant for years with turning points because the Census II-X11 applies the asymmetric weights of the Henderson filters which can adequately estimate only a linear trend.
6. Finally, by adding one more year of preliminary data (with no extremes since they are forecasts) a better estimate of the variance of the irregulars is obtained. The latter allows a significant improvement in the identification and replacement of outliers which, as we already know, can severely distort the estimates obtained with linear moving filters.

For concurrent seasonal factors, the same observations are valid except that the seasonal filters are closer to the two-sided filters than those corresponding to the seasonal factor forecasts. For this reason, the number of revisions in the seasonal factor estimates is also significantly reduced.

4.4.2   The Forecasting Filters of ARIMA Models and Their Properties

We will here analyze the forecasting filters of the ARIMA models. The ARIMA models applied to seasonal series belong to the general multiplicative Box and Jenkins type, that is,

$$\phi_p(B)\Phi_P(B^s)(1-B)^d(1-B^s)^D y_t = \theta_q(B)\Theta_Q(B^s)a_t, \tag{4.3}$$

where s denotes the periodicity of the seasonal component (equal to 12 for monthly series and 4 for quarterly data), B is the backshift operator, such that $By_t = y_{t-1}$, $\phi_p(B) = (1-\phi_1 B-\cdots-\phi_p B^p)$ is the nonseasonal Autoregressive (AR) operator of order p, $\Phi_P(B^s) = (1-\Phi_1 B^s -\cdots-\Phi_P B^{Ps})$ is the seasonal AR operator of order P, $\theta_q(B) = (1-\theta_1 B-\cdots-\theta_q B^q)$ is the nonseasonal Moving Average (MA) operator of order q, $\Theta_Q(B^s) = (1 - \Theta_1 B^s - \cdots - \Theta_Q B^{Qs})$ is the seasonal MA operator of order Q, and the a_t's are i.i.d. with mean zero and constant variance σ_a^2. The $(1-B)^d(1-B^s)^D$ term implies nonseasonal differencing of order d and seasonal differencing of order D. If $d = D = 0$ (no differencing), it is common to replace y_t in (4.3) by deviations from its mean.

The general multiplicative model (4.3) is said to be of order $(p, d, q)(P, D, Q)_s$. Its forecasting function can be expressed in different forms. For computational purpose, the difference equation form is the most useful. Thus, at time $t+l$ the ARIMA model may be written as

$$y_{t+l} = \psi_1 y_{t+l-1} + \cdots + \psi_f y_{t+l-f} - a_{t+l} - \pi_1 a_{t+l-1} - \cdots - \pi_g a_{t+l-g} \qquad (4.4)$$

where $f = p + sP + d + sD$ and $g = q + sQ$; $\psi(B) = \phi_p(B)\Phi_P(B^s)(1-B)^d(1-B^s)^D$ is the general autoregressive operator and $\pi(B) = \theta_q(B)\Theta_Q(B^s)$ is the general moving average operator.

Standing at origin t, to make a forecast l steps ahead of y_{t+l}, denoted by $\hat{y}_t(l)$, the conditional expectation of (4.4) is taken at time t with the following assumptions:

$$E_t(y_{t+l}) = y_{t+l}, \quad \text{if} \quad l < 0; \quad E_t(y_{t+l}) = \hat{y}_t(l), \quad \text{if} \quad l > 0 \qquad (4.5)$$

$$E_t(a_{t+l}) = a_{t+l}, \quad \text{if} \quad l < 0; \quad E_t(a_{t+l}) = 0, \quad \text{if} \quad l > 0, \qquad (4.6)$$

where $E_t(y_{t+l})$ is the conditional expectation of y_{t+l} taken at origin t. Thus, the forecasts $\hat{y}_t(l)$ for each lead time are computed from previous observed y's, previous forecast of y's, and current and previous random shocks a's. The unknown a's are replaced by zeros. In other words, they are weighted averages of previous observations and forecasts made at previous lead times from the same time origin.

In general, if the moving average operator $\pi(B) = \theta_q(B)\Theta_Q(B^s)$ is of degree $q + sQ$, the forecast equations for $\hat{y}_t(1), \hat{y}_t(2), \ldots, \hat{y}_t(q+sQ)$ will depend directly on the a's, but forecasts at longer lead times will not. The latter will receive indirectly the impact of the a's by means of the previous forecasts. In fact, $\hat{y}_t(q + sQ + 1)$ will depend on the $q + sQ$ previous $\hat{y}_t$ which in turn will depend on the a's.

In the case of X11ARIMA, the lead time of the forecasts for monthly series is often equal to 12 and the identification of the ARIMA models has a Q generally equal to 1. Therefore, the forecasts depend directly on the a's.

Finally, it is important to keep in mind that it is the general autoregressive operator $\psi(B)$ defined above that determines whether the forecasting function is to be a polynomial, a mixture of sines and cosines, a mixture of exponentials, or some combinations of these functions. The one-step ARIMA forecasts are minimum mean square error forecasts and are easily updated as new raw values become available.

The ARIMA models for X11ARIMA must fulfill the double condition of (1) fitting the data well, and (2) generating acceptable forecasts, where by acceptable it is understood forecasts that have a mean absolute error smaller than 5 % for well behaved series (e.g., employment males 20 years and over) and smaller than 12 % for more irregular series (e.g., unemployment females 16–19 years).

In the automated version of this method, the user has the choice of: (a) submitting his/her own ARIMA model, or (b) letting the program to select automatically an ARIMA model that passes the above guidelines.

The automatic selection is made from three ARIMA models that have been successfully chosen from a set of 12 ARIMA models testing out of sample extrapolated values for the 4 last years. The sample consisted of 174 monthly and quarterly economic time series of 15 years of length. The three models passed the guidelines in about 80 % of all the cases. The 12 models originally tested were:

(1)	$(1, 1, 1) (1, 1, 1)_s$	(7)	$\log (2, 1, 1) (0, 1, 2)_s$
(2)	$(2, 1, 2) (0, 1, 1)_s$	(8)	$\log (0, 1, 2) (1, 1, 2)_s$
(3)	$(2, 0, 1) (0, 1, 2)_s$	(9)	$\log (0, 1, 1) (0, 1, 1)_s$
(4)	$(2, 0, 1) (1, 2, 1)_s$	(10)	$\log (0, 1, 1) (0, 2, 2)_s$
(5)	$(2, 0, 0) (0, 1, 1)_s$	(11)	$\log (0, 2, 2) (0, 1, 1)_s$
(6)	$(0, 0, 2) (0, 1, 1)_s$	(12)	$\log (0, 2, 2) (0, 2, 2)_s$

The automated option chooses for a multiplicative decomposition from the following ARIMA models:

(1)	$\log(0,1,1)(0,1,1)_s$
(2)	$\log(0,2,2)(0,1,1)_s$
(3)	$(2,1,2)(0,1,1)_s$

and without the log transformation for an additive decomposition. It prints a message indicating the failure of the three models, in which case the user can either identify his own model or try again the automatic option submitting the same series with prior modifications for outliers. The latter is strongly recommended because the guidelines for acceptance or rejection of an ARIMA model are highly sensitive to outliers.

The ARIMA models are robust and once well identified, they are generally adequate for several years. It is advisable, however, to check the adequacy of the model when an extra year of data becomes available to ensure that the most recent movements of the series are taken into account.

Because the correct identification and estimation of ARIMA models have to be done on series not affected by trading day variations and/or moving holidays, these effects are estimated in a first round of the software without the ARIMA identification. Once trading day variations and/or moving holidays are estimated, they are removed from the original series and a second run is done using the above three automated ARIMA models for extrapolation. The ARIMA model that gives the smallest forecasting error for the last 3 years of data and passes a chi-square probability test for the null hypothesis of randomness of the residual is chosen.

4.4.3 Other Main Improvements Incorporated into the Automated Version of X11ARIMA

A set of new statistical tests, tables, and graphs has been incorporated into the automated version of the X11ARIMA. These tests are used to assess the quality of the original series and the reliability of the seasonal adjustment. A brief description of these main improvements follows.

4.4.3.1 An F-Test for the Presence of Seasonality in Table B1

This test is based on a one-way analysis of variance on the seasonal–irregular (SI) ratios (differences), similar to the one already available in Census Method II-X11 variant for the presence of stable seasonality in Table D8. It differs only in that the estimate of the trend-cycle is made directly from the original series, by a centered 12-term moving average.

The estimate of the trend-cycle is removed from the original series by dividing into (subtracting from) the raw data for a multiplicative (additive) model.

The value of the F-ratio is printed in the computer program Table B1. It is the quotient of two variances: (1) the "*between months or quarters*" variance, which is due to seasonality, and (2) the "*residual*" variance, which is due to the irregular component.

Because several of the basic assumptions in the F-test are likely to be violated, the value of the F-ratio to be used, for rejecting the null hypothesis of no significant seasonality being present, is not the one obtained from the tabulated F-distribution. From experimentation with a large number of real and simulated series, it was concluded that, for a monthly series of about 10 years, an F-value less than 10 indicates that there is not enough seasonality to justify using the Census II-X11 filters.

4.4.3.2 A Test for the Presence of Moving Seasonality in Table D8

The moving seasonality test is based on a two-way analysis of variance performed on the SI ratios (differences) from Table D8 and was developed by Higginson [15]. It tests for the presence of moving seasonality characterized by gradual changes in the seasonal amplitude but not in the phase.

The total variance of the SI ratios (differences) is considered as the sum of:

1. The "*between months or quarters*" variance, denoted by σ_m^2, which mainly measures the magnitude of the seasonality. It is equal to the sum of squares of the difference between the average for each month of the SI and the overall average, corrected by the corresponding degrees of freedom.
2. The "*between years*" variance, denoted by σ_y^2, which mainly measures the year-to-year movement of seasonality. It is equal to the sum of squares of the

differences between the annual average of the SI, for each year, and the average of the SI for the whole table, corrected by the corresponding degrees of freedom.
3. The "*residual*" variance, denoted by σ_r^2, equal to the total variance minus the sum of the "*between months or quarters*" variance and the "*between years*" variance.

The F-ratio for the presence of moving seasonality is the quotient between the "*between years*" variance and the "*residual*" variance.

To calculate the variances in an additive model, the $(S + I)$ data are taken in absolute value, otherwise the annual average is always equal to zero. For a multiplicative model, the SI ratios are replaced by absolute deviations from 100, i.e., by $| SI - 100 |$. Contrary to the previous test, for which a high value of the F-statistic is a good indication of the presence of identifiable seasonality, a high value of the F-statistic corresponding to moving seasonality is a bad sign. The F-test, printed in Table D8, indicates whether moving seasonality is present or not.

The presence of moving seasonality could be taken as an indication of residual seasonality when Census II-X11 is used, because this program forces stable or nearly stable seasonality on the first and last 3 years of data. This is partially corrected by the ARIMA option.

4.4.3.3 A Test for the Presence of Identifiable Seasonality in Table D8

This test combines the previous test for the presence of moving seasonality with the Census II-X11 test for the presence of stable seasonality and another nonparametric test for the presence of stable seasonality, the Kruskal–Wallis chi-squared test, whose value is also printed in Table D8.

The main purpose of this test is to determine whether the seasonality of the series is "identifiable" or not. For example, if there is little stable seasonality and most of the process is dominated by highly moving seasonality, the chances are that the seasonal factors will not be well estimated because they will not be properly identified by the Census II-X11 program.

This test basically consists of combining the F-values obtained from the two previous tests as follows:

1. If the test for the presence of stable seasonality, denoted by F_S, fails at the 1 % level of significance, the null hypothesis (that seasonality is not identifiable) is accepted.
2. If step 1 passes but the test for the presence of moving seasonality, indicated by F_M, fails at the 5 % level of significance, then this F_M value is combined with the F_S value from step 1 to give

$$T_1 = \frac{7}{F_M - F_S} \quad \text{and} \quad T_2 = \frac{3F_M}{F_S},$$

and an average of these two statistics T_1 and T_2, $\bar{T} = (T_1 + T_2)/2$, is taken. If $\bar{T}$ is greater than or equal to one, the null hypothesis of identifiable seasonality not being present is accepted.

3. If the F_M-test passes, but either $\bar{T} < 1$ or the Kruskal–Wallis test fails at the 1 % level, then the program prints out that identifiable seasonality is "probably" present.

4. If the F_S, F_M, and the Kruskal–Wallis chi-square value pass, then the null hypothesis of identifiable seasonality not being present is rejected, and the program prints "identifiable seasonality present."

This test has been developed by Lothian and Morry [20], and the program automatically prints out the messages mentioned above at the end of Table D8.

4.4.3.4 A Test for the Presence of Residual Seasonality in Table D11

This is an F-test applied to the values of Table D11 and calculated for the whole length of the series, and for only the last 3 years. The effect of the trend is removed by a first order difference of lag three, that is, $(\hat{y}_t - \hat{y}_{t-3})$ where $\hat{y}_t$ are the values of Table D11. Two F-ratios are printed at the end of the table, as well as a message saying whether residual seasonality is present or not for the last 3 years.

4.4.3.5 A Test for the Randomness of the Residuals

The Census II-X11 program uses the Average Duration of Run (ADR) statistic to test for autocorrelation in the final estimated residuals obtained from Table D13. This nonparametric test was developed by Wallis and Moore [27] and is based on the number of turning points. It is only efficient for testing the randomness of the residuals against the alternative hypothesis that the errors follow a first order autoregressive process of the form $I_t = \rho I_{t-1} + e_t$, where ρ is the autocorrelation coefficient and $\{e_t\}$ is a white noise process. This test, however, is not efficient for detecting the existence of periodicity in the residuals, which can occur when relatively long series are seasonally adjusted, or when the relative variation of the seasonal component is small compared to that of the irregulars.

To test the independence of the residuals against the alternative hypothesis of being a periodic process, the normalized cumulative periodogram as developed by Box and Jenkins [3] has been incorporated.

The normalized cumulative periodogram values are given in a table and also as a graph. By visual inspection, it is possible to determine whether or not components with certain periodicity are present in the residuals. If the residuals are a sample realization of a purely random process, and if the size of the sample tends to infinity, then the normalized cumulative periodogram tends to coincide with the diagonal of the square in which it is drawn.

Deviations of the periodogram from those expected, assuming purely random residuals, can be assessed by using the Kolmogorov–Smirnov test. This test is useful to determine the nature of hidden periodicities left in the residuals, whether of seasonal or cyclical character, and complements the information provided by the test for the presence of residual seasonality.

4.4.3.6 A New Table D11A

This new table, D11A, produces a modified seasonally adjusted series, where the annual totals of the seasonally adjusted values are made equal to those of the raw data. The discrepancy between the annual totals is distributed over the seasonally adjusted values of Table D11 in such a way as to minimize the distortion on the month-to-month or quarter-to-quarter movements of the originally seasonally adjusted series. The procedure is based on a quadratic minimization of the first differences of the annual discrepancies, expressed as differences or ratios. This procedure was originally developed by Hout [16] and improved by Cholette [5].

4.4.3.7 A Set of Guidelines Summarized in One Statistic that Helps to Assess the Reliability of the Seasonal Adjustment

The Statistics Canada Census II-X11 version, as developed in 1975, had two statistics called Q_1 and Q_2 that provided an indication of the size and nature of the irregular and seasonal components, respectively. A description of these statistics and their basic assumptions are discussed by Hout and De Fontenay [17].

Considerable research has been carried out since the first set of guidelines was developed and there is now only one Q statistic, which results from a combination of several other measures developed by Lothian and Morry [21]. Most of these are obtained from the summary measures printed in Table F2 of the X11ARIMA computer program. Their values vary between 0 and 3, and only values less than 1 are considered as acceptable. The set of statistics to produce the final Q statistic is:

1. the relative contribution of the irregular component over spans of 3 months, as obtained from Table F2, denoted by M_1;
2. the relative contribution of the irregular component to the stationary portion of the variance, denoted by M_2;
3. the value of the I/C ratio between the average absolute month-to-month (or quarter-to-quarter) percentage change in the irregular and that in the trend-cycle, from Table D7 and denoted by M_3;
4. the value of the average duration of run for the irregulars from Table F2 and denoted by M_4;
5. the MCD or QCD (the number of months or quarters it takes the average absolute change in the trend-cycle to dominate that in the irregular), from Table F2 and denoted by M_5;

6. the global I/S moving seasonality ratio, obtained as an average of the monthly moving seasonality ratios from Table D9A and denoted by M_6 (it is the ratio of the average absolute year-to-year percentage change in the irregular factors to that in the seasonal factors.);
7. the amount of stable seasonality in relation to the amount of moving seasonality from the tests of Table D8 and denoted by M_7;
8. a measure of the variation of the seasonal component for the whole series from Table F2 and denoted by M_8;
9. the average linear movement of the seasonal component for the whole series, denoted by M_9;
10. same as 8, but calculated for recent years only and denoted by M_{10}; and
11. same as 9, but calculated for recent years only and denoted by M_{11}.

Furthermore, there are two new tables that show the outliers, and a D16 table that shows the total effect of the trading day factors and the seasonal effects. Several other modifications were introduced to correct for some limitations of the Census II-X11 variant [7].

4.5   The X12ARIMA Method

4.5.1   General Outline

The X12ARIMA is today the most often applied seasonal adjustment method by statistical agencies. It was developed by Findley et al. [9] and is an enhanced version of the X11ARIMA method.

The major modifications concern: (1) extending the automatic identification and estimation of ARIMA models for the extrapolation option to many more than the three models available in X11ARIMA, and (2) estimating trading day variations, moving holidays, and outliers in what is called regARIMA. The latter consists of regression models with ARIMA (AutoRegressive Integrated Moving Average) errors. More precisely, they are models in which the mean function of the time series (or its logs) is described by a linear combination of regressors, and the covariance structure of the series is that of an ARIMA process. If no regressors are used, indicating that the mean is assumed to be zero, the regARIMA model reduces to an ARIMA model.

Whether or not special problems requiring the use of regressors are present in the series to be adjusted, a fundamentally important use of regARIMA models is to extend the series with forecasts (and backcasts) in order to improve the seasonal adjustments of the most recent (and the earliest) data. Doing this reduces problems inherent in the trend estimation and asymmetric seasonal averaging processes of the type used by the Census II-X11 method near the ends of the series. The provision of this extension was the most important improvement offered by the X11ARIMA program. Its theoretical and empirical benefits have been documented in many publications, such as Dagum [7], Bobbit and Otto [2], and references therein.

The X12ARIMA method has all the seasonal adjustment capabilities of the X11ARIMA variant. The same seasonal and trend moving averages are available, and the program still offers the Census II-X11 calendar and holiday adjustment routines incorporated in X11ARIMA. Several new options have been included, namely:

1. sliding spans diagnostic procedures, developed by Findley et al. [10];
2. capability to produce the revision history of a given seasonal adjustment;
3. a new Henderson trend filter routine which allows the user to choose any odd number for the length of the Henderson symmetric filter;
4. new options for seasonal filters;
5. several new outlier detection options for the irregular component of the seasonal adjustment;
6. a new table of trading day factors by type of day; and
7. a pseudo-additive seasonal adjustment mode.

The modeling module is designed for regARIMA model building with seasonal socioeconomic time series. To this end, several categories of predetermined regression variables are available, including trend constants or overall means, fixed seasonal effects, trading day effects, holiday effects, additive outliers, level shifts, temporary change outliers, and ramp effects. User-specified regression variables can also be included in the models.

The specification of a regARIMA model requires specification of both the regression variables to be included in the model and the type of ARIMA model for the regression errors (i.e., the order $(p, d, q)(P, D, Q)_s$). Specification of the regression variables depends on user knowledge about the series being modeled. Identification of the ARIMA model for the regression errors follows well-established procedures based on examination of various sample autocorrelation and partial autocorrelation functions produced by the X12ARIMA program.

Once a regARIMA model has been specified, X12ARIMA estimates its parameters by maximum likelihood using an Iterative Generalized Least Squares (IGLS) algorithm. Diagnostic checking involves examination of residuals from the fitted model for signs of model inadequacy. X12ARIMA produces several standard residual diagnostics for model checking, as well as provides sophisticated methods for detecting additive outliers and level shifts. Finally, X12ARIMA can produce point forecasts, forecast standard errors, and prediction intervals from the fitted regARIMA model. In addition to these modeling features, X12ARIMA has an automatic model selection procedure based mainly on the automatic model selection procedure of TRAMO [12]. There are also options that use AICC, a version of Akaike's AIC that adjusts for the length of the series being modeled, to determine if user-specified regression variables (such as trading day or Easter regressors) should be included into a particular series. Also, histories can be generated for likelihood statistics (such as AICC) and forecasts to facilitate comparisons between competing models.

4.5.2 The General RegARIMA Model

ARIMA models are frequently used for seasonal time series. A general multi-plicative seasonal ARIMA model for a time series $y_t, t = 1, \ldots, n$, is specified in Eq. (4.3). A useful extension of ARIMA models results from the use of a time-varying mean function modeled via linear regression effects. More explicitly, suppose we write a linear regression equation for the time series y_t as

$$y_t = \sum_{i=1}^{r} \beta_i x_{it} + z_t \qquad (4.7)$$

where $x_{it}, i = 1, \ldots, r$, are regression variables observed concurrently with y_t, being the $\beta_i, i = 1, \ldots, r$ are the corresponding regression parameters, such that $z_t = y_t - \sum_{i=1}^{r} \beta_i x_{it}$, the time series of regression errors, is assumed to follow the ARIMA model defined in (4.3). Modeling z_t as ARIMA addresses the fundamental problem of applying standard regression methodology to time series data, that is, the standard regression assumes that the regression errors (z_t in (4.7)) are uncorrelated over time. In fact, for time series data, the errors in (4.7) will usually be autocorrelated, and, moreover, will often require differencing. Assuming that z_t is uncorrelated will typically lead to invalid results.

The expressions (4.3) and (4.7) taken together define the general regARIMA model allowed by the X12ARIMA program, that can be written in a single equation as

$$\phi_p(B)\Phi_P(B^s)(1 - B)^d(1 - B^s)^D \left(y_t - \sum_{i=1}^{r} \beta_i x_{it} \right) = \theta_q(B)\Theta_Q(B^s)a_t. \qquad (4.8)$$

The regARIMA model (4.8) can be thought of either as generalising the pure ARIMA model (4.3) to allow for a regression mean function ($\sum_{i=1}^{r} \beta_i x_{it}$), or as generalising the regression model (4.7) to allow the errors z_t to follow the ARIMA model (4.3). In any case, notice that the regARIMA model implies that first the regression effects are subtracted from y_t to get the zero mean series z_t, then the error series z_t is differenced to get a stationary series, say w_t, assumed to follow a stationary ARMA model, $\phi(B)\Phi(B^s)w_t = \theta(B)\Theta(B^s)a_t$.

Another way to write the regARIMA model (4.8) is

$$(1 - B)^d(1 - B^s)^D y_t = \sum_i \beta_i(1 - B)^d(1 - B^s)^D x_{it} + w_t, \qquad (4.9)$$

where w_t follows the stationary ARMA model just given before. Equation (4.9) emphasizes that the regression variables x_{it} in the regARIMA model, as well as in the series y_t, are differenced by $(1 - B)^d(1 - B^s)^D$.

Notice that the regARIMA model as written in (4.8) assumes that the regression variables $x_{it}, i = 1, \ldots, r$, affect the dependent series y_t only at concurrent time points, i.e., model (4.8) does not explicitly provide for lagged regression effects such as $\beta_i x_{i,t-1}$. Lagged effects can be included by the X12ARIMA program, however, by reading appropriate user-defined lagged regression variables.

The X12ARIMA program provides additional flexibility in the specification of the ARIMA part of the regARIMA model by permitting (1) more than two multiplicative ARIMA factors, (2) missing lags within the AR and MA polynomials, (3) the fixing of individual AR and MA parameters at user-specified values when the model is estimated, and (4) inclusion of a trend constant, which is a nonzero overall mean for the differenced series $(1 - B)^d (1 - B^s)^D y_t$.

The most basic regression variable is the constant term. If the ARIMA model does not involve differencing, this is the usual regression intercept, which, if there are no other regression variables in the model, represents the mean of the series. If the ARIMA model does involve differencing, X12ARIMA uses a regression variable such that, when it is differenced according to the ARIMA model [see Eq. (4.9)], a column of ones is produced. The corresponding parameter is then called a trend constant, since it provides for a polynomial trend of the same degree as the order of differencing in the model. For example, with $d > 0$ and $D = 0$, the (undifferenced) trend constant regression variable is proportional to t^d. Notice that the lower order polynomial terms, t^j for $0 \leq j < d$, are not included among the regression variables because they would be differenced to zero by $(1 - B)^d$, hence their coefficients cannot be estimated. With or without the trend constant, the model (4.8) (or 4.9) implicitly allows for these lower order polynomial terms through the differencing. If seasonal differencing is requested $(D > 0)$, the nature of the undifferenced trend constant regression variable is more complicated, though the trend constant can be thought of as allowing for a polynomial of degree $d + sD$. Without a trend constant, model (4.8) implicitly allows for a polynomial of degree $d + sD - 1$.

Fixed seasonal effects in a monthly series can be modeled using 12 indicator variables, one for each calendar month. Since these 12 variables always add to one, they are confounded with an overall level effect. This leads to one of two singularity problems: collinearity with the usual constant term in a model with no differencing; or a singularity in a model with differencing since the 12 variables, when differenced, always sum to 0. One appropriate reparameterization instead uses 11 contrasts in the 12 indicator variables. An alternative reparameterization uses 11 variables taken from the Fourier (trigonometric) series representation of a fixed monthly pattern. X12ARIMA allows either of these options, and also allows specifying the trigonometric terms only for selected frequencies. For quarterly series, or for series with other seasonal periods, X12ARIMA constructs the appropriate versions of these variables. Notice that these variables cannot be used in a model with seasonal differencing, as they would all be differenced to zero.

Trading day effects occur when a series is affected by the different day-of-the-week compositions of the same calendar month in different years. Trading day effects can be modeled with seven variables that represent (*no. of Mondays*), ..., (*no. of Sundays*) in month t. Bell and Hillmer [1] proposed a better parameterization of the same effects using six variables defined as (*no. of Mondays*) − (*no. of*

Sundays), ..., (*no. of Saturdays*) − (*no. of Sundays*), along with a seventh variable
for Length of Month (LOM) or its deseasonalized version, the leap-year regressor
(lpyear). In X12ARIMA the six variables are called the *tdnolpyear* variables.
Instead of using a seventh regressor, a simpler and often better way to handle
multiplicative leap-year effects is to rescale the February values of the original
time series before transformation to $m_{\text{Feb}}y_t/m_t$, where y_t is the original time series
before transformation, m_t is the length of month t (28 or 29), and $\bar{m}_{\text{Feb}} = 28.25$ is
the average length of February. If the regARIMA model includes seasonal effects,
these can account for the length-of-month effect except in Februaries, so the trading
day model only has to deal with the leap-year effect. When this is done, only the
tdnolpyear variables need to be included in the model. X12ARIMA allows explicit
choice of either approach, as well as an option (td) that makes a default choice
of how to handle length-of-month effects. When the time series being modeled
represents the aggregation of some daily series (typically unobserved) over calendar
months they are called *monthly flow series*. If the series instead represents the value
of some daily series at the end of the month, called a *monthly stock series*, then
different regression variables are appropriate. Trading day effects at the end-of-
month stock series can be modeled using seven indicator variables for the day of the
week that the months end on. Since the sum of these variables is always one, this
leads to a singularity problem, so six variables are used instead. X12ARIMA also
allows specification of regression variables appropriate for stock series defined as of
some other day of the month. Trading day effects in quarterly series are relatively
rare because the calendar composition of quarters does not vary as much over time,
on a percentage basis, as that of months do. Trading day variables are not provided
for flow time series with seasonal periods other than monthly or quarterly, or for
stock series other than monthly.

X12ARIMA also provides a simplified model for trading day variation of
monthly or quarterly series that uses only one regressor, a weekday–weekend
variable:

$$D_t = (\text{no. of Weekdays}) - \frac{5}{2}(\text{no. of Saturdays and Sundays}).$$

The underlying assumption for this model is that all weekdays (Monday–Friday)
have identical effects, and Saturday and Sunday have identical effects.

Holiday effects in a monthly flow series arise from holidays whose dates vary
over time if: (1) the activity measured by the series regularly increases or decreases
around the date of the holiday, and (2) this affects 2 (or more) months depending on
the date the holiday occurs each year. Effects of holidays with a fixed date, such as
Christmas, are indistinguishable from fixed seasonal effects. Easter effects are the
most frequently found holiday effects in American and European economic time
series, since the date of Easter Sunday varies between March 22 and April 25. Labor
Day and Thanksgiving also are found in American and Canadian time series. The
basic model used by X12ARIMA for Easter and Labor Day effects assumes that
the level of activity changes on the k-th day before the holiday for a specified k,

and remains at the new level until the day before the holiday. For Thanksgiving the model used assumes that the level of activity changes on the day that is a specified number of days before or after Thanksgiving and remains at the new level until December 24. The regression variable constructed for the holiday effect is, for a given month t, the proportion of the affected time period that falls in month t. Actually, these regressors are deseasonalized by subtracting their long-run monthly means. Essentially the same Easter effect variable applies also to quarterly flow time series, but Labor Day and Thanksgiving effects are not present in quarterly series.

X12ARIMA provides four other types of regression variables to deal with abrupt changes in the level of a series of a temporary or permanent nature: *additive outliers* (AO), *level shifts* (LS), *temporary changes* (TC), and *ramps*. AOs affect only one observation in the time series, LSs increase or decrease all observations from a certain time point onward by some constant amount, TCs allow for an abrupt increase or decrease in the level of the series that returns to its previous level exponentially rapidly, and ramps allow for a linear increase or decrease in the level of the series over a specified time interval. LS regression variables are defined as -1 and then 0, in preference to an equivalent 0 and then 1, to make the overall level of the regression mean function of any forecasts consistent with the most recent level of the time series. Similar considerations dictate the definition of ramp variables. Often, however, large seasonal movements make it difficult to identify where such changes in level have occurred. Identifying the location and nature of potential outliers is the object of the outlier detection methodology implemented . This methodology can be used to detect AOs, TCs, and LSs (not ramps); any that are detected are automatically added to the model as regression variables. Prespecified AOs, LSs, TCs, and ramps are actually simple forms of interventions as discussed by Box and Tiao [4]. While X12ARIMA does not provide the full range of dynamic intervention effects discussed by Box and Tiao [4], often a short sequence of suitably chosen AO, LS, TC, and/or ramp variables can produce reasonable approximations to more complex dynamic intervention effects, although with one or two additional parameters.

4.6   Illustrative Example: X12ARIMA Seasonal Adjustment of the US NODG Series

The X12ARIMA seasonal adjustment program is an enhanced version of the X11ARIMA method [7]. The enhancements include a more self-explanatory and versatile user interface and a variety of new diagnostics to help the user detect and remedy any inadequacies in the seasonal and calendar adjustments obtained under the program options selected. As described in the previous sections, the program also includes a variety of new options to overcome adjustment problems and thereby enlarge the range of economic time series that can be adequately seasonally adjusted.

The main source of these new options is an extensive set of time series model building facilities for fitting what are called regARIMA models, that is, regression models with ARIMA errors. A fundamental use of regARIMA models is to extend the series with forecasts (and backcasts) in order to improve the seasonal adjustments of the most recent (and the earliest) data. Doing this mitigates problems inherent in the trend estimation and asymmetric seasonal averaging processes of the type used by the Census Method II-X11 near the ends of the series. The introduction of the ARIMA extension was a major improvement introduced by Estela Bee Dagum in the X11ARIMA. Its benefits, both theoretical and empirical, have been documented in many publications, see Dagum [7], Bobbitt and Otto [2], and references therein.

X12ARIMA is available as an executable program for computers running DOS (version 3.0 or higher), Sun 4 UNIX workstations, and VAX/VMS computers. When released, the X12ARIMA program is in the public domain, and can be copied or transferred. The corresponding documentation and examples are available online at http://www.census.gov/srd/www/x12a/. X12ARIMA is also implemented in OxMetrics, in the open source software Gretl, and in the Eurostat software Demetra+, and its recent extension JDemetra+. Since the 1990s, in the European countries, Eurostat has been playing a leading role in the promotion, development, and maintenance of source available software solution for seasonal adjustment of time series in line with established best practices. In this regard, Eurostat developed Demetra to provide a convenient and flexible tool for seasonal adjustment using either TRAMO/SEATS and X12ARIMA methods. Subsequently, Demetra+ was developed by the National Bank of Belgium, but due to constraints implied by the use of FORTRAN components, Eurostat created a new seasonal adjustment software which is platform independent and extensible, that is, JDemetra+ [13]. It is highly influenced by the output of TRAMO/SEATS, structural models, and X12ARIMA. JDemetra+ runs on operating systems that support the Java VM (Virtual Machine), such as Microsoft Windows XP SP3/Vista SP1/Windows 7, Ubuntu 9.10, various other Linux distributions, Solaris OS version 11 Express (SPARC and x86/x64 Platform Edition), and Macintosh OS X 10.6 Intel.

In this section, we make use of the X12ARIMA program as developed by the US Bureau of Census to perform the seasonal adjustment of the US New Orders for Durable Goods (NODG) series, observed from February 1992 to December 2013. We do not provide a detailed discussion of all the tables and results produced by the software, but we focus on the most important outputs for the interpretation of the seasonal adjustment of the NODG series.

4.6.1 Input: Specification File

To apply X12ARIMA to any particular time series, an input file, called specification file, must be created. This ASCII (or text) file contains a set of specifications that X12ARIMA reads to obtain the information it needs about the time series data, the time series model to be used, the analysis to be performed, and the output desired. This specification file has the extension .spc. It is common practice to allow the software to test automatically for the logarithmic or level specification, for the presence of trading days effects as well as Easter and leap-year effects. Furthermore, an automatic detection of several types of outliers [Additive Outliers (AO) and Level Shift (LS)] is generally required by the software. The most appropriate ARIMA$(p, d, q)(P, D, Q)_s$ model can be automatically selected by the software. Based on these considerations, we have written the specification file presented in Table 4.1. In detail, we have selected the following options:

- **series** describes the time series data (*data*), start date (*start*), seasonal period (*period*), and the series title (*title*);
- **transform** illustrates that we allow the software to automatically select the most appropriate transformation (level or log) for the data;
- **regression** specifies the regression variables used to form the regression part of the regARIMA model. In particular, we allow the estimation of the trading day effects (*td*), Easter effect over a span of 9 days before and after Easter (*easter[9]*), as well as of the leap-year effect if the log transformation is selected;
- **outlier** allows for an automatic detection of additive outliers and/or level shifts using the estimated model;
- **pickmdl** specifies an automatic selection procedure based on X11ARIMA [7]. It differs from the option **automdl** according to which an automatic model selection procedure is performed based on TRAMO [12]. The latter specification is implemented by default in both Gretl and JDemetra+. However, it should be noticed that we have found that this option generally selects less parsimonious models than needed. The main reason be the fact that it is automatically used together with the simplified trading day model with only one regressor. This generally produces wrong estimates that are left in the residuals and must be picked up by the ARIMA model which is then more complex than necessary. Using this option the model chosen for the log transformed data of the US New Orders for Durable Goods time series is the ARIMA$(1,1,2)(0,1,1)_{12}$, but using the trading

Table 4.1 Specification file to perform the X12ARIMA seasonal adjustment of the NODG series

```
series{
period=12
title="nodg"
start=
data=(...)
}
transform{function=auto}
regression{variables = (td easter[9])}
outlier{}
pickmdl{}
check{print=all}
x11{save=( d11 d12 d13 ) print=alltables}
```

day with six regressors and the automatic selected ARIMA$(0,1,1)(0,1,1)_{12}$ we obtained better results. As discussed by the main X11ARIMA guidelines [7] and by Fischer and Planas [11], the airline model, ARIMA$(0,1,1)(0,1,1)_s$, is generally found to be appropriate in the seasonal adjustment of the major economic and social indicators. In this regard, we recommend to use the trading day model based on six regressors together with the pickmdl option in X12ARIMA;

- **check** produces statistics useful to perform a diagnostic checking of the estimated model. Based on the *print=all* specification, the software provides summary statistics of the residuals, corresponding sample ACF and PACF with associated standard errors, Ljung–Box Q statistics, normality test statistics, and a spectral plot of the model residuals, as well as an histogram of the standardized residuals;
- **x11** describes the seasonal adjustment options, including mode of adjustment, seasonal and trend filters, and Easter holiday adjustment option, and some seasonal adjustment diagnostics. Based on the selected option *save=(d11 d12 d13)* the seasonally adjusted series, the estimated trend-cycle, and irregular components are saved in the workspace. Furthermore, based on *print=alltables*, all the tables produced by the X12ARIMA software are displayed in the output.

4.6.2   Testing for the Presence of Identifiable Seasonality

Before going through the X12ARIMA decomposition, it is necessary to check if the original series presents identifiable seasonality. In this regard, Table D8 provides *combined seasonality tests* that are useful to check for the presence of seasonality in the original series. This test combines the test for the presence of moving seasonality with the *F*-test for the presence of stable seasonality and the Kruskal–Wallis chi-squared test, another nonparametric test for the presence of stable seasonality. The main purpose of this test is to determine whether the seasonality of the series is identifiable or not. For example, if there is little stable seasonality and most of the process is dominated by rapidly moving seasonals, chances are that the seasonals will not be accurately estimated for they will not be properly identified by the X12ARIMA method. Table 4.2 illustrates the output for the NODG series.

The *nonparametric Friedman test* is used to determine if stable seasonality is present in the series. This test uses a preliminary estimate of the unmodified Seasonal–Irregular (SI) component. Indeed, the seasonal component includes the intra-year variation that repeats each year (stable seasonality) or evolves from year to year (moving seasonality). A high test statistic and low *p*-value indicate that a significant amount of variation in the SI ratios is due to months (or quarters), which in turn is evidence of seasonality. If the *p*-value is lower than 1 %, the null hypothesis of no seasonal effect is rejected. Conversely, a small value of the *F*-test and great *p*-value (close to 1.0) is evidence that variation due to months or quarters could be determined by the random error and the null hypothesis of no month/quarter effect is not rejected. As shown in Table 4.2, in our specific case, the null hypothesis is rejected, such that seasonality is present.

Table 4.2 Combined seasonality test performed on the NODG series

D8.A F-tests for seasonality

Test for the presence of seasonality assuming stability.

	Sum of squares	Dgrs. freedom	Mean square	F-value
Between months	9305.6408	11	845.96735	111.033 $**$
Residual	1912.3856	251	7.61907	
Total	11,218.0264	262		

**Seasonality present at the 0.1 % level.

Nonparametric test for the presence of seasonality assuming stability

Kruskal–Wallis statistic	Dgrs. freedom	Probability level
213.9990	11	0.000 %

Seasonality present at the 1 % level.

Moving Seasonality Test

	Sum of squares	Dgrs. freedom	Mean square	F-value
Between years	60.1151	20	3.005756	0.495
Error	1335.2196	220	6.069180	

No evidence of moving seasonality at the 5 % level.

COMBINED TEST FOR THE PRESENCE OF IDENTIFIABLE SEASONALITY

IDENTIFIABLE SEASONALITY PRESENT

Kruskal–Wallis test is the second test for stable seasonality provided by the software. The test is calculated for the final estimation of the unmodified SI component from which k samples are derived ($k = 12$ for monthly series and $k = 4$ for quarterly series) of size $n_1, n_2, \ldots, n_k$, respectively. The test is based on the statistic:

$$W = \frac{12}{n(n+1)} \sum_{j=1}^{k} \frac{S_j}{n_j} - 3(n+1),$$

where S_j is the sum of the ranks of the observations from the j-th sample within the whole sample of $n = \sum_{j=1}^{k} n_j$ observations. Under the null hypothesis, the test statistic follows a chi-square distribution with $k - 1$ degrees of freedom. Table 4.2 shows that, for the NODG series, the outcome of the test is *stable seasonality present*.

A more detailed description of these two tests is provided in Sect. 4.4.3, and we refer the reader to that section for the theoretical description of the *moving seasonality test* as well as of the *combined seasonality test*. As shown in Table 4.2, all the tests confirm the presence of stable seasonality and the absence of moving seasonality in the US NODG series.

4.6.3 Pre-processing

As discussed in the previous sections, the X12ARIMA method consists of two linked parts: the regARIMA model for estimation of the deterministic components, and the decomposition part of the linearized series for the stochastic components performed using the Census II-X11 filters combined with those of the ARIMA extrapolation. Part A of the X12ARIMA output provides the results for the user selected options to correct a priori the series by introducing adjustment factors. For the NODG series, these latter include correction for the effect of Easter holiday, of changes in the level of the series (e.g., effect of a strike), and also introduce seven weights for each day of the week to take into account the variations due to the trading day composition of the months.

The first information provided by the software is about the automatic selection between the level and logarithmic transformation for the data as illustrated in Table 4.3. In particular, likelihood statistics for model fit to untransformed and log transformed series are displayed. It can be noticed that for the NODG series the logarithmic transformation is applied and a multiplicative seasonal adjustment model is selected.

4.6.3.1 Regressors

Based on our specification, the X12ARIMA provides the estimated coefficients of the trading days variables to evaluate their significant effect on the NODG series. Table 4.4 illustrates that the variables related to Wednesday and Sunday (derived) are characterized by coefficients that are significantly different from one. Specifically, Fig. 4.1 displays the monthly trading days component that ranges from 98.7 to 103.99 over the observed period February 1992–December 2013, being on average equal to 100. The Easter effect is also evaluated in the model specification, and the value of the estimated corresponding coefficient is reported with its standard error and t-statistic. As shown in Table 4.4, there is a significant effect lasting over 9 days around Easter. In particular, Fig. 4.2 displays that the Easter effect is felt 9 days before Easter and on Easter Sunday but not after Easter. This is evidenced by years 1997, 1999, 2002, 2005, 2008, 2010, and 2013, where Easters fall very early in April or in March. Indeed, orders are generally affected by Easter, because customs do not operate from Good Friday to Easter Monday.

The software automatically detects the presence of three different outliers: an additive outlier at June 2000, and two level shift outliers on October 2008 and January 2009, respectively.

Table 4.3 Automatic selection between level and logarithmic transformation of the NODG series

Likelihood statistics for model fit to untransformed series.

Likelihood statistics	
Number of observations (nobs)	263
Effective number of observations (nefobs)	250
Number of parameters estimated (np)	11
AIC	5258.6176
AICC (*F*-corrected-AIC)	5259.7268
Hannan–Quinn	5274.2077
BIC	5297.3537

Likelihood statistics for model fit to log transformed series.

Likelihood statistics	
Number of observations (nobs)	263
Effective number of observations (nefobs)	250
Number of parameters estimated (np)	10
Log likelihood	428.0586
Transformation Adjustment	−3029.8684
Adjusted Log likelihood (L)	−2601.8098
AIC	5223.6195
AICC (*F*-corrected-AIC)	5224.5400
Hannan–Quinn	5237.7924
BIC	5258.8342

AICC (with aicdiff= −2.00) prefers log transformation
Multiplicative seasonal adjustment will be performed.

4.6.3.2 ARIMA Model

For the log transformed NODG series under study, the following ARIMA(0,1,1) $(0,1,1)_{12}$ (airline) model has been automatically selected and estimated by the software:

$$(1 - B)(1 - B^{12})Y_t = (1 - 0.5357B)(1 - 0.7018B^{12})a_t,$$

where a_t is assumed to follow a white noise process with zero mean and estimated variance equal to 0.0124. All the coefficients result significantly different from zero and satisfy the invertibility conditions (Table 4.5).

Similar to X11ARIMA the ARIMA model to be used in the context of the X12ARIMA method must fulfill the double condition of fitting the data well and generating "acceptable" projections for the last 3 years of observed data. By "acceptable" projection is meant a projection with a mean absolute error smaller than 5 % for well-behaved series (e.g., unemployment adult males) and smaller than 12 % for highly irregular series (e.g., unemployment teenage males). In this regard,

Table 4.4 Estimates of calendar effects and outliers detection

Regression model

Variable	Parameter estimate	Standard error	t-Value
Trading day			
Mon	0.0048	0.00427	1.12
Tue	−0.0076	0.00426	−1.78
Wed	0.0129	0.00424	3.05
Thu	−0.0034	0.00423	−0.81
Fri	0.0059	0.00428	1.38
Sat	−0.0025	0.00425	−0.59
*Sun (derived)	−0.0101	0.00427	−2.37
Easter [9]	−0.0250	0.00898	−2.79
Automatically identified outliers			
AO2000.Jun	0.1547	0.02904	5.32
LS2008.Oct	−0.1808	0.02803	−6.45
LS2009.Jan	−0.2358	0.02808	−8.40

*For full trading day and stable seasonal effects, the derived parameter estimate is obtained indirectly as minus the sum of the directly estimated parameters that define the effect.

Chi-squared tests for groups of regressors

Regression effect	df	Chi-square	p-value
Trading day	6	28.88	0.00

it is recommended the identification of the ARIMA model be performed on data in which extreme values have been previously modified. This recommendation is even more relevant if the outliers fall in the most recent years, in order to avoid the rejection of good models simply because the outliers have inflated the absolute average extrapolation error above the acceptance level of the guidelines. It can be noticed in Table 4.6 that these requirements are satisfied by the estimated ARIMA model.

To determine whether or not a model fits the data well, the Portmanteau test of fit developed by Box and Pierce [3] with the variance correction for small samples introduced by Ljung and Box [19] is used. As shown in Table 4.6, the null hypothesis of randomness of the residuals is tested and not rejected at a 10 % level.

4.6.3.3   Residuals

Beside the Portmanteau test previously introduced, a more complete analysis of the residuals of the estimated ARIMA model has been carried out by specifying the option *check* in the specification file. As suggested by the Portmanteau test, the residuals satisfy the assumption of being generated by a purely random process

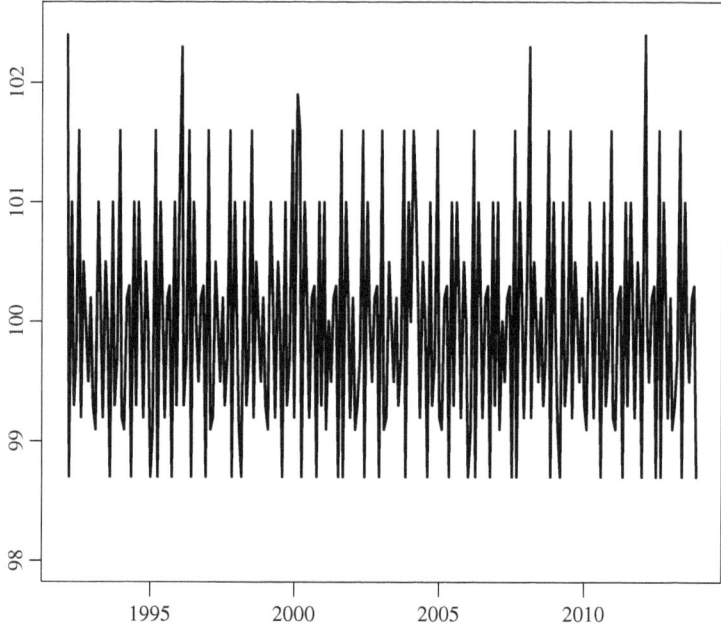

Fig. 4.1 Trading day estimates of the US New Orders for Durable Goods

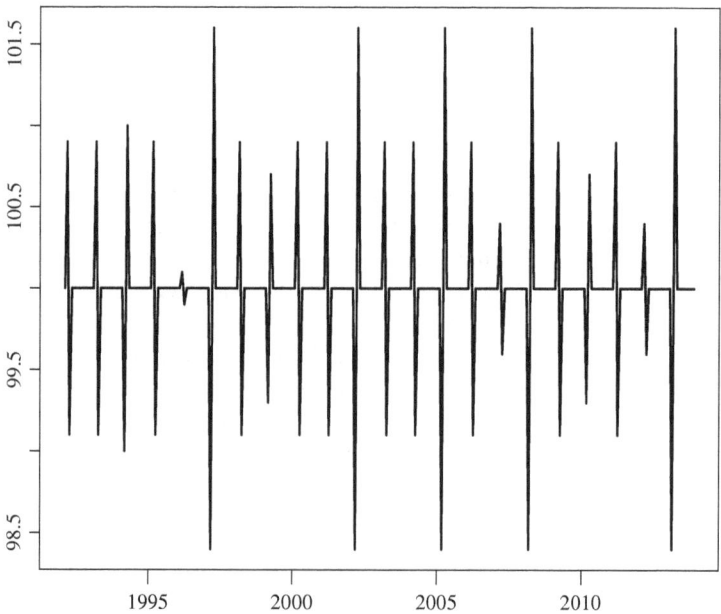

Fig. 4.2 Moving holiday component of the US NODG

Table 4.5 Estimated ARIMA$(0,1,1)(0,1,1)_{12}$ model

ARIMA model: (0 1 1)(0 1 1)
Nonseasonal differences: 1
Seasonal differences: 1

Parameter	Estimate	Standard errors
Nonseasonal MA		
Lag 1	0.5357	0.05229
Seasonal MA		
Lag 12	0.7018	0.04671
Variance	0.12400E−02	
SE of Var	0.11091E−03	

Likelihood statistics	
Number of observations (nobs)	263
Effective number of observations (nefobs)	250
Number of parameters estimated (np)	13
Log likelihood	477.6115
Transformation Adjustment	−3029.8684
Adjusted Log likelihood (L)	−2552.2569
AIC	5130.5139
AICC (*F*-corrected-AIC)	5132.0562
Hannan–Quinn	5148.9386
BIC	5176.2928

Table 4.6 Average absolute percentage error in forecasts and goodness of fit for the estimated ARIMA$(0,1,1)(0,1,1)_{12}$ model

Average absolute	percentage error in	within-sample forecasts:	
Last year:	3.27	Last 1 year:	6.93
Last 2 years:	2.09	Last 3 years:	4.10
Chi-square	Probability: 17.34 %	($Q = 28.0679$, 22 df)	

with constant mean and variance. However, to verify this assumption in more detail, several statistics are computed on these residuals. Furthermore, the normality assumption on the residual distribution is not rejected.

4.6.4 Decomposition

The *Decomposition* part includes tables with results from consecutive iterations of the Census II-X11 filters and quality measures. As discussed in Sects. 4.4 and 4.5, to estimate the different components of the series, while taking into account the possible presence of extreme observations, X12ARIMA proceeds iteratively through four processing stages, denoted as A, B, C, and D, plus two stages, E, and Quality measures, that propose statistics and charts and are not part of the decomposition per se. Stage A has been widely discussed in the previous section. On stages B, C, and D the basic algorithm is used. Specifically,

B. *First automatic correction of the series*. This stage consists in a first estimation and downweighting of the extreme observations and a first estimation of the working day effects. This stage is performed by applying the basic algorithm detailed earlier. These operations lead to Table B20, adjustment values for extreme observations, used to correct the unadjusted series and result in the series from Table C1.

C. *Second automatic correction of the series*. Still applying basic algorithm once again, this part leads to a more precise estimation of replacement values of the extreme observations (Table C20). The series, finally "cleaned up", is shown in Table D1 of the printouts.

D. *Seasonal adjustment*. This part, at which the basic algorithm is applied for the last time, is that of the seasonal adjustment per se, as it leads to final estimates:

  • of the seasonal component (Table D10);
  • of the seasonally adjusted series (Table D11);
  • of the trend-cycle component (Table D12); and
  • of the irregular component (Table D13).

  X12ARIMA selects the filters automatically, taking into account the global moving seasonality ratio, which is computed on preliminary estimates of the irregular component and of the seasonal. For the NODG series, a 3×5 m.a. has been selected to estimate the seasonal component, whereas the trend-cycle has been estimated using a 13-term Henderson filter, being the signal-to-noise ratio equal to 2.47. The seasonally adjusted NODG series is shown in Fig. 4.3 together with the original series, and the irregular component in Fig. 4.4.

E. *Components modified for large extreme values*. Part E includes: (1) components modified for large extreme values, (2) a comparison between the annual totals of the raw time series and seasonally adjusted time series, (3) changes in the final seasonally adjusted series, (4) changes in the final trend-cycle, and (5) robust estimation of the final seasonally adjusted series. The results from part E are used in the subsequent part to calculate the quality measures.

F. *Quality measures*. The final part contains statistics for judging the quality of the seasonal adjustment, as described in Sects. 4.4 and 4.5. The M statistics are used to assess the quality of the seasonal adjustment. They account for the relative contribution of the various components (irregular, trend-cycle, seasonal,

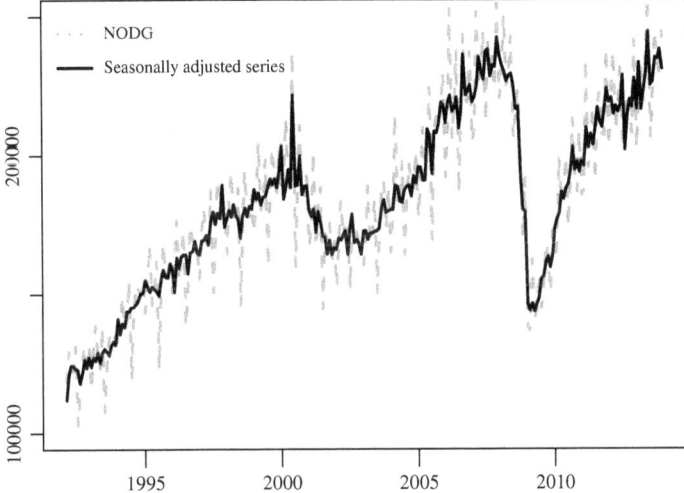

Fig. 4.3 NODG original and seasonally adjusted series

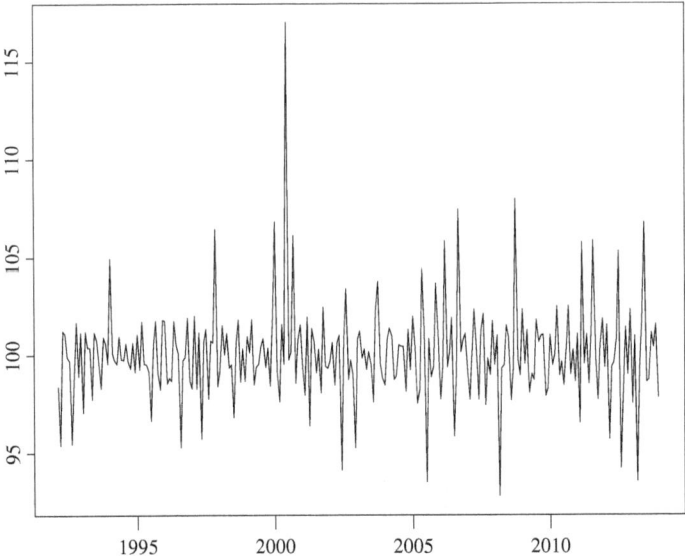

Fig. 4.4 Irregular component for the US NODG series

preliminary factors, trading days, and holidays effects) to the variance of the stationary part of the original time series. These statistics vary between 0 and 3, but only values smaller than 1 are acceptable. Table 4.7 illustrates these statistics for the components estimated in the X12ARIMA seasonal adjustment of the NODG series. It can be noticed that all the measures range from 0 to 3, indicating a good seasonal adjustment for the series.

Table 4.7 Diagnostic checking on the residuals of the estimated ARIMA$(0,1,1)(0,1,1)_{12}$ model

F3. Monitoring and quality assessment statistics

The measures below are between 0 and 3; acceptance region from 0 to 1.

1. The relative contribution of the irregular over 3 months span (from Table F2.B). $M1 = 0.827$

2. The relative contribution of the irregular component to the stationary portion of the variance (from Table F 2.F). $M2 = 0.939$

3. The amount of month-to-month change in the irregular component as compared to the amount of month-to-month change in the trend-cycle (from Table F2.H). $M3 = 0.736$

4. The amount of autocorrelation in the irregular as described by the average duration of run (Table F 2.D). $M4 = 0.285$

5. The number of months it takes the change in the trend-cycle to surpass the amount of change in the irregular (from Table F 2.E). $M5 = 0.625$

6. The amount of year-to-year change in the irregular as compared to the amount of year-to-year change in the seasonal (from Table F 2.H). $M6 = 0.295$

7. The amount of moving seasonality present relative to the amount of stable seasonality (from Table F 2.I). $M7 = 0.195$

8. The size of the fluctuations in the seasonal component throughout the whole series. $M8 = 0.518$

9. The average linear movement in the seasonal component throughout the whole series. $M9 = 0.236$

10. Same as 8, calculated for recent years only. $M10 = 0.454$

11. Same as 9, calculated for recent years only. $M11 = 0.332$

ACCEPTED at the level 0.50

Q (without M2) = 0.45 ACCEPTED.

A composite indicator calculated from M statistics is the Q statistic, defined as follows:

$$Q = \frac{10M_1 + 12M_2 + 10M_3 + 8M_4 + 11M_5 + 10M_6 + 18M_7 + 7M_8 + 7M_9 + 4M_{10} + 4M_{11}}{100}$$

Q without M_2 (also called Q_2) is the Q statistic without the M_2 statistics. If time series does not cover at least 6 years, statistics M_8, M_9, M_{10}, and M_{11} cannot be calculated. In this case, the Q statistics is computed as:

$$Q = \frac{14M_1 + 15M_2 + 10M_3 + 8M_4 + 11M_5 + 10M_6 + 32M_7 + 0M_8 + 0M_9 + 0M_{10} + 0M_{11}}{100}.$$

The model has satisfactory quality if the Q statistic is less than 1. In this regard, it can be noticed that the X12ARIMA seasonal adjustment of the NODG series is satisfactory.

References

1. Bell, W. H., & Hillmer, S. C. (1984). Issues involved with the seasonal adjustment of economic time series. *Journal of Business and Economic Statistics, 2*, 291–320.
2. Bobbit, L., & Otto, M. C. (1990). Effects of forecasts on the revisions of seasonally adjusted values using the X11 seasonal adjustment procedure. In *Proceedings of the American Statistical Association Business and Economic Statistics Session* (pp. 449–453).
3. Box, G. E. P., & Jenkins, G. M. (1970). *Time series analysis: Forecasting and control*. Holden-Day: San Francisco (2nd ed., 1976).
4. Box, G. E. P., & Tiao, G. C. (1975). Intervention analysis with applications to economics and environmental problems. *Journal of the American Statistical Association, 70*, 70–79.
5. Cholette, P. A. (1979). A comparison and assessment of various adjustment methods of sub-annual series to yearly benchmarks. Research paper. Seasonal Adjustment and Time Series Staff, Statistics Canada, Ottawa.
6. Dagum, E. B. (1978). Modelling, forecasting and seasonally adjusting economic time series with the X-11 ARIMA method. *The Statistician, 27*(3), 203–216.
7. Dagum, E. B. (1988). *The X-11 ARIMA/88 seasonal adjustment method-foundations and user's manual*. Statistics Canada, Ottawa: Time Series Research and Analysis Centre.
8. Dagum, E. B., Chhab, N., & Chiu, K. (1990). Derivation and properties of the X11ARIMA and census X11 linear filters. *Journal of Official Statistics, 12*(4), 329–347.
9. Findley, D. F., Monsell, B. C., Bell, W. R., Otto, M.C., & Chen B. C. (1998). New capabilities and methods of the X12ARIMA seasonal adjustment program. *Journal of Business and Economic Statistics, 16*(2), 127–152.
10. Findley, D. F., Monsell, B. C., Shulman, H. B., & Pugh, M. G. (1990). Sliding spans diagnostics for seasonal and related adjustments. *Journal of the American Statistical Association, 85*, 345–355.
11. Fischer, B., & Planas, C. (2000). Large scale fitting of regression models with ARIMA errors. *Journal of Official Statistics, 16*(2), 173–184.
12. Gomez, V., & Maravall, A. (1996). *Program TRAMO and SEATS*. Madrid: Bank of Spain.
13. Grudkowska, S. (2015). *JDemetra+ Reference manual*. Warsaw: National Bank of Poland.
14. Henderson, R. (1916). Note on graduation by adjusted average. *Transaction of Actuarial Society of America, 17*, 43–48.

15. Higginson, J. (1975). *An F-test for the presence of moving seasonality when using Census Method II X-11 variant*. Research paper, Seasonal Adjustment and Time Series Staff, Statistics Canada, Ottawa.

16. Huot, G. (1975). Quadratic minimization adjustment of monthly or quarterly series to annual totals. Research paper. Seasonal Adjustment and Time Series Staff, Statistics Canada, Ottawa.

17. Huot, G., & De Fontenay, A. (1973). General seasonal adjustment guidelines. Part 1, Statistics Canada's version of the U.S. Bureau of the Census monthly multiplicative and additive X-11 program. Research paper. Seasonal Adjustment and Time Series Staff, Statistics Canada, Ottawa

18. Ladiray, D., & Quenneville, B. (2001). *Seasonal adjustment with the X11 method*. Lecture notes in statistics (Vol. 158). New York: Springer.

19. Ljung, G., & Box, G. (1979). The likelihood function of stationary autoregressive-moving average models. *Biometrika, 66*, 265–270.

20. Lothian, J., & Morry, M. (1978a). A test for the presence of identifiable seasonality when using the X-11 program. Research paper. Seasonal Adjustment and Time Series Analysis Staff, Statistics Canada, Ottawa.

21. Lothian, J., & Morry, M. (1978b). A set of guidelines to assess the reliability of seasonal adjustment by the X-11 program. Research paper. Seasonal Adjustment and Time Series Staff, Statistics Canada, Ottawa.

22. Macaulay, F. R. (1931). *The smoothing of time series*. New York: National Bureau of Economic Research.

23. Musgrave, J. (1964). A set of end weights to end all end weights. Working paper. U.S. Bureau of Census, Washington, DC.

24. Newbold, P., & Granger, C.W. J. (1974). Experience with forecasting univariate time series and the combination of forecasts (with discussion). *Journal of the Royal Statistical Society. Series A, 137*, 131–165.

25. Reid, D. J. (1975). A review of short-term projection techniques. In H. A. Gorden (Ed.), *Practical aspects of forecasting* (pp. 8–25). London: Operational Research Society.

26. Shiskin, J., Young, A. H. & Musgrave, J. C. (1967). The X-11 variant of the Census Method II seasonal adjustment program. Technical Paper 15 (revised). US Department of Commerce, Bureau of the Census, Washington, DC.

27. Wallis, W. A., & Moore, G. H. (1941). A significance test for time series analysis. *Journal of the American Statistical Association, 36*, 401–409.

28. Young, A. H. (1968). A linear approximation to the Census and BLS seasonal adjustment methods. *Journal of the American Statistical Association, 63*, 445–457.

Chapter 5
Seasonal Adjustment Based on ARIMA Model Decomposition: TRAMO-SEATS

Abstract TRAMO-SEATS is a seasonal adjustment method based on ARIMA modeling. TRAMO estimates via regression dummy variables and direct ARIMA modeling (regARIMA) the deterministic components, trading days, moving holidays, and outliers, which are later removed from the input data. In a second round, SEATS estimates the stochastic components, seasonality, and trend-cycle, from an ARIMA model fitted to the data where the deterministic components are removed. SEATS uses the filters derived from the ARIMA model that describes the behavior of the time series. By imposing certain conditions, a unique canonical decomposition is performed to obtain the ARIMA models for each component. This chapter discusses with details the estimation methods used by TRAMO and SEATS as well as the basic assumptions on the derivation of ARIMA models for each component. An illustrative example of the seasonal adjustment with the TRAMO-SEATS software default option is shown with the US New Orders for Durable Goods series. The illustrative example concentrates on the most important tables of this software that enable to assess the quality of the seasonal adjustment.

Peter Burman [4] was the first to develop a seasonal adjustment method based on ARIMA model decomposition that he named SIGEX (Signal Extraction). Later, working on the same topic, Hillmer and Tiao [16] developed what these authors called ARIMA model-based seasonal adjustment, largely discussed in Bell and Hillmer [2]. An ARIMA model is identified from the observed data and, by imposing certain restrictions, models for each component are derived. Since the components are unknown, to obtain a unique solution Hillmer and Tiao proposed a canonical decomposition which has the property of maximizing the variance of the irregulars and minimizing the variance of the estimated components. Because ARIMA model identification and estimation are not robust to outliers or extreme values and cannot deal with deterministic components such as trading days and moving holidays, further changes were made by combining dummy variables regression models with ARIMA models. In this regard, Gomez and Maravall [12] produced at the Bank of Spain a seasonal adjustment software called TRAMO-SEATS which is currently applied mainly by European statistical agencies. TRAMO

© Springer International Publishing Switzerland 2016
E. Bee Dagum, S. Bianconcini, *Seasonal Adjustment Methods and Real Time Trend-Cycle Estimation*, Statistics for Social and Behavioral Sciences,
DOI 10.1007/978-3-319-31822-6_5

stands for "Time series Regression with ARIMA noise, Missing observations and Outliers," and SEATS for "Signal Extraction in ARIMA Time Series." First, TRAMO estimates via regression the deterministic components, which are later removed from the input data. In a second round, SEATS estimates the seasonal and trend-cycle components from the ARIMA model fitted to the data where the deterministic components are removed. SEATS uses the filters derived from the ARIMA model that describes the stochastic behavior of the linearized time series.

It should be mentioned that Eurostat, in collaboration with the National Bank of Belgium, the US Bureau of the Census, the Bank of Spain, and the European Central Bank, has developed an interface of TRAMO-SEATS and X12ARIMA called Demetra+. In the Bank of Spain and Eurostat websites it is also possible to find a considerable number of papers relevant to TRAMO-SEATS as well as in the European Statistical System (ESS) Guidelines. Next sections briefly describe the supporting methodology of TRAMO-SEATS.

5.1 TRAMO: Time Series Regression with ARIMA Noise, Missing Observations, and Outliers

TRAMO (Time Series Regression with ARIMA Noise, Missing Observations, and Outliers) is a regression method that performs the estimation, forecasting, and interpolation of missing observations and ARIMA errors, in the presence of possibly several types of outliers. The ARIMA model can be identified automatically or by the user. Given the vector of observations $\mathbf{y} = (y_1, \ldots, y_n)'$, the program fits the regression model

$$y_t = \mathbf{x}_t' \boldsymbol{\beta} + v_t, \tag{5.1}$$

where $\boldsymbol{\beta} = (\beta_1, \ldots, \beta_p)'$ is a vector of regression coefficients, $\mathbf{x}_t = (x_{1t}, \ldots, x_{pt})'$ are p regression variables that define the deterministic part of the model. On the other hand, v_t represents the stochastic part of the model, that is assumed to follow the general ARIMA model

$$\Psi(B)v_t = \pi(B)a_t, \tag{5.2}$$

where B is the backshift operator, $\Psi(B)$ and $\pi(B)$ are finite polynomials in B, and a_t is assumed to be normally identically distributed, that is, $\mathrm{NID}(0, \sigma_a^2)$.

The polynomial $\Psi(B)$ contains the unit roots associated with regular and seasonal differencing, as well as the polynomial with stationary autoregressive roots (and complex unit roots, if present). $\pi(B)$ denotes the invertible moving average polynomial. In TRAMO, these polynomials assume the following multiplicative form:

$$\Psi(B) = \phi_p(B)\Phi_P(B^s)(1 - B)^d(1 - B^s)^D$$
$$= (1 + \phi_1 B + \cdots + \phi_p B^p)(1 + \Phi_1 B^s + \cdots + \Phi_P B^{sP})(1 - B)^d(1 - B^s)^D,$$
$$\pi(B) = \theta_q(B)\Theta_Q(B^s)$$
$$= (1 + \theta_1 B + \cdots + \theta_q B^q)(1 + \Theta_1 B^s + \cdots + \Theta_Q B^{sQ}),$$

where s denotes the number of observations per year, e.g., 12 for monthly and 4 for quarterly data. The model may contain a constant c related to the mean of the differenced series $(1 - B)^d(1 - B^s)^D y_t$. In practice, this parameter is estimated as one of the regression parameters in (5.1). Initial estimates of the parameters can be input by the user, set to the default values, or computed by the program.

The regression variables $\mathbf{x}_t$ can be given by the user or generated by the program. In the latter case, the variables correspond to trading day variations, Easter effects, and outliers.

Trading day variations are linked to the calendar of each year. Traditionally, six variables have been used to model the trading day effects. These are: (*no. of Mondays*) − (*no. of Sundays*), ..., (*no. of Saturdays*) − (*no. of Sundays*). The sum of the effects of each day of the week cancels out. Mathematically, this can be expressed by the requirement that the trading day coefficients $\beta_j, j = 1, \ldots, 7$, verify $\sum_{j=1}^{7} \beta_j = 0$, which implies $\beta_7 = -\sum_{j=1}^{6} \beta_j$. There is the possibility to consider a more parsimonious modeling of the trading day effects using one variable instead of six. In this case, the days of the week are first divided into two categories: working days and nonworking days. Then the variable is defined as (*no. of Mondays, Tuesdays, Wednesdays, Thursdays, and Fridays*) − 5/2(*no. of Saturdays and Sundays*). Again, the motivation is that it is desirable that the trading day coefficients $\beta_j, j = 1, \ldots, 7$, verify $\sum_{j=1}^{7} \beta_j = 0$. Since $\beta_1 = \beta_2 = \cdots = \beta_5$ and $\beta_6 = \beta_7$, we have $5\beta_1 = -2\beta_6$. The set of trading day variables may also include the *leap-year effect*. The leap-year variable assumes value 0 for months that are not February, −0.25 for Februaries with 28 days, and 0.75 for Februaries with 29 days (leap year). Over a 4-year period, this effect cancels out. Trading day effects typically characterize series reflecting economic activity, and if significant, most often adding Easter effect variables improves significantly the results.

Easter effects. This variable models a constant change in the level of daily activity during the d days before Easter, and its typical value ranges between three and six. The value of d can also be supplied by the user. The variable has zeros for all months different from March and April. The value assigned to March is equal to $p_M - m_M$, where p_M is the proportion of the d days that fall on that month and m_M is the mean value of the proportions of the d days that fall on March over a long period of time. The value assigned to April is $p_A - m_A$, where p_A and m_A are defined analogously. Usually, a value of $m_M = m_A = 1/2$ is a good approximation. Since $p_A - m_A = 1 - p_M - (1 - m_M) = -(p_m - m_M)$, a desirable feature is that the sum of the effects of both months, March and April, cancels out. If forecast of the series is desired, the Easter effect variable has to be extended over the forecasting horizon (2 years when TRAMO is used with SEATS). Furthermore, given that the holidays vary among countries, it is strongly recommended that each country builds its own (set of) holiday regression variables, extended over the forecast period.

Outliers reflect the effect of some special, nonregularly repetitive events, such as implementation of a new regulation, major political or economic changes, modifications in the way the variable is measured, occurrence of a strike or natural disaster, etc. Consequently, discordant observations and various types of abrupt changes are often present in times series data. The location and type of outliers are "a priori" unknown. TRAMO uses an improved Chen and Liu [6] type procedure for outlier detection and correction. The effect of an outlier is modeled by $y_t = \omega \xi(B) I_t(t_0) + v_t$, where $\xi(B)$ is a quotient of polynomials in B that models the type of outlier (its effect over time), $I_t(t_0)$ is an indicative variable of the occurrence of the outlier, that is,

$$I_t(t_0) = \begin{cases} 1 \text{ if } & t = t_0 \\ 0 \text{ otherwise,} \end{cases}$$

whereas ω represents the impact of the outlier at time t_0 and v_t is the outlier free series which follows the model specified in Eq. (5.2). In the automatic detection and correction, by default, three types of outliers can be considered:

(1) *Additive Outlier (AO)*: $\xi(B) = 1$;
(2) *Transitory Change (TC)*: $\xi(B) = 1/(1 - \delta B)$, where, by default, $\delta = 0.7$; and
(3) *Level Shift (LS)*: $\xi(B) = 1/(1 - B)$.

One can also include a fourth type of outlier, that is *Innovational Outlier (IO)* for which $\xi(B) = \theta(B)/\phi(B)\delta(B)$ that resembles a shock in the innovations a_t.

The procedure followed by TRAMO consists of:

1. *Pretest for the log/level specification.* A trimmed range mean regression test is performed to select whether the original data will be transformed into log or maintain the level. The former is selected if the slope is positive, whereas the latter is chosen if it is negative. When the slope is close to zero, a selection is made based on the Bayesian Information Criterion (BIC) when applied to the default model using both specifications. A second test consists of direct comparison of the BICs of the default model in levels and in logs (with a proper correction).
2. *Pretest for trading days and Easter effects.* These pretests are made with regressions using the default model for the noise and, if the model is subsequently changed, the test is redone. Thus, the output file of TRAMO may say at the beginning "Trading day is not significant," but the final model estimated may contain trading day variables (or vice versa).
3. *Automatic detection and correction of outliers.* The program has a facility for detecting outliers and for removing their effects. The outliers can be entered by the user or they can be automatically identified by the program, using an original

approach based on those of Tsay [29] and Chen and Liu [6]. The outliers are identified one by one, as proposed by Tsay [29], and multiple regressions are used, as in [6], to detect spurious outliers. The procedure used to incorporate or reject outliers is similar to the stepwise regression procedure for selecting the "best" regression equation. This results in a more robust procedure than that of Chen and Liu [6], which uses "backward elimination" and may therefore detect too many outliers in the first step of the procedure.

4. *Automatic model selection.* The program further performs an automatic identi-fication of the ARIMA model. This is done in two steps. The first one yields the nonstationary polynomial $(1 - B)^d (1 - B^s)^D$ of model (5.2). This is done by iterating on a sequence of AR and ARMA models (with constant) which have a multiplicative structure when the data is seasonal. The procedure is based on the results of Tiao and Tsay [27], and Tsay [28]. Regular and seasonal differences are obtained, up to a maximum order of $(1 - B)^2 (1 - B^s)$.

The second step identifies an ARMA model for the stationary series (modified for outliers and regression-type effects) following the Hannan–Rissanen proce-dure [15], with an improvement which consists of using the Kalman filter to calculate the first residuals in the computation of the estimator of the variance of the innovations of model (5.2). For the general multiplicative model

$$\phi_p(B)\Phi_P(B^s)(1 - B)^d(1 - B^s)^D y_t = \theta_q(B)\Theta_Q(B^s)a_t, \qquad a_t \sim WN(0, \sigma_a^2),$$

the search is made over the range $0 \le (p, q) \le 3$, and $0 \le (P, Q) \le 2$. This is done sequentially (for fixed regular polynomials, the seasonal ones are obtained, and vice versa), and the final orders of the polynomials are chosen according to the BIC criterion, with some possible constraints aimed at increasing parsimony and favoring "balanced" models (similar AR and MA orders).

Finally, the program combines the facilities for automatic identification and correction of outliers and automatic ARIMA model identification just described so that it has an option for automatic model identification of a nonstationary series in the presence of outliers.

5. *Estimation of the regARIMA model.* The basic methodology is described by Gomez and Maravall [11], and additional documentation is contained in [8, 9, 13]. Estimation of the regression parameters, including outliers, and the missing observations among the initial values of the series, plus the ARIMA model parameters, can be made by concentrating the former out of the likelihood, or by joint estimation. Several algorithms are available for computing the nonlinear sum of squares to be minimized. When the differenced series can be used, the algorithm of Morf et al. [26], with a simplification similar to that of Mélard [25], is used.

For the nonstationary series, it is possible to use the ordinary Kalman filter (default option), or its square root version (see [1]). The latter is adequate when numerical difficulties arise. By default, the exact maximum likelihood method is employed, and the unconditional and conditional least squares methods are available as options. Nonlinear maximization of the likelihood function and computation of the parameter estimates standard errors are made using the Marquardt's method and first numerical derivatives.

Estimation of regression parameters is made by using first the Cholesky decomposition of the inverse error covariance matrix to transform the regression equation. In particular, the Kalman filter provides an efficient algorithm to compute the variables in this transformed regression. Then, the resulting least squares problem is solved by applying the QR algorithm, where the Householder orthogonal transformation is used. This procedure yields an efficient and numerically stable method to compute Generalized Least Square (GLS) estimators of the regression parameters, which avoids matrix inversion. For forecasting, the ordinary Kalman filter or the square root filter options are available. These algorithms are applied to the original series. See Gomez and Maravall [10] for a more detailed discussion on how to build initial conditions on a nonstationary case.

Missing observations can be handled in two equivalent ways. The first one is an extension to nonstationary models of the skipping approach by Jones [17]. In this case, interpolation of missing values is made by a simplified Fixed Point Smoother, and yields identical results to those of Kohn and Ansley [19]. The second one consists of assigning a tentative value and specifying an additive outlier to each missing observation. If this option is used, the interpolator is the difference between the tentative value and the estimated regression parameter. If the ARIMA parameters are the same, it coincides with the interpolator obtained with the skipping approach. Also, the likelihood can be corrected so that it coincides with that of the skipping approach. When concentrating the regression parameters out of the likelihood, the mean squared errors of the forecasts and interpolations are obtained following the approach of Kohn and Ansley [18]. When some of the initial missing values cannot be estimated (free parameters), the program detects them, and flags the forecasts or interpolations that depend on these free parameters. The user can then assign arbitrary values (typically, very large or very small) to the free parameters and rerun the program. Proceeding in this way, all parameters of the ARIMA model can be estimated because the function to minimize does not depend on the free parameters. Moreover, it will be evident which forecasts and interpolations are affected by these arbitrary values because they will strongly deviate from the rest of the estimates. However, if all unknown parameters are jointly estimated, the program may not flag all free parameters. It may happen that there is convergence to a valid arbitrary set of solutions, i.e., some linear combinations of the initial missing observations, including the free parameters, can be estimated.

In brief, regression parameters are initialized by OLS and the ARMA model parameters are first estimated with two regressions, as in Hannan and Rissanen [15]. Next, the Kalman filter and the QR algorithm provide new regression parameter estimates and regression residuals. For each observation, t-tests are computed for four types of outliers, namely additive, innovational, level shift, and transitory. If there are outliers whose absolute t-values are greater than a preselected critical level C, the one with the greatest absolute t-value is selected. Otherwise, the series is free from outlier effects and the algorithm stops.

   If outliers are identified, the series is corrected and the ARMA model parameters are reestimated. Then, a multiple regression is performed using the Kalman filter and the QR algorithm. If some outliers have absolute t-values below the critical value C, the one with the lowest absolute value is removed and the multiple regression is reestimated. In the next step, using the residuals from the last multiple regression model, t-tests are calculated for the four types of outliers and for each observation. If there are outliers with absolute t-values greater than the critical level C, the one with the greatest absolute t value is chosen and the algorithm goes on to the estimation of the ARMA model to iterate, otherwise it stops.

6. *Diagnostic checks.* The main diagnostic tests are on the residuals. The estimated residuals $\hat{a}_t$ are analyzed to test the hypothesis that a_t are independent and identically normally distributed with zero mean and constant variance σ_a^2. Besides inspection of the residual graph, the corresponding sample autocorrelation function is examined. The lack of residual autocorrelation is tested using the Ljung–Box test statistics, and skewness and kurtosis tests are applied to test for normality of the residuals. Specific out-of-sample forecast tests are also performed to evaluate if forecasts behave in agreement with the model.

7. *Optimal forecasts.* If the diagnostics are satisfied, the model is used to compute optimal forecasts for the series, together with their Mean Square Error (MSE). These are obtained using the Kalman filter applied to the original series.

   TRAMO is to be used as a preadjustment process to eliminate all the deterministic components such that the residual is a linearized series that can be modeled with an ARIMA process. The latter is decomposed by SEATS in stochastic trend-cycle, seasonality, and irregulars. Both programs can handle routine applications to a large number of series and provide a complete model-based solution to the problems of forecasting, interpolation, and signal extraction for nonstationary time series.

5.2 SEATS: Signal Extraction in ARIMA Time Series

SEATS stands for "Signal Extraction in ARIMA Time Series." It belongs to the class of procedures based on ARIMA models for the decomposition of time series into unobserved components. The method was mainly developed for economic series, and the corresponding computer program is based on a seasonal adjustment method developed by Burman [4] at the Bank of England. In particular, the SEATS procedure consists of the following steps.

1. *ARIMA model estimation.* SEATS starts by fitting an ARIMA model to a series not affected by deterministic components, such as trading day variations, moving holidays, and outliers. Let y_t denote this linearized series, and consider an additive decomposition model (multiplicative if applied to the log transformation

of y_t), such that

$$z_t = (1 - B)^d (1 - B^s)^D y_t, \tag{5.3}$$

represent the "differenced" series. The model for the differenced linearized series z_t can be written as

$$\phi_p(B)\Phi_P(B^s)(z_t - \bar{z}) = \theta_q(B)\Theta_Q(B^s)a_t, \tag{5.4}$$

where $\bar{z}$ is the mean of $z_t, t = 1, \ldots, n$, a_t is a series of innovations, normally distributed with zero mean and variance σ_a^2, $\phi_p(B)\Phi_P(B^s)$ and $\theta_q(B)\Theta_Q(B^s)$ are Autoregressive (AR) and Moving Average (MA) polynomials in B, respectively, which are expressed in multiplicative form as the product of a regular polynomial in B and a seasonal polynomial in B^s. The complete model can be written in detailed form as

$$\phi_p(B)\Phi_P(B^s)(1 - B)^d (1 - B^s)^D y_t = \theta_q(B)\Theta_Q(B^s)a_t + c, \tag{5.5}$$

and, in concise form, as

$$\Psi(B)y_t = \pi(B)a_t + c, \tag{5.6}$$

where c is equal to $\Psi(B)\bar{y}$, being $\bar{y}$ the mean of the linearized series y_t. In words, the model that SEATS assumes is that of a linear time series with Gaussian innovations. When used with TRAMO, estimation of the ARIMA model is made by the exact maximum likelihood method described in Gomez and Maravall [11]. When used by itself, SEATS applies the quasi-maximum likelihood method described by Burman [4]. In both cases, a (faster) least squares algorithm is also available. SEATS contains an option where the automatic model identification procedure is replaced by a faster one. This simplified procedure starts directly by fitting the default model and, if found inappropriate, tries a sequence of models according to the results.

The program starts with the estimation of the ARIMA model. The (inverse) roots of the AR and MA polynomials are always constrained to remain in or inside the unit circle. When the modulus of a root converges within a preset interval around 1 (by default (0.98,1]), the program automatically fixes the root. If it is an AR root, the modulus is made 1. If it is an MA root, it is fixed to the lower limit. This simple feature makes the program very robust to over- and underdifferencing. In the standard case in which SEATS and TRAMO are used jointly, SEATS controls the AR and MA roots mentioned above, uses the ARIMA model to filter the linearized series, obtains in this way new residuals, and produces a detailed analysis of them.

2. *Derivation of the ARIMA models for each component.* The program proceeds by decomposing the series that follows model (5.6) into several components. The decomposition can be multiplicative or additive. Next, we shall discuss the

additive model, since the multiplicative relation can be taken care with the log transformation of the data. That is,

$$y_t = T_t + C_t + S_t + I_t, \tag{5.7}$$

where T_t denotes the trend component, C_t the cycle, S_t represents the seasonal component, and I_t the irregulars.

The decomposition is done in the frequency domain. The spectrum (or pseudospectrum) is partitioned into additive spectra, associated with the different components which are determined, mostly, from the AR roots of the model. The trend component represents the long-term evolution of the series and displays a spectral peak at frequency 0, whereas the seasonal component captures the spectral peaks at seasonal frequencies (e.g., for monthly data these are 0.524, 1.047, 1.571, 2.094, 2.618, and 3.142). The cyclical component captures periodic fluctuations with period longer than a year, associated with a spectral peak for a frequency between 0 and $(2\pi/s)$, and short-term variation associated with low order MA components and AR roots with small moduli. Finally, the irregular component captures white noise behavior, and hence has a flat spectrum. The components are determined and fully derived from the structure of the (aggregate) ARIMA model (5.6) for the linearized series directly identified from the data. The program is aimed at monthly or quarterly frequency data and the maximum number of observations that can be processed is 600.

One important assumption is that of orthogonality among the components, and each one will be described through an ARIMA model. In order to identify the components, the *canonical decomposition* is used, such that the variance of the irregulars is maximized, whereas the trend, seasonal, and cycle are as stable as possible.

The canonical condition on the trend, seasonal, and cyclical components identifies a unique decomposition, from which the ARIMA models for the components are obtained (including the component innovation variances). This is achieved as follows.

Let the total AR polynomial $\Psi(B)$ of the ARIMA model (5.6) be factorized as

$$\Psi(B) = \phi_p(B)\Phi_P(B^s)(1 - B)^d(1 - B^s)^D.$$

The roots of $\Psi(B)$ are assigned to the unobserved components as follows:

- the roots of $(1 - B)^d = 0$ are assigned to trend component;
- the roots of $(1 - B^s)^D = 0$ are factorized such that $((1 - B)(1 + B + \cdots + B^{s-1}))^D = 0$. In particular, the root of $(1 - B) = 0$ goes to the trend, whereas those of $(1 + B + \cdots + B^{s-1}) = 0$ go to the seasonal component;
- the positive real roots of $\phi_p(B) = 0$ are assigned to the trend if they are in modulus greater than or equal to k, and assigned to cycle if in modulus smaller than k;

- the negative real roots of $\phi_p(B) = 0$ are assigned to the seasonal component if $s \neq 1$, and to the cycle if $s = 1$;
- the complex roots of $\phi_p(B) = 0$ are assigned to the seasonal if $\omega \in$ [a seasonal frequency $\pm\epsilon$], being ω the frequency of the root. Otherwise, they are assigned to the cycle; and
- the positive real roots of $\Phi_P(B^s) = 0$, if they are greater than or equal to k they are assigned to the trend. On the other hand, when they are smaller than k they are assigned to the cycle.

The parameters k and ϵ are automatically set equal to 0.5 and -0.4, respectively, but they can be entered by the user as well.

The factorization of $\Psi(B)$ can be rewritten as

$$\Psi(B) = \Psi_T(B)\Psi_S(B)\Psi_C(B),$$

where $\Psi_T(B)$, $\Psi_S(B)$, and $\Psi_C(B)$ are the AR polynomials with the trend, seasonal, and cyclical roots, respectively. Let P and Q denote the orders of the polynomials $\Psi(B)$ and $\pi(B)$ in (5.6).

Considering the case in which $P \geq Q$, a polynomial division of the spectrum (or pseudospectrum) of model (5.6) yields a first decomposition of the type

$$\frac{\pi(B)}{\Psi(B)}a_t = \frac{\tilde{\pi}(B)}{\Psi(B)}a_{1t} + v_1,$$

where the order of $\tilde{\pi}(B)$ is $\min(Q, P-1)$, and v_1 is a constant (0 if $P > Q$).

A partial fraction expansion of the spectrum of $[\tilde{\pi}(B)/\Psi(B)]a_{1t}$ yields the decomposition

$$\frac{\tilde{\pi}(B)}{\Psi(B)}a_{1t} = \frac{\tilde{\pi}_T(B)}{\Psi_T(B)}\tilde{a}_{T_t} + \frac{\tilde{\pi}_S(B)}{\Psi_s(B)}\tilde{a}_{S_t} + \frac{\tilde{\pi}_C(B)}{\Psi_C(B)}\tilde{a}_{C_t},$$

where, given $j = T, S, C$, we have order$(\tilde{\pi}_j) \leq$ order(Ψ_j). Let $\tilde{g}_j(\omega)$ denote the spectrum of $[\tilde{\pi}_j(B)/\Psi_j(B)]\tilde{a}_{j_t}$, with $v_j = \min\{\tilde{g}_j(\omega) : 0 \leq \omega \leq \pi\}$. Imposing the canonical condition

$$g_j(\omega) = \tilde{g}_j(\omega) - v_j, \quad j = T, S, C,$$

$$v = v_1 + \sum_j v_j,$$

the spectrum of the final components are obtained, which give the models for the components

$$\Psi_T(B)T_t = \pi_T(B)a_{T_t}$$

$$\Psi_S(B)S_t = \pi_S(B)a_{S_t}$$

$$\Psi_C(B)C_t = \pi_C(B)a_{C_t}$$

$$I_t \sim \text{WN}(0, v).$$

All components have balanced models, in the sense that the order of the AR polynomial equals that of the MA one. On the other hand, when $Q > P$, the decomposition proceeds as follows. First a decomposition is performed, whereby

$$\text{ARIMA}(P, Q) = \text{ARIMA}(P, P - 1) + \text{MA}(Q - P).$$

The first component falls under the previous case of $P \geq Q$, and hence can be decomposed in the previous way. In general,

$$\text{ARIMA}(P, P - 1) = T_t + S_t + C_t + I_t,$$

where T_t, S_t, C_t, and I_t denote the trend, seasonal, cyclical, and irregular component. The $\text{MA}(Q - P)$ component, which represents stationary short-term deviations, is added to the cyclical component. The series is decomposed then into a balanced trend model, a balanced seasonal model, a top heavy cycle model, and a white noise irregular. The first three components are made canonical (i.e., noise free).

As a general rule, it is recommended that balanced models be favored because the excess of MA structure often produces nonadmissible decompositions.

3. *Estimation of the components.* To perform the series decomposition, SEATS selects a different ARIMA model from the one estimated by TRAMO if the latter does not admit an admissible decomposition. In this regard, the forecasts of the original series done by SEATS differ from the ones obtained by TRAMO. This can also happen when the ARIMA models are the same. Indeed, whereas TRAMO uses white noise residuals, SEATS uses residuals obtained by the filtering of the linearized series. The latter is obtained using an ARIMA model plus maximum likelihood estimates of the residuals deleted using the differentiation process, such that the number of residuals equals the sample size. Finally, the components estimated by SEATS with their forecasts are modified to reincorporate the deterministic effects that have been previously removed by TRAMO.

For a particular time series $(y_1, y_2, \ldots, y_n)'$, the program yields the Minimum Mean Square Error (MMSE) estimators of the components, computed with a Wiener–Kolmogorov type filter applied to the finite series by extending the latter with forecasts and backcasts (see [4]). For $i = 1, \ldots, n$, the estimate $\hat{y}_{t|n}$, equal to the conditional expectation $E(y_t | y_1, \ldots, y_n)$, is obtained for all components. When $n \to \infty$, the estimate $\hat{y}_{t|n}$ becomes the "final" or "historical" estimate, which is denoted by $\hat{y}_t$. For $t = n$, the concurrent estimate, $\hat{y}_{n|n}$, is obtained, i.e., the estimate for the last observation of the series. The final and concurrent estimates are the ones of most interest. When $n - k < t < n$, $\hat{y}_{t|n}$ yields a preliminary estimator, and, for $t > n$, a forecast. Besides their estimates, the program produces several years of forecasts of the components, as well as corresponding Standard Errors (SE).

4. *Diagnostic checks.* For the last two and the next two forecasted years, the SE of the revision of the preliminary estimates and of the forecast are provided. The program further computes MMSE estimates of the innovations in each one of the components. The joint distributions of the components and of their MMSE estimators are obtained; they are characterized by the variances and auto- and cross-correlations. The comparison between the theoretical moments for the MMSE estimators and the corresponding estimates provides additional elements for diagnosis (see [21]). The program also presents the Wiener–Kolmogorov filter for each component, and the filter which expresses the weights with which the different innovations a_j in the observed series contribute to the estimate $\hat{y}_{t|n}$. These weights directly provide the moving average expressions for the revisions.

Next, an analysis of the estimation errors for the trend and for the seasonally adjusted series (and for the cycle, if present) is performed. Let denote with

$$d_t = y_t - \hat{y}_t,$$
$$d_{t|n} = y_t - \hat{y}_{t|n},$$
$$r_{t|n} = \hat{y}_t - \hat{y}_{t|n}.$$

for $t = 1, \ldots, n$, the final estimation error, the preliminary estimation error, and the revision error in the preliminary estimator. The variances and autocorrelation functions for $d_t, d_{t|n}, r_{t|n}$ are displayed. The program shows how the variance of the revision error in the concurrent estimator $r_{n|n}$ decreases as more observations are added, and hence the time it takes in practice to converge to the final estimator. Similarly, the program computes the decrease in precision as the forecast moves away from the concurrent estimate. Finally, the SE of the estimates of the linearized rates of growth for the concurrent estimate of the rate and its successive revisions are presented, both for the trend and seasonally adjusted series. All SE computed are exact given that the ARIMA model for the observed series is correct. Further details can be found in [22, 23].

5.3 Illustrative Example: TRAMO-SEATS Seasonal Adjustment of the US NODG Series

TRAMO-SEATS is a DOS program, but its Windows version with some modifications and additions, called TSW, has been developed by Caporello and Maravall [5]. TRAMO-SEATS is also available, along with X12ARIMA, in the open source software Gretl and in the commercial software Eviews. The two seasonal adjustment procedures are also implemented in the Eurostat software Demetra+, recently updated in JDemetra+ [14], both developed by the National Bank of Belgium.

The main feature of Demetra+ and JDemetra+ is to normalize the two different methods and provide ease of use for national statistics agencies. This is also the aim declared in the European Statistical System (ESS) Guidelines for Seasonal Adjustment [7] that want to harmonize and provide guidelines for seasonal adjustment. From the viewpoint of a typical user there is little difference between JDemetra+ and Demetra+. The actual added value of JDemetra+ to the development of the seasonal adjustment tools lies in the rewriting of the original FORTRAN codes of X12ARIMA and TRAMO-SEATS in Java. This solution is a crucial factor that enables the long-term maintenance of the tool, integration of the libraries in the IT environments of many institutions, and reuse of the modules and algorithms for other purposes. It runs on operating systems that support the Java VM (Virtual Machine), such as Microsoft Windows XP SP3/Vista SP1/Windows 7, Ubuntu 9.10, various other Linux distributions, Solaris OS version 11 Express (SPARC and x86/x64 Platform Edition), and Macintosh OS X 10.6 Intel.

Eurostat uses the TRAMO-SEATS method in the JDemetra+ interface for seasonal adjustment, and this is the common practice in several European countries. Hence, in this section, JDemetra+ is applied to perform the seasonal adjustment of the US New Orders for Durable Goods (NODG) series, observed from February 1992 to December 2013, using the TRAMO-SEATS method. The results of the seasonal adjustment provided by JDemetra+ are divided into six parts: input, main results, pre-processing, decomposition, benchmarking, and diagnostics. A detailed discussion of all the tables, charts, and results produced by the software is not provided. We focus on the most important outputs for the interpretation of the seasonal adjustment of the NODG series performed using the TRAMO-SEATS method.

5.3.1 Input: Specifications

The first part in the JDemetra+ output, called *Input*, presents information about the specification used for seasonal adjustment including most important settings. Several predefined specifications are available, but also user-defined specifications can be applied. The predefined specifications refer to automatically test whether a logarithmic transformation is made of the series or just left as is (*log/level*), to test for the presence of working days (*working days*), or trading days effects (*trading days*), as well as Easter (*Easter*) and leap-year effects. Furthermore, an automatic detection of several types of outliers (AO, LS, and TC) can be performed by the software. For some predefined specifications, the airline model ARIMA$(0,1,1)(0,1,1)_s$ is used as default, but also there are cases in which JDemetra+ is required to automatically identify and estimate the best ARIMA model for the series. In detail, the predefined TRAMO-SEATS specifications available in JDemetra+ are the following:

- **RSA0**: level, airline model.

- **RSA1**: log/level, outliers detection (AO, TC, and LS), airline model.
- **RSA2**: log/level, working days, Easter, outliers detection, airline model.
- **RSA3**: log/level, outliers detection, automatic model identification.
- **RSA4**: log/level, working days, Easter, outliers detection, automatic model identification.
- **RSA5/RSAfull**: log/level, trading days, Easter, outliers detection, automatic model identification.

The default option in JDemetra+ is RSAfull, in agreement with recommendation of Maravall and Perez [24], who suggested to start with the automatic option that pretests for trading day and Easter effects, and for the presence of outliers. Hence, we select this predefined specification in JDemetra+ for the seasonal adjustment of the NODG series.

5.3.2   Testing for the Presence of Identifiable Seasonality

Before going through the TRAMO-SEATS decomposition, as suggested in Sect. 4.6.2. for the X12ARIMA seasonal adjustment, it is necessary to check if the original series presents identifiable seasonality. The *Diagnostics* part includes *Combined seasonality tests* that are useful to check for the presence of seasonality in the NODG series. The corresponding output is illustrated in Table 5.1. The tests are the same discussed for the X12ARIMA decomposition, and we remind the reader to Sect. 4.6.2 for a detail description. The nonparametric Friedman and Kruskal–Wallis tests confirm the presence of stable seasonality in the series, whereas the test for the presence of moving seasonality rejects such a hypothesis. It turns out that the combined test confirms the presence of identifiable seasonality in the NODG series.

5.3.3   Pre-processing

The *Main Results* part includes basic information about pre-processing (TRAMO) and the quality of the decomposition (SEATS). Information about the TRAMO part of the process includes the estimation span, the number of observations, transformation (if any), and deterministic effects. On the other hand, the second part of the *Main results* summaries the results from the decomposition of the variance of the white noise innovation for each component extracted by SEATS from the linearized time series where the regression effects estimated by TRAMO are removed. The general interest is on a rather stable seasonal component and hence, when competing models are compared, the preference is given to the model that minimizes the innovation variance of the seasonal component. For **RSA5** and **RSAfull** options the trading day and leap-year effects are pretested. If the trading

Table 5.1 Combined seasonality test performed on the NODG series

Non parametric tests for stable seasonality

Friedman test

Friedman statistic = 187.9744
Distribution: Chi2(11)
p-value: **0.0000**

Stable seasonality present at the 1 % level

Kruskall–Wallis test

Kruskall–Wallis statistic =216.7785474118
Distribution: Chi2(11)
p-value: **0.0000**
Stable seasonality present at the 1 % level

Test for the presence of seasonality assuming stability

	Sum of squares	Degrees of freedom	Mean square
Between months	0.9073464537171304	11.0	0.08248604124701185
Residual	0.16251917583422604	251.0	6.474867563116575E-4
Total	1.0698656295513564	262.0	0.004083456601341055

Value: 127.39417515948156
Distribution: *F* with 11 degrees of freedom in the numerator and 251 degrees of freedom in the denominator
p-**value: 0.0000**
Seasonality present at the 1 % level

Evolutive seasonality test

	Sum of squares	Degrees of freedom	Mean square
Between years	0.00594528505243273	20.0	2.972642526216365E-4
Error	0.11407660206322308	220.0	5.1853Q0093782867E-4

Value: 0.573282639857342
Distribution: *F* with 20 degrees of freedom in the numerator and 220 degrees of freedom in the denominator
p-**value: 0.9283**
No evidence of moving seasonality at the 20 % level

Combined seasonality test

Identifiable seasonality present

day effects have been detected, the message "Trading days effect (6 regressors)" or "Trading days effect (7 regressors)" is displayed, depending on whether the leap-year effect has been detected or not. If the Easter effect is statistically significant, Easter effect detected is displayed. The additional information, i.e., type, location, and coefficients of every outlier, can be found in the *Pre-processing* node.

As shown in Table 5.2, it also includes information about the data, such as the estimation span, the number of observations, the data transformation, the correction for leap years, and the number of regression variables (calendar-related variables and outliers). The notation of the estimation span varies according to the frequency of the observations, in this case monthly, such that [2-1992:12-2013]. Based on the predefined (default) specification **RSAfull**, the message "Series has been log-transformed" is displayed since the logarithmic transformation has been applied as result of the test performed by the software. Based on the selected predefined specification, trading days effects, leap-year and the Easter effects are pretested and estimated if present. It can be noticed that all these effects, except the leap year, result significant in the series under analysis. Likelihood statistics related to the

Table 5.2 Pre-processing summary information

Summary

Estimation span: [2-1992–12-2013]
263 observations
Series has been log-transformed
Trading days effects (6 variables)
Easter(6) effect detected
3 detected outliers

Final model

Likelihood statistics
Number of effective observations = 250
Number of estimated parameters = 13

Loglikelihood = 460.5213592108896
Transformation adjustment = −3029.8684210569045
Adjusted loglikelihood = −2549.347061846015

Standard error of the regression (ML estimate) = 0.034824441988354465
AIC = 5124.69412369203
AICC = 5126.236496573386
BIC (corrected for length)= −6.449841441444706

Scores at the solution
−0.012825 0,001446

final model are also shown. The number of effective observations is the number of observations used to estimate the model, i.e., the number of observations of regular and seasonal differenced series. The number of estimated parameters is the sum of the regular and seasonal parameters for both AR and MA, mean effect, trading/working days effect, outliers, regressors, and standard error of model. In the pre-processing part, the model is estimated using the exact Maximum Likelihood Estimation (MLE) method. The *standard error of the regression (ML estimate)* is the standard error of the regression from MLE. JDemetra+ displays the maximized value of the likelihood function, that is used for the selection of the proper ARIMA model, that is, AIC, AICC, and BIC [14]. The model with the smallest values of these three criteria is preferred.

5.3.3.1 Regressors

Based on the **RSAfull** specification, JDemetra+ provides the estimated coefficients of the trading day variables to evaluate their significant effect on the NODG series. Table 5.3 illustrates that only the Wednesday variable is characterized by a coefficient that is significantly different from zero. This implies that the outcome of the joint F-test indicates that the trading day regressors are jointly significant (the

Table 5.3 Estimates of calendar effects and outliers detection

Regression model

Trading days

	Coefficients	t-Stat	p-value
Monday	0.0043	1.01	0.3128
Tuesday	−0.0072	−1.69	0.0926
Wednesday	0.0125	2.94	0.0036
Thursday	−0.0032	−0.76	0.4464
Friday	0.0065	1.51	0.1319
Saturday	−0.0034	−0.79	0.4309

Joint F-Test = 4.68(0.0002)

Easter [6]

	Coefficients	t-Stat	p-value
	−0.0246	−2.68	0.0078

Outliers

	Coefficients	t-Stat	p-value
LS (1-2009)	−0.2379	−8.32	0.0000
LS (10-2008)	−0.1788	6.26	0.0000
AO (6-2000)	0.1528	5.22	0.0000

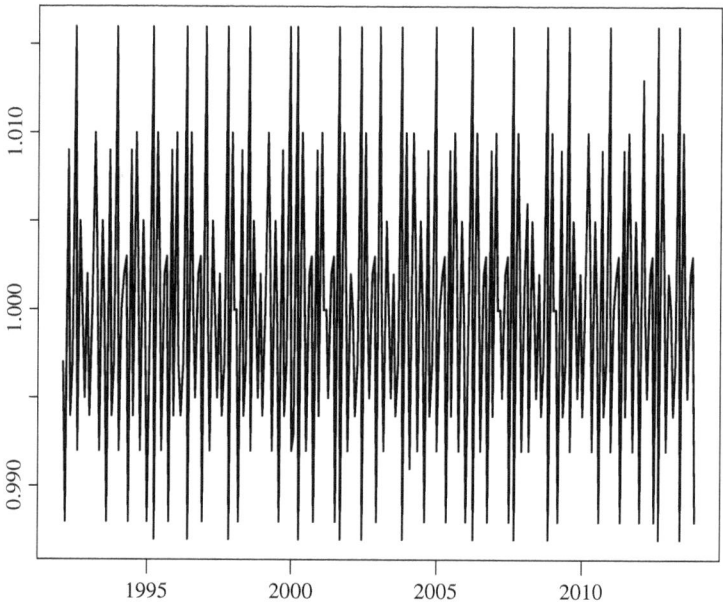

Fig. 5.1 Trading day estimates of the US New Orders for Durable Goods

p-value associated to the F-test statistic is lower than 5 %). Figure 5.1 displays the monthly trading days component that ranges from 0.988 to 1.016 over the period February 1992–December 2013.

The Easter effect is also estimated, and the value of the corresponding coefficient is reported with its standard error, t-statistics, and the related p-value. As shown in Table 5.3, there is a significant effect lasting over 6 days around Easter. In particular, Fig. 5.2 displays that the Easter effect is felt 6 days before Easter and on Easter Sunday but not after Easter.

JDemetra+ presents also the results of the outliers detection. This table includes the type of outlier, its date, the value of the coefficient, and corresponding t-statistics and p-values. For the series under study, the software automatically detects the presence of three outliers: an additive outlier at June 2000, and two level shift outliers on October 2008 and January 2009, respectively.

5.3.3.2 ARIMA Model

The *Arima* section shows that, for the log transformed NODG series under study, the ARIMA$(0,1,1)(0,1,1)_{12}$ model is chosen and estimated. It should be noticed that we have reversed the sign of the parameters given by JDemetra+ following the notation

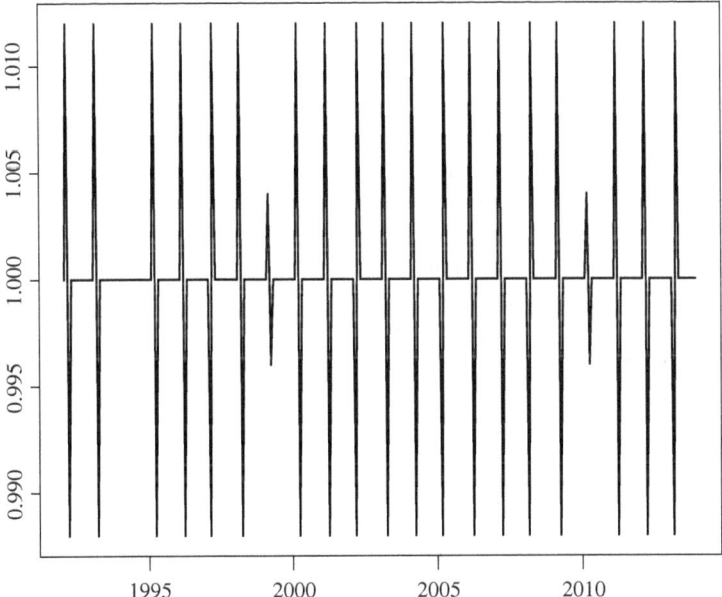

Fig. 5.2 Moving holiday component of the US NODG

Table 5.4 Estimated values of the ARIMA$(0,1,1)(0,1,1)_{12}$ model parameters

Arima model

[(0,1,1)(0,1,1)]

	Coefficients	t-stat	p-value
Theta(1)	0.5235	9.47	0.0000
BTheta(1)	0.6940	14.07	0.0000

adopted by Box and Jenkins [3] (see Table 5.4):

$$(1 - B)(1 - B^{12})Y_t = (1 - 0.5235B)(1 - 0.6940B^{12})a_t, \qquad (5.8)$$

where a_t is assumed to follow a white noise process.

All the coefficients result significantly different from zero. Even if this is not the case, when autoregressive components are estimated, for each regular AR root (i.e., the solution of the corresponding characteristic equation) the argument and modulus are also reported (if present) to inform to which component the regular roots would be assigned if the SEATS decomposition is used. Rules used by SEATS to assign the roots to the components have been described in Sect. 5.3, and they are detailed for the NODG series in the following subsections. The roots with seasonal frequency belong to the seasonal component while real and positive roots with a high modulus value are assigned to the trend.

5.3.3.3 Residuals

The examination of residuals from the fitted model is a crucial step of model validation. JDemetra+ produces several residual diagnostics and graphs for model checking. They are presented in the *Residuals* section and its subcategories. Specifically, Fig. 5.3 illustrates the residuals from the ARIMA$(0,1,1)(0,1,1)_{12}$ fitted to the NODG series, after deterministic effects have been removed.

Their pattern seems to be in agreement with the assumption that they have been generated by a purely random process with constant mean and variance. However, to verify this assumption in more detail, several tests are computed on these residuals, and the results are presented both in summary and detail. *Summary* statistics report for each test the corresponding p-value. Bold p-values (ranging from 0.1 to 1) mean "good," italics (p-value ranging from 0.05 to 0.1) means "uncertain," and red (p-value ranging from 0 to 0.05) means "bad" (the latter are not present in our illustrative example). The four tests considered evaluate the normality, no correlation, randomness, and linearity of the residuals. Table 5.5 shows that, for the NODG series, all the tests do not reject the null hypothesis.

The *Details* sections, not reported here, correspond to the appropriate tables from the *Summary* diagnostics. Specifically, detailed information are provided for the

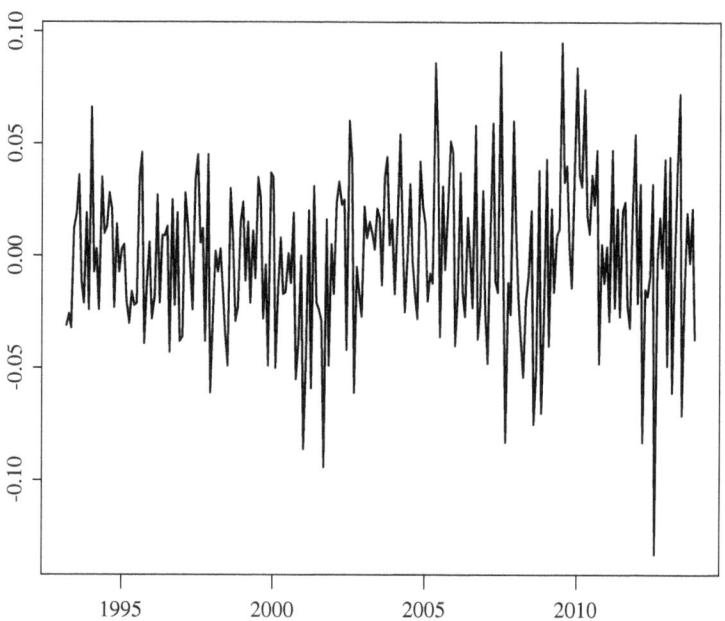

Fig. 5.3 Residuals from the ARIMA$(0,1,1)(0,1,1)_{12}$ fitted to the NODG series

Table 5.5 Summary statistics for residuals from the regARIMA$(0,1,1)(0,1,1)_{12}$ model

Analysis of the residuals

Summary

1. *Normality of the residuals*

	p-value
Mean	**0.9231**
Skewness	**0.2516**
Kurtosis	*0.0948*
Normality	**0.1151**

2. *Independence of the residuals*

	p-value
Ljung-Box(24)	**0.3738**
Box-Pierce(24)	**0.4324**
Ljung-Box on seasonality (2)	**0.8426**
Box-Pierce on seasonality (2)	**0.8507**

Durbin-Watson statistic: 2.0673

3. *Randomness of the residuals*

	P-value
Runs around the mean: number	*0.0992*
Runs around the mean: length	**1.0000**
Up and Down runs: number	**1.0000**
Up and Down runs: length	**1.0000**

4. *Linearity of the residuals*

	P-value
Ljung-Box on squared residuals(24)	**0.4171**
Box-Pierce on squared residuals(24)	**0.4742**

analysis of the:

1. *Distribution* of the residuals. For the NODG series, the estimated mean of the residuals results not significantly different from zero. Furthermore, the normality assumption is tested in terms of skewness and kurtosis, separately, as well as jointly using the Doornik–Hansen test which combines both the skewness and kurtosis tests [14]. In this specific case, the residuals can be considered as realization of a normal random process.
2. The *independence test* provided by JDemetra+ is the Ljung–Box test [20]. Denoting with ρ_j the sample autocorrelation at rank j of the n residuals, the

statistics is given by

$$\text{LB}(k) = n(n-2) \sum_{j=1}^{k} \frac{\rho_j^2}{(n-j)}$$

which is distributed as a chi-square variable with $k - p$ degrees of freedom if the residuals are independent. In particular, k depends on the frequency of the series ($k = 24$ for monthly series, equal to 8 for quarterly series, and $4s$ for other frequencies, where s is a frequency of the time series), and p is the number of parameters in the ARIMA model with deterministic effects from which the residuals are derived.

3. *Randomness* of the residuals. The Wald–Wolfowitz tests, also known as *Run tests*, are used to account for the randomness of the residuals. Generally, it examines the hypothesis that a series of numbers is random. For data centered around the mean, the test calculates the number and length of runs. A run is defined as a set of sequential values that are either all above or below the mean. An up run is a sequence of numbers each of which is above the mean; a down run is a sequence of numbers each of which is below the mean. The test checks if the number of up and down runs is distributed equally in time. Both too many runs and too few runs are unlikely a real random sequence. The null hypothesis is that the values of the series have been independently drawn from the same distribution. The test also verifies the hypothesis that the length of runs is random. In the case under investigation, the randomness assumption is satisfied.

4. The *linearity* residuals test provides an evidence of autocorrelation in residuals. In this regard, the Ljung–Box test is performed on the square of the residuals, and, although not shown, the linearity condition is satisfied by the residuals of the estimated ARIMA model.

5.3.4 Decomposition

SEATS processes the linearized series obtained with TRAMO. The decomposition made by SEATS assumes that all the components in the time series, trend, seasonal, transitory effects (if present), and irregular, are orthogonal and can be expressed by ARIMA models. The identification of the components requires that only irregular components include noise.

5.3.4.1 Wiener–Kolmogorov (WK) Analysis

The spectrum of the ARIMA$(0,1,1)(0,1,1)_{12}$ model estimated by TRAMO is partitioned into additive spectrums for each component. We recall that the trend component captures low frequency variations of the series and displays a spectral peak at frequency 0. By contrast, the seasonal component picks up the spectral peaks at seasonal frequencies and the irregular component captures a white noise behavior.

In some cases, the ARIMA model chosen by TRAMO is changed by SEATS. This is done when it is nonadmissible for its decomposition. The decomposition is admissible when the variance of the seasonal component $V(a_{St}) = \sigma_S^2$ ranges in a given interval $[\underline{\sigma}_S^2, \bar{\sigma}_S^2]$ (see for details Bell and Hillmer [2]), under the assumption that the series can be decomposed into the sum of a seasonal and nonseasonal component, that are independent of each other. Furthermore, the decomposition is defined to be canonical when this variance σ_S^2 is exactly equal to $\bar{\sigma}_S^2$.

For the NODG series, the ARIMA$(0,1,1)(0,1,1)_{12}$ model estimated by TRAMO is kept by SEATS, and, from that, ARIMA representations of the several components are derived as described in Sect. 5.3. In the *Decomposition* part of the JDemetra+ output, each model is presented in closed form (i.e., using the backshift operator B) as illustrated in Table 5.6.

Specifically, the theoretical model for the trend results

$$(1 - B)^2 T_t = (1 + 0.0299B - 0.9701B^2)a_{Tt} \quad \text{with} \quad a_{Tt} \sim \text{WN}(0, 0.041), \quad (5.9)$$

whereas the model for the seasonal component is

$$(1 + B + \cdots + B^{11})S_t = \Theta_{11}(B)a_{St}, \quad (5.10)$$

Table 5.6 The ARIMA models for original series, seasonally adjusted series and components

Model

D: $1.00000 - B - B^{12} + B^{13}$

MA: $1.00000 - 0.523460\, B - 0.694005\, B^{12} + 0.363284\, B^{13}$

Sa

D: $1.00000 - 2.00000\, B + B^2$

MA: $1.00000 - 1.49745\, B + 0.511615\, B^2$

Innovation variance: **0.73560**

Trend

D: $1.00000 - 2.00000\, B + B^2$

MA: $1.00000 + 0.0299450\, B - 0.970055\, B^2$

Innovation variance: **0.04117**

Seasonal

D: $1.00000 + B + B^2 + B^3 + B^4 + B^5 + B^6 + B^7 + B^8 + B^9 + B^{10} + B^{11}$

MA: $1.00000 + 1.23274\, B + 1.17869\, B^2 + 1.02437\, B^3 + 0.790409\, B^4 + 0.539837\, B^5 + 0.294997\, B^6 + 0.0730214\, B^7 - 0.100683\, B^8 - 0.252590\, B^9 - 0.346686\, B^{10} - 0.571634\, B^{11}$

Innovation variance: **0.02285**

Irregular

Innovation variance: **0.41628**

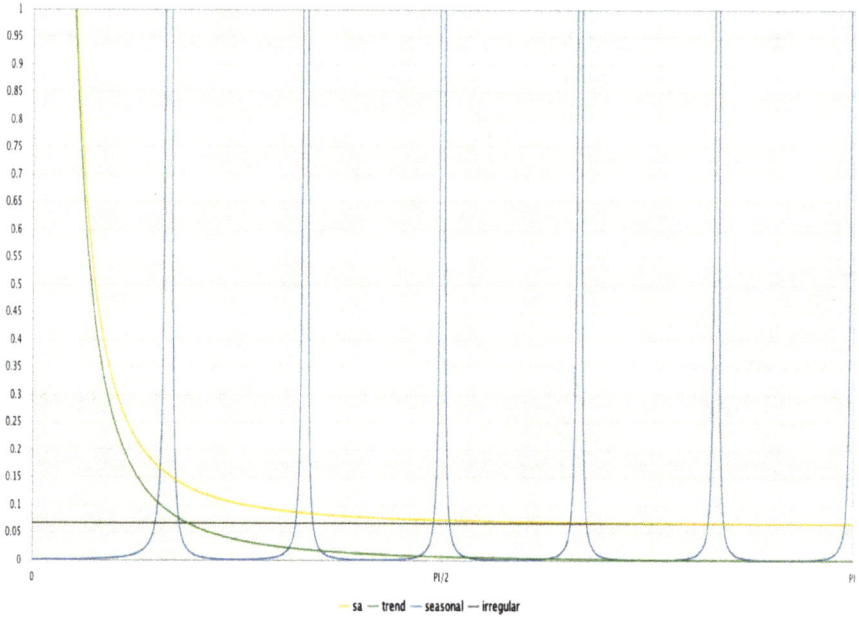

Fig. 5.4 The theoretical spectra of the components and seasonally adjusted function

where $a_{S_t} \sim WN(0, 0.023)$, and $\Theta_{11}(B)$ is a polynomial of order 11 in the backshift operator B.

Expressions (5.9) and (5.10) theoretically describe the components and at these equations JDemetra+ refers with the term *Components*. The corresponding spectra, as well as the one of the seasonally adjusted series, are presented in the section *WK analysis* in the subsection *Components*. The sum of the components spectra should be equal to the spectrum for the linearized time series, presented in the *Arima* subsection in the *Pre-processing* part. Figure 5.4 shows the standard spectra for trend (green), seasonal (blue), and irregular (grey) components defined in Table 5.6. The seasonally adjusted series spectrum (yellow) is the sum of trend component spectrum (green) and the irregular component spectrum (grey).

As discussed in Sect. 5.3, the estimation of the components is performed using symmetric and bi-infinite Wiener–Kolmogorov (WK) filters that define Minimum Mean Square Error (MMSE) estimators. As example, in the case of the seasonal component, the estimator is given by

$$\hat{S}_t = W_S(B)y_t = \sum_{j=-\infty}^{\infty} w_{S,j} B^j y_t \qquad t = 1, \ldots, n,$$

where $W_S(B)$ is the Wiener–Kolmogorov filter obtained as

$$W_S(B) = \sigma_S^2 \frac{\psi_S(B)\psi_S(F)}{\psi(B)\psi(F)} y_t,$$

where $F = B^{-1}$ is the forward operator. $\psi_S(B) = \frac{\theta_S(B)}{\phi_S(B)}$, with $\phi_S(B)$ and $\theta_S(B)$ are the autoregressive and moving average polynomials, respectively, in the ARIMA model (5.10) defining the seasonal component, whereas $\psi(B)$ refers to the ARIMA model (5.8) specified for the linearized NODG series. Replacing in $W_S(B)$ at y_t its ARIMA specification (5.8), and based on the theoretical components specified in Eqs. (5.9) and (5.10), we obtain

$$(1 + B + \cdots + B^{11})\hat{S}_t = \sigma_S^2 \left[\theta_S(B) \frac{\theta_S(F)(1-F)^2}{\theta(F)} \right] a_t. \tag{5.11}$$

This equation defines what JDemetra+ indicates as *Final estimator*, or simply *Estimator*, for the seasonal component, from which the variance and autocorrelations of the (theoretical) MMSE estimator can be obtained. Similarly, we can derive the theoretical estimator for the trend component defined in Eq. (5.9).

The behavior of the WK filters is generally studied in the frequency domain, such that the spectrum of the seasonal estimator is derived as

$$W_S(e^{-i\lambda}) = \sum_{j=-\infty}^{\infty} w_{S,j} e^{-i\lambda j} = \frac{f_S(\lambda)}{f_Y(\lambda)},$$

where $f_S(\lambda)$ and $f_Y(\lambda)$ are the spectral densities of the theoretical seasonal component and of the original series, respectively. These spectral densities are generally unknown, but they are approximated based on the ARIMA specifications given in Eqs. (5.8) and (5.10), respectively. The transfer functions of the WK estimators of the other components are similarly derived. They are represented graphically in the subsection *Final estimators* in the *WK analysis* part of the JDemetra+ output. Beyond the spectra of the theoretical estimators, JDemetra+ also provides the corresponding squared gain functions and WK filter weights, for a maximum of 71 weights. In particular, the spectra of the final estimators are shown in Fig. 5.5, and they result similar to those of the theoretical components shown in Fig. 5.4, although the spectra of the final estimators show spectral zeros at the frequencies where the corresponding theoretical components have spectral peaks. However, the estimator adapts to the structure of the analyzed series, that is, the spectral holes in seasonally adjusted series (yellow line) are wider as more stochastic is the seasonal component.

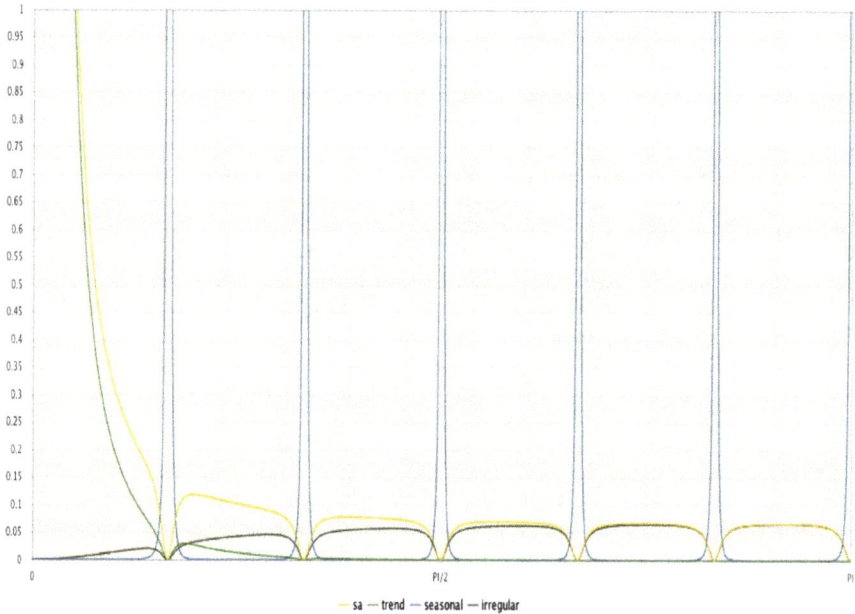

Fig. 5.5 Final WK estimators spectra

5.3.4.2   Estimated Components

The *Stochastic series* part presents the table containing the estimated components produced by SEATS, that is, the seasonally adjusted series, the trend, the seasonal component, and the irregular component. The seasonally adjusted NODG series is shown in Fig. 5.6 together with the original series.

The estimated components, denoted by JDemetra+ as *Estimates*, are derived by applying the WK filter to the observations extended at both ends with forecasts and backcasts from the ARIMA model specified for the linearized NODG series. Alternatively, the spectrum of the component is estimated by applying the spectrum (transfer function) of the filter to the spectral density of the ARIMA model (5.8) [2].

5.3.4.3   Diagnostic and Quality Assessment

The *Model based tests* part analyzes the theoretical distribution of estimators making use of the estimated components. Specifically, the tests concern the variance and the autocorrelation function of each estimator, as well as the cross-correlation among them. Hence, the output is organized in three main sections.

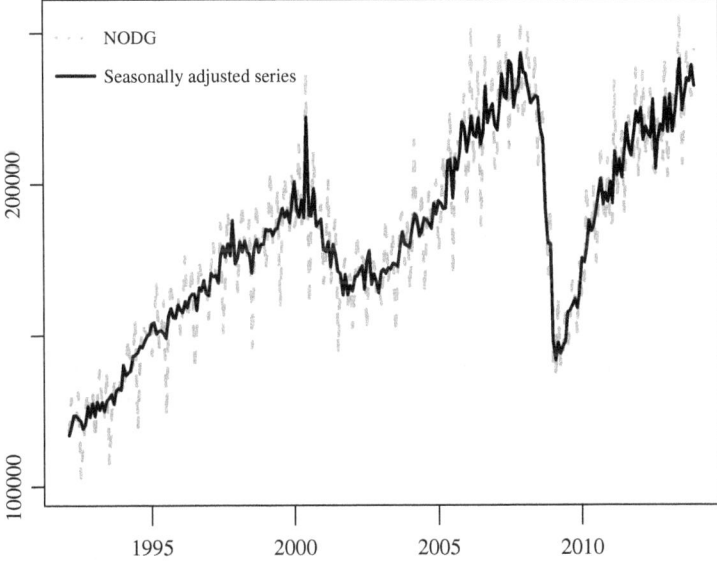

Fig. 5.6 The US NODG original and seasonally adjusted series

Table 5.7 Cross-correlation between the components estimators and the components sample estimates

Cross-correlation

	Estimator	Estimate	p-value
Trend/seasonal	**−0.1011**	−0.1708	**0.3719**
Trend/irregular	**−0.2041**	−0.1173	**0.2919**
Seasona/irregular	**0.0513**	0.1131	**0.2324**

The most important one is on the *Cross-correlation function*. The decomposition made by SEATS assumes that the components are orthogonal. To test this assumption, Table 5.7 contains the cross-correlations between pairs of final components estimators (denoted by "Estimator") that are compared with the cross-correlations computed on the estimated components (indicated as "Estimate"). The cross-correlations among the theoretical components are not reported since, by assumption, they are uncorrelated, whereas the MMSE estimators could be correlated. The p-values in Table 5.7 are related to the test statistics used to analyze if there is a statistically significant difference among the expected cross-correlations between the final estimators and the cross-correlations computed between the estimated components. It is evident that, for the NODG series, there is no significant difference between them.

Table 5.8 Variance of the components, of their theoretical WK estimators and of empirical estimates

Variance

	Component	Estimator	Estimate	p-value
Trend	0.0799	0.0096	0.0074	**0.0655**
Seasonally adjusted	2.5776	2.1820	2.0913	**0.7398**
Seasonal	0.1482	0.0162	0.0080	**0.1238**
Irregular	0.4163	0.2686	0.2458	**0.3849**

SEATS identifies the components assuming that, except for the irregular, they are clean of noise. This implies that the variance of the irregular is maximized in contrast to that of the trend-cycle and seasonal components which are as stable as possible. In this regard, JDemetra+ compares the variance expected for the theoretical estimator with the variance empirically computed on the corresponding estimated component. In the NODG example, the variance of the WK estimator of the seasonal component is derived using its theoretical specification (5.11). For all the components, Table 5.8 shows that, for the NODG series, there are no statistically significant differences between the variances of the theoretical WK estimators and their estimated values.

Finally, for each component, JDemetra+ exhibits autocorrelations corresponding to the ARIMA model selected for each component, the corresponding WK estimators and the estimated components. They are calculated from the first lag up to the seasonal lag.

For each table, the user should check whether the empirical estimates agree with the model, i.e., if their ACFs are close to those of the WK estimator. Special attention should be given to first and/or seasonal order autocorrelations. The autocorrelation functions for the trend and seasonal components of the NODG series are provided in Tables 5.9 and 5.10, respectively. It can be noticed that for both components there is evidence of difference in the theoretical and estimated autocorrelations in the first lags, but not at the seasonal lag.

It has to be noticed that the coefficients of the autocorrelation function of the irregular component are always null ("Component" column) as shown in Table 5.11 since the theoretical model for irregular component is always a white noise. As outlined by the p-values column, the estimated irregular component can be thought as realization of an uncorrelated process.

Table 5.9 Autocorrelations of the stationary transformation of trend, the estimators and the sample estimates

Trend

Lag	Component	Estimator	Estimate	p-value
1	0.0005	0.4622	0.3619	*0.0078*
2	−0.4995	−0.2542	−0,3984	*0.0261*
3	0.0000	−0.3292	−0.3102	**0.7656**
4	0.0000	−0.1708	−0.0062	**0.0525**
5	0.0000	−0.0865	0.1200	*0.0216*
6	0.0000	−0.0399	0.1204	*0.0397*
7	0.0000	−0.0108	0.0038	**0.8717**
8	0.0000	0.0137	−0.0983	**0.2171**
9	0.0000	0.0440	−0.0451	**0.3054**
10	0.0000	0.0357	0.0515	**0.8594**
11	0.0000	−0.0723	−0.0125	**0.4977**
12	0.0000	−0.1536	−0.1440	**0.9009**

Table 5.10 Autocorrelations of the stationary transformation of the seasonal component, the estimators and the sample estimates

Seasonal

Lag	Component	Estimator	Estimate	p-value
1	0.8654	0.7040	0.5605	*0.0188*
2	0.6674	0.3311	0.1542	*0. 0216*
3	0.4467	−0.0428	−0.1614	*0.0374*
4	0.2340	−0.3588	−0.3517	**0.8991**
5	0.0510	0.5750	−0.3705	*0.0491*
6	−0.0886	−0.6661	−0.4018	*0.0452*
7	−0.1781	−0.6233	−0.3492	*0.0073*
8	−0.2168	−0.4537	−0.3495	**0.1107**
9	−0.2087	−0.1798	−0.1658	**0.8765**
10	−0.1621	0.1604	0.1012	**0.6234**
11	−0.0881	0.5139	0.4329	**0.4704**
12	0.0000	0.8129	0.7992	**0.8198**

Table 5.11 Autocorrelations of the stationary transformation of the irregular component, the estimators and the sample estimates

Irregular

Lag	Component	Estimator	Estimate	p-value
1	−0.0000	−0.2382	−0.2980	**0.2870**
2	−0.0000	−0.1246	−0.1918	**0.3210**
3	−0.0000	−0.0651	0.0111	**0.2715**
4	−0.0000	−0.0338	−0.0292	**0.9476**
5	−0.0000	−0.0171	0.0672	**0.2300**
6	−0.0000	−0.0079	0.0110	**0.7568**
7	−0.0000	−0.0022	0.0511	**0.4486**
8	−0.0000	0.0027	−0.1111	**0.1004**
9	−0.0000	0.0086	0.0321	**0.7347**
10	−0.0000	0.0184	−0.0040	**0.7465**
11	−0.0000	0.0361	0.0732	**0.5936**
12	−0.0000	−0.1532	−0.1196	**0.5818**

References

1. Anderson, B., & Moore, J. (1979). *Optimal filtering*. New Jersey: Prentice Hall.
2. Bell, W. R., & Hillmer, S. C. (1984). Issues involved with the seasonal adjustment of economic time series. *Journal of Business and Economic Statistics, 2*, 291–320.
3. Box, G. E. P., & Jenkins, G. M. (1970). *Time series analysis: Forecasting and control*. San Francisco: Holden-Day (Second Edition 1976).
4. Burman, J. P. (1980). Seasonal adjustment by signal extraction. *Journal of the Royal Statistical Society A, 143*, 321–337.
5. Caporello, G., Maravall, A., & Sanchez, F. J. (2001). Program TSW reference manual. Working Paper 0112, Servicio de Estudios, Banco de España.
6. Chen, C., & Liu, L. M. (1993). Joint estimation of model parameters and outlier effects in time series. *Journal of the American Statistical Association, 88*, 284–297.
7. Eurostat. (2009). *ESS guidelines on seasonal adjustment*. Luxembourg: Office for Official Publications of the European Communities.
8. Gomez, V. (1997). Automatic model identification in the presence of missing observations and outliers. *Mimeo*. Ministerio de Economia y Hacienda, Direccion General de Analisis y Programacion Presupuestaria, June 1997
9. Gomez, V., & Maravall. A. (1992). Time series regression with ARIMA noise and missing observations - Program TRAM. EUI Working Paper Eco No. 92/81, Department of Economics, European University Institute.
10. Gomez, V., & Maravall, A. (1993). Initializing the Kalman filter with incompletely specified initial conditions. In G. R. Chen (Ed.), *Approximate Kalman filtering*. Series on approximation and decomposition. London: World Scientific Publishing Company.
11. Gomez, V., & Maravall, A. (1994). Estimation, prediction and interpolation for nonstationary series with the Kalman filter. *Journal of the American Statistical Association, 89*, 611–624.
12. Gomez, V., & Maravall, A. (1996). Program TRAMO and SEATS: Instructions for users. Working Paper 9628, Service de Estudios, Banco de Espana.
13. Gomez, V., Maravall, A., & Pena, D. (1999). Missing observations in ARIMA models: Skipping strategy versus additive outlier approach. *Journal of Econometrics, 88*, 341–364.
14. Grudkowska, S. (2015). *JDemetra+ reference manual*. Warsaw: National Bank of Poland.

15. Hannan, E. J., & Rissanen, J. (1982). Recursive estimation of mixed autoregressive moving average order. *Biometrika, 69*, 81–94.
16. Hillmer, S. C., & Tiao, G. C. (1982). An ARIMA-model based approach to seasonal adjustment. *Journal of the American Statistical Association, 77*, 63–70.
17. Jones, R. (1980). Maximum likelihood fitting of ARMA models to time series with missing observations. *Technometrics, 22*, 389–395.
18. Kohn, R., & Ansley, C. F. (1985). Efficient estimation and prediction in time series regression models. *Biometrika, 72*, 694–697.
19. Kohn, R., & Ansley, C. F. (1986). Estimation, prediction and interpolation for ARIMA models with missing data. *Journal of the American Statistical Association, 81*, 751–761.
20. Ljung, G., & Box, G. (1979). The likelihood function of stationary autoregressive-moving average models. *Biometrika, 66*, 265–270.
21. Maravall, A. (1987). On minimum mean squared error estimation of the noise in unobserved component models. *Journal of Business and Economic Statistics, 5*, 115–120.
22. Maravall, A. (1988). The use of ARIMA models in unobserved components estimation. In W. Barnett, E. Berndt, & H. White (Eds.), *Dynamic econometric modeling*. Cambridge: Cambridge University Press.
23. Maravall, A. (1995). Unobserved components in economic time series. In H. Pesaran, P. Schmidt, & M. Wickens (Eds.), *The handbook of applied econometrics* (Vol. 1). Oxford: Basil Blackwell.
24. Maravall, A., & Pérez, D. (2012). Applying and interpreting model-based seasonal adjustment - The Euro-area industrial production series. In W. R. Bell, S. H. Holan, & T. S. McElroy (Eds.), *Economic time series: Modeling and seasonality*. London: Chapman and Hall.
25. Melard, G. (1984). A fast algorithm for the exact likelihood of autoregressive moving average models. *Applied Statistics, 35*, 104–114.
26. Morf, M., Sidhu, G. S., & Kailath, T. (1974). Some new algorithms for recursive estimation on constant, linear, discrete-time systems. *IEEE Transactions on Automatic Control, 19*, 315–323.
27. Tiao, G. C., & Tsay, R. S. (1983). Consistency properties of least squares estimates of autoregressive parameters in ARMA Models. *The Annals of Statistics, 11*, 856–871.
28. Tsay, R. S. (1984). Regression models with time series errors. *Journal of the American Statistical Association, 79*, 118–124.
29. Tsay, R. S. (1986). Time series model specification in the presence of outliers. *Journal of the American Statistical Association, 81*, 132–141.

Chapter 6
Seasonal Adjustment Based on Structural Time Series Models

Abstract The structural model decomposition method starts directly with an observation equation (sometimes called measurement equation) that relates the observed time series to the unobserved components. Simple ARIMA or stochastic trigonometric models are a priori assumed for each unobserved component. Structural Time series Analyzer, Modeler, and Predictor (STAMP) is the main software and includes several types of models for each component. This chapter discusses in detail the basic structural time series model with explicit specifications for each component. It deals also with the estimation of the parameters which is carried out by the method of maximum likelihood where the maximization is done by means of a numerical optimization method. Based on the parameter estimates, the components can be estimated using the observed time series. Model adequacy is generally diagnosed using classical test statistics applied to the standardized one-step ahead prediction errors. An illustrative example of the seasonal adjustment performed using the default option of the STAMP software is shown with the US Unemployment Rate for Males (16 years and over) series.

The structural time series approach involves decomposing a series into components which have a direct interpretation. A structural model, also called unobserved component (UC) model, consists of a number of stochastic linear processes that stand for the trend, cycle, seasonality, and remaining stationary dynamic features in an observed time series. The trend component typically represents the longer term movement of the series and is often specified as a smooth function of time. The recurring but persistently changing patterns within the years are captured by the seasonal component. In economic time series, the cycle component can represent the dynamic features associated with the business cycle.

The key to handling structural time series models is the *state space form* with the state of the system representing the various unobserved components such as trend, cycle, and seasonality. The estimate of the unobservable state can be updated by means of a *filtering* procedure as new observations become available. Predictions are made by *extrapolating* these estimated components into the future, while *smoothing* algorithms give the best estimate of the state at any point within the sample. The statistical treatment can therefore be based on the Kalman filter and its related methods. A detailed discussion of the methodological and technical

© Springer International Publishing Switzerland 2016

E. Bee Dagum, S. Bianconcini, *Seasonal Adjustment Methods and Real Time Trend-Cycle Estimation*, Statistics for Social and Behavioral Sciences,
DOI 10.1007/978-3-319-31822-6_6

concepts underlying structural time series models is contained in the monographs by Harvey [11], Kitagawa and Gersh [17], and Kim and Nelson [16]. Durbin and Koopman [8] also provide an exhaustive overview on state space methods for time series, whereas an introduction is given by Commandeur and Koopman [5].

In this chapter, we first introduce the basic structural time series model with explicit specifications for each component. The estimation of parameters is carried out by the method of maximum likelihood in which the likelihood is evaluated via the Kalman filter. The likelihood is maximized by means of a numerical optimization method. Based on the parameter estimates, the components can be estimated using the observed time series. Model adequacy is generally diagnosed using classical test statistics applied to the standardized one-step ahead prediction errors.

6.1 Structural Time Series Models

The structural time series model for quarterly or monthly observations $y_t, t = 1, \ldots, n$, is given by

$$y_t = T_t + C_t + S_t + I_t, \qquad (6.1)$$

where T_t stands for the trend, C_t for the cycle, S_t for seasonality, and I_t is the irregular component generally assumed to be $NID(0, \sigma_I^2)$. All four components are stochastic and the disturbances driving them are assumed to be mutually uncorrelated. The definitions of the components are given below, but a full explanation of the underlying rationale can be found in Harvey [11]. The effectiveness of structural time series models compared to ARIMA-type models is discussed in Harvey et al. [12].

6.1.1 Trend Models

The trend component can be specified in many different ways. The most common specifications are listed below.

Local Level Model The trend component can be simply modeled as a random walk process, that is

$$T_{t+1} = T_t + a_{T_t}, \qquad (6.2)$$

where the disturbance series a_{T_t} is normally independently distributed with mean zero and variance σ_T^2, that is $NID(0, \sigma_T^2)$. It is also mutually independent on all other disturbance series related to y_t in (6.1). The initial trend T_1 is generally treated as an unknown coefficient that has to be estimated together with the variance σ_T^2.

In the specification (6.2), the trend component is an integrated process of order 1, $I(1)$. When this trend is included in the decomposition of y_t, the time series y_t is at least $I(1)$ as well. In particular, when T_t is given by Eq. (6.2), the model $y_t = T_t + I_t$ is called *local level model*.

Local Linear Trend An extension of the random walk trend is obtained by including a stochastic drift component

$$T_{t+1} = T_t + \beta_t + a_{T_t} \tag{6.3}$$

$$\beta_{t+1} = \beta_t + \zeta_t \qquad \zeta_t \sim NID(0, \sigma_\zeta^2), \tag{6.4}$$

where the disturbance series a_{T_t} is as in (6.2). The initial values T_1 and β_1 are treated as unknown coefficients. When T_t is given by (6.3), Harvey [10] defines $y_t = T_t + I_t$ as the *local linear trend model*. In case $\sigma_\zeta^2 = 0$, the trend (6.3) reduces to an $I(1)$ process given by $T_{t+1} = T_t + \beta_1 + a_{T_t}$, where the drift β_1 is fixed. This specification is referred as a *random walk plus drift* process. If in addition $\sigma_T^2 = 0$, the trend reduces to the deterministic linear trend $T_{t+1} = T_1 + \beta_1 t$. When $\sigma_T^2 = 0$ and $\sigma_\zeta^2 > 0$, the trend T_t in (6.3) remains an $I(2)$ process.

Trend with Stationary Drift An extension of the previous model is obtained by including a stationary stochastic drift component:

$$T_{t+1} = T_t + \beta_t + a_{T_t}, \tag{6.5}$$

$$\beta_{t+1} = \varphi_\beta \beta_t + \zeta_t,$$

with the autoregressive coefficient $0 < \varphi_\beta < 1$, and where the disturbance series a_{T_t} and ζ_t are as in Eqs. (6.3) and (6.4). The restriction for φ_β is necessary to have a stationary process for the drift β_t. In this case, the initial variable T_1 is treated as an unknown coefficient, while the initial drift is specified as $\beta_1 \sim N(0, \sigma_\zeta^2/1 - \varphi_\beta^2)$. The stationary drift process for β_t can be generalized to a higher order autoregressive process and can include moving average terms. However, in practice it may be difficult to empirically identify such drift processes without very large data samples.

Higher Order Smooth Trend The local linear trend (6.3) with $\sigma_T^2 = 0$ is a smooth $I(2)$ process. It can be alternatively specified as $\Delta^2 T_{t+2} = \zeta_t$, where the initial variables T_1 and T_2 are treated as unknown coefficients. To enforce more smoothness in the trend component, it can be generalized as $\Delta^k T_{t+k} = \zeta_t$, where the initial variables $T_1, \ldots, T_k$ are treated as unknown coefficients for $k = 1, 2, \ldots$. In the usual way, we can specify the higher order smooth trend component by $T_t = T_t^{(k)}$, where

$$T_{t+1}^{(j)} = T_t^{(j)} + T_t^{(j-1)}, \quad T_t^{(0)} = \zeta_t, \tag{6.6}$$

for $j = k, k-1, \ldots, 1$, and where the disturbance series ζ_t is as in (6.4). In case $k = 2$, we obtain the trend model (6.3) with $\sigma_T^2 = 0$, where $T_t = T_t^{(2)}$ and $\beta_t = T_t^{(1)}$. This trend specification is discussed in more detail by Gomez [9].

Trend with Smooth Stationary Drift Although the smoothness of a trend is a desirable feature for many economic time series, the fact that the smooth trend is defined by an $I(k)$ process can be less convincing. Therefore, Koopman and Ooms [20] propose a smooth $I(1)$ trend given by

$$T_{t+1} = T_t + \beta_t^{(m)}, \tag{6.7}$$

$$\beta_{t+1}^{(j)} = \varphi_\beta \beta_t^{(j)} + \beta_t^{(j-1)}, \quad \beta_t^{(0)} = \zeta_t,$$

for $j = m, m-1, \ldots, 1$, and where the disturbance series ζ_t is as in (6.3). In case $m = 1$, the trend with stationary drift model (6.5) is derived with $\sigma_T^2 = 0$ and $\beta_t = \beta_t^{(1)}$. The autoregressive coefficient $0 < \varphi_\beta < 1$ is the same for each $\beta_{t+1}^{(j)}$ with $j = m, m-1, \ldots, 1$. This restriction can be lifted by having different autoregressive coefficients for each j, but generally the parsimonious specification (6.7) is preferred.

6.1.2   The Cyclical Component

To capture the business cycle features of a time series, various stochastic specifications of the cycle component can be considered.

Autoregressive Moving Average Process The cycle component C_t can be formulated as a stationary autoregressive moving average (ARMA) process as

$$\varphi_C(B)C_{t+1} = \theta_C(B)a_{C_t}, \quad a_{C_t} \sim NID(0, \sigma_C^2), \tag{6.8}$$

where $\varphi_C(B)$ is the autoregressive polynomial of order p in the lag operator B with coefficients $\varphi_{C,1}, \ldots, \varphi_{C,p}$ and $\theta_C(B)$ is the moving average polynomial of order q with coefficients $\theta_{C,1}, \ldots, \theta_{C,q}$. The requirement of stationarity applies to the autoregressive polynomial $\varphi_C(B)$, and states that the roots $|\varphi_C(B)| = 0$ lie outside the unit circle. The theoretical autocorrelation function of an ARMA process has cyclical properties when the roots of $|\varphi_C(B)| = 0$ are within the complex range. It requires $p > 1$. In this case, the autocorrelations converge to zero with increasing lags, but the convergence pattern is cyclical. It implies that the time series itself has cyclical dynamic properties. Once the autoregressive coefficients are estimated, it can be established whether the empirical model with C_t as in (6.8) has detected cyclical dynamics in the time series.

Time-Varying Trigonometric Cycle Another stochastic formulation of the cycle component is based on a time-varying trigonometric process as follows:

$$\begin{pmatrix} C_{t+1} \\ C_{t+1}^* \end{pmatrix} = \rho \begin{pmatrix} \cos \lambda_c & \sin \lambda_c \\ -\sin \lambda_c & \cos \lambda_c \end{pmatrix} \begin{pmatrix} C_t \\ C_t^* \end{pmatrix} + \begin{pmatrix} a_{C_t} \\ a_{C_t^*} \end{pmatrix} \quad t = 1, \ldots, n, \tag{6.9}$$

with frequency λ_c associated with the typical length of a business cycle, say between 1.5 and 8 years according to Burns and Mitchell [4]. The factor $0 < \rho < 1$ is

introduced to enforce a stationary process for the stochastic cycle component. The
disturbance and the initial conditions for the cycle variables are given by

$$
\begin{pmatrix} a_{C_t} \\ a_{C_t^*} \end{pmatrix} \sim NID\left(\mathbf{0}, \sigma_C^2 I_2\right) \qquad \begin{pmatrix} C_1 \\ C_1^* \end{pmatrix} \sim NID\left(\mathbf{0}, \frac{\sigma_C^2}{\rho^2} I_2\right) \tag{6.10}
$$

where the cyclical disturbance series a_{C_t} and $a_{C_t^*}$ are serially and mutually independent, also with respect to all other disturbance series. The coefficients ρ, λ_c,
and σ_C^2 are unknown and need to be estimated together with the other parameters.
This stochastic cycle specification is discussed by Harvey [11], that shows that the
process (6.9) to be the same as and ARMA process (6.8) with complex roots where
$p = 2$ and $q = 1$.

Smooth Time-Varying Trigonometric Cycle To enforce smoothness on the cycle
component, Harvey and Trimbur [13] propose the specification $C_t = C_t^{(m)}$, where

$$
\begin{pmatrix} C_{t+1}^{(j)} \\ C_{t+1}^{(j)*} \end{pmatrix} = \rho \begin{pmatrix} \cos \lambda_c & \sin \lambda_c \\ -\sin \lambda_c & \cos \lambda_c \end{pmatrix} \begin{pmatrix} C_t^{(j)} \\ C_t^{(j)*} \end{pmatrix} + \begin{pmatrix} C_t^{(j-1)} \\ C_t^{(j-1)*} \end{pmatrix}, \tag{6.11}
$$

for $j = m, m - 1, \ldots, 1$, and

$$
\begin{pmatrix} C_{t+1}^{(0)} \\ C_{t+1}^{(0)*} \end{pmatrix} = \begin{pmatrix} a_{C_t} \\ a_{C_t^*} \end{pmatrix} \sim NID(0, \sigma_C^2 I_2), \tag{6.12}
$$

for $t = 1, \ldots, n$. The initial conditions for this stationary process are provided by
Trimbur [27].

6.1.3 Seasonality

As discussed in Chap. 2, to account for the seasonal variation in a time series, the
component S_t is included in model (6.1). Common specifications for the seasonal
component S_t are provided in the following:

Dummy Seasonal Since the seasonal effects should sum to zero over a year, a basic
model for this component is given by

$$
S_t = -\sum_{j=1}^{s-1} S_{t-j} + \omega_t \qquad t = s, \ldots, n, \tag{6.13}
$$

where s denotes the number of "seasons" in a year. In words, the seasonal effects are
allowed to change over time by letting their sum over the previous year be equal to a
random disturbance term ω_t with mean zero and variance σ_ω^2. Writing out Eq. (6.13)

in terms of the lag operator B gives

$$(1 + B + \cdots + B^{s-1})S_t = \gamma(B)S_t = \omega_t \quad t = s, \ldots, n. \qquad (6.14)$$

However, since $(1 - B^s) = (1 + B + \cdots + B^{s-1})(1 - B) = \gamma(B)(1 - B)$ the model can also be expressed in terms of the seasonal difference operator as $(1 - B^s)S_t = (1 - B)\omega_t$. The normally distributed disturbance ω_t drives the changes in the seasonal effect over time and is serially and mutually uncorrelated with all other disturbances and for all time periods. In the limiting case where $\sigma_\omega^2 = 0$ for all t, the seasonal effects are fixed over time and are specified as a set of unknown fixed dummy coefficients that sum up to zero.

Trigonometric Seasonal Alternatively, a seasonal pattern can also be modeled by a set of trigonometric terms at the seasonal frequencies, $\lambda_j = 2\pi j/s, j = 1, \ldots, [s/2]$, where $[s/2]$ is equal to $s/2$ if s is even, and $(s - 1)/2$ if s is odd. The seasonal effect at time t is then described as

$$S_t = \sum_{j=1}^{[s/2]} (S_j \cos \lambda_j t + S_j^* \sin \lambda_j t). \qquad (6.15)$$

When s is even, the sine term disappears for $j = s/2$, and so the number of trigonometric parameters, the S_j and S_j^*, is always $(s - 1)/2$, which is the same as the number of coefficients in the seasonal dummy formulation. A seasonal pattern based on (6.15) is the sum of $[s/2]$ cyclical components, each with $\rho = 1$, and it may be allowed to evolve over time in exactly the same way as the cycle was allowed to move in Eq. (6.10). The model is

$$S_t = \sum_{j=1}^{[s/2]} S_{j,t}, \qquad (6.16)$$

where, following Eq. (6.10),

$$\begin{pmatrix} S_{j,t} \\ S_{j,t}^* \end{pmatrix} = \rho \begin{pmatrix} \cos \lambda_j & \sin \lambda_j \\ -\sin \lambda_j & \cos \lambda_j \end{pmatrix} \begin{pmatrix} S_{j,t-1} \\ S_{j,t-1}^* \end{pmatrix} + \begin{pmatrix} \omega_{j,t} \\ \omega_{j,t}^* \end{pmatrix}, \qquad (6.17)$$

with $\omega_{j,t}$ and $\omega_{j,t}^*, j = 1, \ldots, [s/2]$, being white noise processes which are uncorrelated with each other and have a common variance σ_ω^2. $S_{j,t}^*$ appears as a matter of construction, and its interpretation is not particularly important. Note that, when s is even, the component at $j = s/2$ collapses to

$$S_{j,t} = S_{j,t-1} \cos \lambda_j + \omega_{j,t}.$$

Other seasonal models that incorporate time-varying seasonal effects in the unobserved components time series model can be considered. We refer the reader to Proietti [25] for a review of other different seasonal specifications and their properties.

6.1.4 Regression Component

The model (6.1) may provide a good description of the time series, although it may sometimes be necessary to include additional components. As discussed in the previous chapters, seasonal economic time series is often affected by trading day effects and holiday effects which can influence the dynamic behavior of the series. Hence, a set of explanatory variables need to be included in the model to capture specific (dynamic) variations in the time series, as well as outliers and breaks. Therefore, Koopman and Ooms [20] suggest to extend model (6.1) as follows:

$$y_t = T_t + C_t + S_t + \mathbf{x}'_t \boldsymbol{\delta} + I_t, \qquad I_t \sim N(0, \sigma_I^2),$$

for $t = 1, \ldots, n$, and where $\mathbf{x}_t$ is a K-dimensional vector of predetermined covariates and $\boldsymbol{\delta}$ is a $K \times 1$ vector of regression coefficients, that can be allowed to change over time.

6.2 Linear State Space Models

The statistical treatment of structural time series models is based on the corresponding state space representation, according to which the observations are assumed to depend linearly on a *state vector* that is unobserved and generated by a stochastic time-varying process. The observations are further assumed to be subject to a measurement error that is independent on the state vector. The state vector can be estimated or identified once a sufficient set of observations becomes available.

The state space form provides a unified representation of a wide range of linear time series models, as discussed by Harvey [11], Kitagawa and Gersch [17], and Durbin and Koopman [8].

The general linear state space model for a sequence of n observations, $y_1, \ldots, y_n$, is specified as follows:

$$y_t = \mathbf{z}_t^T \boldsymbol{\alpha}_t + I_t, \qquad t = 1, \ldots, n \qquad (6.18)$$

$$\boldsymbol{\alpha}_{t+1} = \boldsymbol{\Gamma}_t \boldsymbol{\alpha}_t + \mathbf{R}_t \boldsymbol{\varepsilon}_t. \qquad (6.19)$$

Equation (6.18) is called the *observation* or *measurement equation* which relates the observations $y_t, t = 1, \ldots, n$, to the state vector $\boldsymbol{\alpha}_t$ through $\mathbf{z}_t$, that is, an $m \times 1$

vector of fixed coefficients. In particular, $\boldsymbol{\alpha}_t$ is the $m \times 1$ state vector that contains the unobserved trend T_t, cycle C_t, and the seasonal component S_t. The irregular component I_t is generally assumed to follow a white noise process with zero mean and variance $\sigma_{I_t}^2$.

On the other hand, Eq. (6.19) is called the *state* or *transition equation*, where the dynamic evolution of the state vector $\boldsymbol{\alpha}_t$ is described through the fixed matrix $\boldsymbol{\Gamma}_t$ of order $m \times m$. $\boldsymbol{\varepsilon}_t$ is a $r \times 1$ vector of disturbances which are assumed to follow a multivariate white noise process with zero mean vector and covariance matrix $\boldsymbol{\Sigma}_{\varepsilon_t}$. It is assumed to be distributed independently on I_t at all time points. The matrix $\mathbf{R}_t$ is an $m \times r$ selection matrix with $r < m$, that in many standard cases is the identity matrix $\mathbf{I}_m$, being generally $r = m$. Indeed, although matrix $\mathbf{R}_t$ can be specified freely, it is often composed of a selection from the first r columns of the identity matrix $\mathbf{I}_m$.

Initial conditions have to be specified for the state vector at the first time point, $\boldsymbol{\alpha}_1$. It is generally assumed to be generated as $\boldsymbol{\alpha}_1 \sim N(\mathbf{a}_1, \mathbf{P}_1)$, and to be independent on the observation and state disturbances, I_t and $\boldsymbol{\varepsilon}_t$. The mean vector $\mathbf{a}_1$ and covariance matrix $\mathbf{P}_1$ can be treated as given and known in almost all stationary processes. For nonstationary processes and in presence of regression effects in the state vector, the associated elements in the initial mean vector $\mathbf{a}_1$ can be treated as unknown and estimated. For an extensive discussion of initialization in state space analysis, we refer the reader to Durbin and Koopman [8].

By appropriate choices of $\boldsymbol{\alpha}_t, I_t$, and $\boldsymbol{\varepsilon}_t$, of the matrices $\mathbf{z}_t, \boldsymbol{\Gamma}_t, \mathbf{R}_t$, and of the scalar $\sigma_{I_t}^2$, a wide range of different structural time series models discussed in Sect. 6.2 can be specified from Eqs. (6.18) and (6.19).

Basic Decomposition Model Consider the model $y_t = T_t + S_t + I_t$ with the trend component specified through Eqs. (6.3) and (6.4), and S_t as in (6.13) in presence of quarterly data $(s = 4)$. A state vector of five elements and a disturbance vector of four elements are required, and they are given by

$$\boldsymbol{\alpha}_t = \begin{pmatrix} T_t \\ \beta_t \\ S_{1,t} \\ S_{2,t} \\ S_{3,t} \end{pmatrix}, \quad \boldsymbol{\varepsilon}_t = \begin{pmatrix} a_{T_t} \\ \zeta_t \\ \omega_t \end{pmatrix}.$$

The state space formulation of the basic decomposition model is given by Eqs. (6.18) and (6.19) with the system matrices

$$\boldsymbol{\Gamma}_t = \begin{pmatrix} 1 & 0 & 0 & 0 & 0 \\ 0 & 1 & 0 & 0 & 0 \\ 0 & 0 & -1 & -1 & -1 \\ 0 & 0 & 1 & 0 & 0 \\ 0 & 0 & 0 & 1 & 0 \end{pmatrix}, \mathbf{z}_t = \begin{pmatrix} 1 \\ 0 \\ 1 \\ 0 \\ 0 \end{pmatrix}, \boldsymbol{\Sigma}_{\varepsilon_t} = \begin{pmatrix} \sigma_T^2 & 0 & 0 \\ 0 & \sigma_\zeta^2 & 0 \\ 0 & 0 & \sigma_\omega^2 \end{pmatrix}, \mathbf{R}_t = \begin{pmatrix} 1 & 0 & 0 \\ 0 & 1 & 0 \\ 0 & 0 & 1 \\ 0 & 0 & 0 \\ 0 & 0 & 0 \end{pmatrix}.$$

Here the system matrices $\boldsymbol{\Gamma}_t$ and $\mathbf{z}_t$ do not depend on time. The variances of the disturbances are unknown and need to be estimated. They are $\sigma_\zeta^2, \sigma_T^2, \sigma_\xi^2$, and σ_ω^2. For the trend component, the initial variables T_1 and β_1 are treated as unknown, whereas for the seasonal component, with $s = 4$, the unknown coefficients are the initial variables S_{-1}, S_0, S_1. All these coefficients form part of the initial state vector $\boldsymbol{\alpha}_1$. This is the *basic structural time series model*. A typical application of this basic model is for the seasonal adjustment of time series, being the seasonally adjusted series given by $y_t - S_t = T_t + I_t$ for $t = 1, \ldots, n$.

Local Trend Plus Cycle Model Another model is given by $y_t = T_t + C_t + I_t$ with T_t specified as in Eq. (6.2) and C_t as in Eq. (6.9). Defining

$$
\boldsymbol{\alpha}_t = \begin{pmatrix} T_t \\ C_t \\ C_t^* \end{pmatrix}, \boldsymbol{\varepsilon}_t = \begin{pmatrix} a_{T_t} \\ a_{C_t} \\ a_{C_t^*} \end{pmatrix}, \boldsymbol{\Gamma}_t = \begin{pmatrix} 1 & 0 & 0 \\ 0 & \rho\cos(\lambda_c) & \rho\sin(\lambda_c) \\ 0 & -\rho\sin(\lambda_c) & \rho\cos(\lambda_c) \end{pmatrix}, \mathbf{z}_t^T = (1,1,0),
$$

$$
\boldsymbol{\Sigma}_{\varepsilon_t} = \begin{pmatrix} \sigma_\eta^2 & 0 & 0 \\ 0 & \sigma_C^2(1-\rho^2) & 0 \\ 0 & 0 & \sigma_C^2(1-\rho^2) \end{pmatrix}, \mathbf{R}_t = \begin{pmatrix} 1 & 0 & 0 \\ 0 & 1 & 0 \\ 0 & 0 & 1 \end{pmatrix}
$$

in (6.18) and (6.19), we obtain the following local level plus cycle model:

$$
\begin{aligned}
y_t &= T_t + C_t + I_t, & I_t &\sim N(0, \sigma_I^2) \\
T_{t+1} &= T_t + \eta_t, & \eta_t &\sim N(0, \sigma_\eta^2) \\
C_{t+1} &= \rho[\cos(\lambda_c)C_t + \sin(\lambda_c)C_t^*] + a_{C_t}, & a_{C_t} &\sim NID(0, \sigma_C^2(1-\rho^2)) \\
C_{t+1}^* &= \rho[-\sin(\lambda_c)C_t + \cos(\lambda_c)C_t^*] + a_{C_t^*}, & a_{C_t^*} &\sim NID(0, \sigma_C^2(1-\rho^2))
\end{aligned}
$$

for $t = 1, \ldots, n$, where $0 < \rho < 1$ is the damping factor and λ_c is the frequency of the cycle in radians so that $2\pi/\lambda_c$ is the period of the cycle. In case $\rho = 1$, the cycle reduces to a fixed sine–cosine wave, but the component is still stochastic since the initial values C_1 and C_1^* are stochastic variables with mean zero and variance σ_C^2. A typical application of this model is for the signal extraction of business cycles from macroeconomic time series.

6.2.1 The Kalman Filter

Given the values of all system matrices and initial conditions $\mathbf{a}_1$ and $\mathbf{P}_1$, the predictive estimator of the state vector $\boldsymbol{\alpha}_{t+1}$ is based on observations $y_1, \ldots, y_n$. The Kalman filter [15] computes the minimum mean square linear estimator (MMSLE) of the state vector $\boldsymbol{\alpha}_{t+1}$ conditional on the observations $y_1, \ldots, y_t$, denoted by $\mathbf{a}_{t+1|t}$, together with its mean square error (MSE) matrix, denoted by $\mathbf{P}_{t+1|t}$. $\mathbf{a}_{t+1|t}$ is generally called the state prediction estimate, whereas $\mathbf{P}_{t+1|t}$ is its state prediction error variance matrix. Specifically, the one-step ahead prediction error $v_t = y_t - E(y_t|y_1,\ldots,y_{t-1})$ and its variance $\mathrm{Var}(v_t) = f_t$ are derived as

$$v_t = y_t - \mathbf{z}_t^T \mathbf{a}_{t|t-1}, \quad f_t = \mathbf{z}_t^T \mathbf{P}_t \mathbf{z}_t + \sigma_I^2,$$

such that the innovations have mean zero and are serially independent by construction, that is, $E(v_t v_s') = 0$ for $t \neq s$, and $t, s = 1,\ldots,n$. For particular initial values $\mathbf{a}_{1|0}$ and $\mathbf{P}_{1|0}$, the MMSLE and corresponding MSE matrix are then determined as follows:

$$\mathbf{a}_{t+1|t} = \boldsymbol{\Gamma}_t \mathbf{a}_{t|t-1} + \mathbf{k}_t v_t$$
$$\mathbf{P}_{t+1} = \boldsymbol{\Gamma}_t \mathbf{P}_t \boldsymbol{\Gamma}_t^T + \mathbf{R}_t \boldsymbol{\Sigma}_{\varepsilon_t} \mathbf{R}_t^T - \mathbf{k}_t \mathbf{M}_t^T$$

where $\mathbf{k}_t = \mathbf{M}_t f_t^{-1}$ is the Kalman gain matrix, being $\mathbf{M}_t = \boldsymbol{\Gamma}_t \mathbf{P}_t \mathbf{z}_t$.

Before the MMSLE $\mathbf{a}_{t+1|t}$ and the MSE $\mathbf{P}_{t+1|t}$ are computed in the Kalman filter, the MMSLE of the state vector $\boldsymbol{\alpha}_t$ conditional on $y_1,\ldots,y_t$, denoted by $\mathbf{a}_{t|t}$, and its corresponding MSE matrix, denoted by $\mathbf{P}_{t|t}$, can be computed as

$$\mathbf{a}_{t|t} = \mathbf{a}_{t|t-1} + \mathbf{P}_{t|t-1} \mathbf{z}_t f_t^{-1} v_t, \qquad \mathbf{P}_{t|t} = \mathbf{P}_{t|t-1} - \mathbf{P}_{t|t-1} \mathbf{z}_t f_t^{-1} \mathbf{z}_t^T \mathbf{P}_{t|t-1}.$$

It follows that

$$\mathbf{a}_{t+1|t} = \boldsymbol{\Gamma}_t \mathbf{a}_{t|t}, \qquad \mathbf{P}_{t+1|t} = \boldsymbol{\Gamma}_t \mathbf{P}_{t|t} \boldsymbol{\Gamma}_t' + \mathbf{R}_t \boldsymbol{\Sigma}_{\varepsilon_t} \mathbf{R}_t^T.$$

Formal proofs of the Kalman filter can be found in Anderson and Moore [1], Harvey [11], and Durbin and Koopman [8].

6.2.2 Likelihood Estimation

The Kalman filter can be used to evaluate the likelihood function via the prediction error decomposition [11, 14, 26], according to which the joint density of $y_1,\ldots,y_n$ is specified as follows:

$$p(y_1,\ldots,y_n) = p(y_1) \prod_{t=2}^{n} p(y_t|y_1,\ldots,y_{t-1}).$$

The predictive density $p(y_t|y_1,\ldots,y_{t-1})$ is Gaussian with mean $E(y_t|y_1,\ldots,y_{t-1}) = \mathbf{z}_t^T \mathbf{a}_{t|t-1}$ and variance $V(y_t|y_1,\ldots,y_{t-1}) = \mathbf{z}_t^T \mathbf{P}_{t|t-1} \mathbf{z}_t + \sigma_I^2 = f_t$. Hence, for a realized time series $y_1, \ldots, y_n$, and denoting with θ the vector of model parameters, the log-likelihood function is given by

$$\ell(\theta) = \log p(y_1, \ldots, y_n) = \sum_{t=1}^{n} \log p(y_t|y_1, \ldots, y_{t-1}) \tag{6.20}$$

$$= -\frac{n}{2} \log(2\pi) - \frac{1}{2} \sum_{t=1}^{n} \log |f_t| - \frac{1}{2} \sum_{t=1}^{n} v_t^2 f_t^{-1}.$$

The one-step ahead prediction errors v_t and their variances f_t are computed by the Kalman filter for a given value of the parameter vector θ.

The log likelihood (6.20) has not closed form and has to be numerically maximized with respect to θ. This can be done using a numerical quasi-Newton method, such as the Broyden–Fletcher–Goldfarb–Shanno (BFGS) method. It is generally regarded as computationally efficient in terms of convergence speed and numerical stability [24]. This iterative optimization method is based on information from the gradient (or score). Analytical and computationally fast methods for computing the score for a current value of θ in a state space analysis are developed by Koopman and Shepard [21].

An alternative method for maximum likelihood estimation is the EM-algorithm. It is based on the joint density $p(y_1, \ldots, y_n, \alpha_1, \ldots, \alpha_n)$ and consists of two step. The expectation (E) step takes the expectation of the components of the joint density conditional on $y_1, \ldots, y_n$. Hence, in the E-step, the estimated state space vector is evaluated using a smoothing algorithm related to the Kalman filter. In the maximization (M) step the resulting expression of the joint density is maximized with respect to θ. It is usually done analytically, and is simpler than maximizing the full likelihood function directly. Given the updated estimate of θ from the M-step, the algorithm returns to the E-step and evaluate the smoothed estimates based on the new parameter estimates. This iterative procedure converges to the maximum likelihood estimate of θ. Details on the properties of the EM algorithm for state space models can be found in Koopman [18].

The nonstationary trend and seasonal components, as discussed in Sect. 6.2, rely on initial variables that are treated as fixed unknown coefficients. A straightforward approach to the estimation of α_1 is to estimate it jointly with θ by the method of maximum likelihood as previously discussed. However, numerical problems may arise when the likelihood function is maximized with respect to a parameter vector of high dimension. Several methods have been used in the literature, and most of them can be embedded in a unified treatment for the initialization of the Kalman filter with respect to the initial variables, and we refer the reader to Ansley and Kohn [2], de Jong [6], and Koopman [18] for a detailed discussion. The latter references also provide a detailed treatment in presence of stationary variables, where in this case the initial conditions for α_1 can be obtained from the theoretical autocovariance function.

6.2.3 Diagnostic Checking

The assumptions underlying model (6.1), and all the component specifications given in Sect. 6.2, are that all disturbances related to the unobserved components are normally distributed and serially and mutually independent with constant variances. Under these assumptions the standardized one-step ahead prediction errors (or *prediction residuals*) are given by

$$e_t = \frac{v_t}{\sqrt{f_t}}, \quad t = 1, \ldots, n,$$

being also normally distributed and serially independent with unit variance.

The *Normality* assumption is analyzed looking at the first four moments of the standardized forecast errors, that are given by

$$m_1 = \frac{1}{n} \sum_{t=1}^{n} e_t, \qquad m_q = \frac{1}{n} \sum_{t=1}^{n} (e_t - m_1)^q, \quad q = 2, 3, 4.$$

When the model assumptions are valid, skewness M_3 and kurtosis M_4 are normally distributed as [3]

$$M_3 = \frac{m_3}{\sqrt{m_2^3}} \sim N\left(0, \frac{6}{n}\right), \quad M_4 = \frac{m_4}{m_2^2} \sim N\left(3, \frac{24}{n}\right).$$

Standard statistical tests can be used to check whether the observed values of M_3 and M_4 are consistent with their asymptotic densities.

A simple test for *heteroscedasticity* is obtained by comparing the sum of squares of two exclusive subsets of the sample. For example, the statistic

$$\frac{\sum_{t=n-h+1}^{n} e_t^2}{\sum_{t=1}^{h} e_t^2}$$

is $F_{h,h}$-distributed under the null hypothesis of homoscedasticity.

Finally, the correlogram of the prediction residuals should not reveal significant serial correlation. A standard Portmanteau test statistic for serial correlation is based on the Ljung–Box statistic [23]. This is given by

$$Q(q) = n(n+2) \sum_{j=1}^{q} \frac{c_j^2}{n-j},$$

for some positive integer q, that depends on the frequency of the series (generally $q = 24$ for monthly series), where c_j is the j-th correlation $c_j = \frac{1}{nm_2} \sum_{t=j+1}^{n} (e_t - m_1)(e_{t-j} - m_1)$. The statistic Q is distributed as a chi-square variable with $q - p$

degrees of freedom if the residuals are independent, being p the number of parameters in the model.

6.3 Illustrative Example: Analysis of the US Unemployment Rate for Males Using STAMP

The implementation and estimation of the structural time series models discussed in the previous sections are made simpler by the availability of a specific software called STAMP (Structural Time series Analyzer, Modeller, and Predictor). It is a graphical user interface (GUI)-based package designed to model and forecast time series using the structural time series approach developed by Koopman et al. [19]. STAMP runs on Windows, Macintosh, and Linux operating systems as part of OxMetrics, a software system for (econometric) data analysis and forecasting developed by Doornik [7]. It provides a powerful, flexible, easy to use, and up-to-date environment to perform state space analysis of time series data.

STAMP enables a large variety of models to be analyzed, including all the structural time series models discussed in the previous sections. The software also provides an extensive array of test statistics, residual, and auxiliary residual graphics after estimation for assessing the goodness of the model fit. Furthermore, OxMetrics integrates STAMP into a larger class of related models, and provides a broad suite of graphics, data transformation, and arithmetic options.

STAMP default model is the *basic structural model* discussed in Sect. 6.3 which includes a stochastic level and slope specified through Eqs. (6.3) and (6.4), as well as a stochastic seasonal component given in (6.13). By default, STAMP does not allow for the treatment of trading day and holiday effects. Hence, it is appropriate for the analysis of stock series. In this regard, we analyze the monthly series of US Unemployment Rate for Males (16 years and over) observed from January 1992 to December 2013. In general, we recommend to select the automatic detection of outliers in the *Select components* window among the several possible options. Once the estimation is complete, some basic information appears in the *Results* window as illustrated in Table 6.1.

The "Estimation report" tells us that the convergence was very strong. Maximum likelihood estimation has been carried out by numerical optimization.

The "Diagnostic summary report" provides some of the basic diagnostic and goodness of fit statistics discussed in Sect. 6.2.3. In Table 6.1, T denotes the sample size (indicated by n in our notation), p the number of model parameters to estimate, and N is the number of analyzed time series. The Ljung–Box test for residual serial correlation, $Q(24, 21)$, is based on $q = 24$ residual autocorrelations and should be tested against a chi-square distribution with $q - p = 21$ degrees of freedom. Based on the observed value of Q equal to 24.628, we can conclude that there is no serial correlations in the residuals, a strong indication that the model is adequately capturing the dynamic structure of the series. The classical Durbin–Watson (DW)

Table 6.1 STAMP output: diagnostic summary report

```
Estimating...
Very strong convergence relative to 1e-07
 - likelihood cvg 7.35241e-15
 - gradient cvg 1.03322e-08
 - parameter cvg 5.02437e-08
- number of bad iterations 0
 Estimation process completed.

UC( 1) Estimation done by Maximum Likelihood (exact score)
 The selection sample is: 1992(1) - 2013(12) (T = 264, N = 1)
 The dependent variable Y is: UNEMP_RATE
 The model is: Y = Trend + Seasonal + Irregular + Interventions
 Steady state ......found without full convergence

Log-Likelihood is 355.342 (-2 LogL = -710.684).
Prediction error variance is 0.041706

Summary statistics
                UNEMP_RATE
 T                 264.00
 p                   3.0000
 std.error           0.2042
 Normality           0.7291
 H(82)               1.2800
 DW                  2.0335
 r(1)               -0.0248
 q                  24.000
 r(q)               -0.0316
 Q(q,q-p)           24.628
 Rs^2                0.2933
```

test is also given. Since the DW statistic is approximately equal to $2(1 - \hat{\rho})$, where $\hat{\rho}$ is the sample first order autocorrelation of the residuals, a value of the DW statistic equal to 2 indicates no autocorrelation. The value of the DW statistics always lies between 0 and 4. If the Durbin–Watson statistic is substantially less than 2, there is evidence of positive serial correlation. As a rough rule of thumb, if Durbin/Watson is less than 1, there may be cause for alarm. Small values of the DW statistic indicate that successive error terms are, on average, close in value to one another, or positively correlated. If the statistic is greater than 2, successive error terms are, on average, much different in value from one another.

The *heteroscedasticity* test statistic $H(82)$, illustrated in Sect. 6.2.3, presents a small value indicating that the residuals are characterized by constant variance over time. Finally, the *Normality* test statistic of Bowman–Shenton discussed in

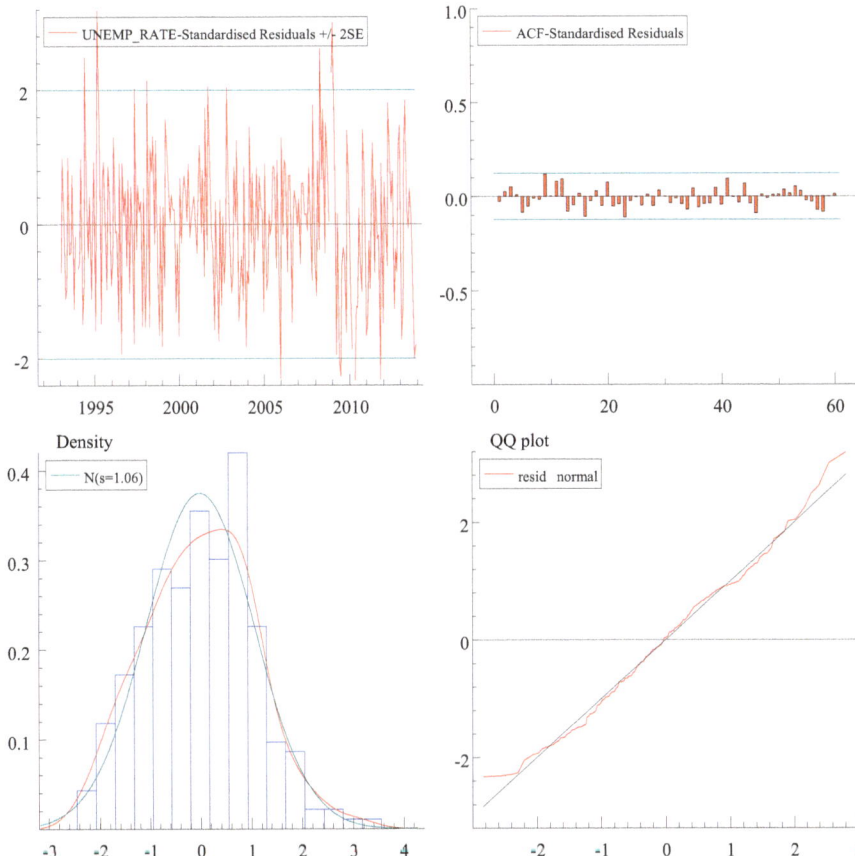

Fig. 6.1 Standardized residual graphics

Sect. 6.2.3, and based on third and fourth moments of the residuals, has a χ^2 distribution with 2 degrees of freedom when the model is correctly specified. The 5 % critical value is 5.99. The small value of the Normality test statistic together with the results shown for the other statistics in Table 6.1 clearly indicates that the basic structural model is adequate for the US Unemployment Rate series. This is also confirmed by the standardized residuals graphs illustrated in Fig. 6.1.

STAMP also provides further information for the estimated model. As shown in Table 6.2, in the *Print parameter* section it reports on the parameters, in this example estimated by maximum likelihood, where the variances and all the parameters are organized by each component. Concerning the variances of the disturbances, a zero parameter estimate would suggest that the corresponding component is fixed. In our specific case, this is true for the seasonal component, whereas the trend appears to be stochastic as assumed. The q-ratios in parenthesis are the ratio of each standard deviation to the standard deviation associated with the largest variance, being the

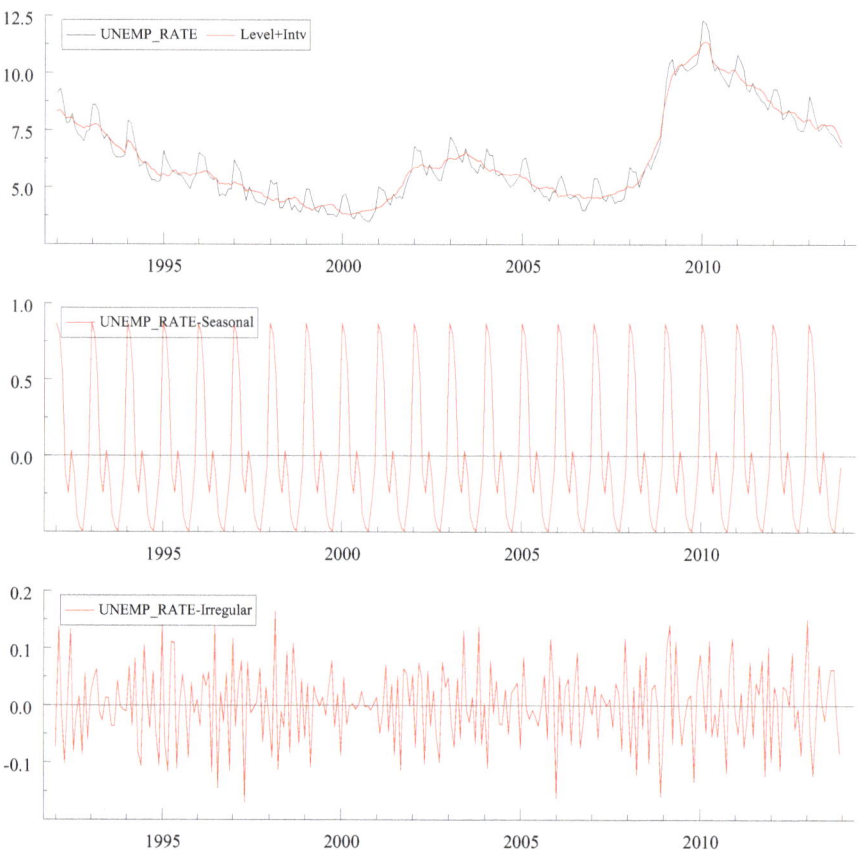

Fig. 6.2   Estimated components in the basic structural model

latter related to the level of the trend component. The original series together with the estimated components are shown in Fig. 6.2.

The final state vector contains information on the values taken by the various components at the end of the sample. Following Koopman et al. [22], the final state is the filtered estimate at time n, that is $a_{n|n}$ as defined in Sect. 6.3. The square roots of the diagonal elements of the corresponding MSE matrix, $\hat{\mathbf{P}}_{n|n}$, are the root mean square errors (RMSEs) of the corresponding state elements. The t-value is the estimate of the state divided by the RMSE, and the corresponding p-value is also reported. The null hypothesis of a zero value is rejected (at 5 % level) for both the level and slope in the trend component, and for the seasonal dummies, except for those related to June and December (at 1 % level).

STAMP automatically detected three outliers. In particular, level breaks are identified in January 1994, December 2008, and April 2010.

Table 6.2 STAMP output: print parameter section

```
Variances of disturbances:
           Value          (q-ratio)
Level      0.0161857      (1.000)
Slope      0.000674890    (0.042)
Seasonal   0.00000        (0.000)
Irregular  0.00937361     (0.579)


State vector analysis at period 2013(12)
                         Value      Prob
Level                   6.13693    [0.00000]
Slope                  -0.13053    [0.03842]
Seasonal chi2 test   1209.31038    [0.00000]
Seasonal effects:
Period                   Value      Prob
1                       0.86587    [0.00000]
2                       0.79596    [0.00000]
3                       0.53084    [0.00000]
4                      -0.11948    [0.00075]
5                      -0.24775    [0.00000]
6                       0.03332    [0.33913]
7                      -0.11718    [0.00087]
8                      -0.39016    [0.00000]
9                      -0.47199    [0.00000]
10                     -0.49448    [0.00000]
11                     -0.31218    [0.00000]
12                     -0.07277    [0.04151]


Regression effects in final state at time 2013(12)
                        Coefficient     RMSE    t-value      Prob
Level break 1994(1)         0.69029   0.19110   3.61224   [0.00037]
Level break 2008(12)        0.84900   0.19090   4.44744   [0.00001]
Level break 2010(4)        -0.72008   0.19089  -3.77215   [0.00020]
```

References

1. Anderson, B. D. O., & Moore, J. B. (1979). *Optimal filtering*. Englewood Cliffs: Prentice-Hall.
2. Ansley, C. F., & Kohn, R. (1985). Estimation, filtering and smoothing in state space models with incompletely specified initial conditions. *Annals of Statistics, 13*, 1286–1316.
3. Bowman, K. O., & Shenton, L. R. (1975). Omnibus test contours for departures from normality based on $\sqrt{b_1}$ and b_2. *Biometrika, 62*, 243–250.
4. Burns, A., & Mitchell, W. (1946). *Measuring business cycles*. Working paper. New York: NBER.

5. Commandeur, J. J. F., & Koopman, S. J. (2007). *An introduction to state space time series analysis*. Oxford: Oxford University Press.
6. de Jong, P. (1991). The diffuse Kalman filter. *Annals of Statistics, 19*, 1073–1083.
7. Doornik, J. A. (2009). *An introduction to OxMetrics 6: A software system for data analysis and forecasting*. London: Timberlake Consultants Ltd.
8. Durbin, J., & Koopman, S. J. (2001). *Time series analysis by state space methods*. Oxford: Oxford University Press.
9. Gomez, V. (2001). The use of Butterworth filters for trend and cycle estimation in economic time series. *Journal of Business and Economic Statistics*, 19, 365–373.
10. Harvey, A. C. (1985). Trends and cycles in macroeconomic time series. *Journal of Business and Economic Statistics, 3*, 216–227.
11. Harvey, A. C. (1989). *Forecasting, structural time series models and the Kalman filter*. Cambridge: Cambridge University Press.
12. Harvey, A. C., Koopman, S. J., & Penzer, J. (1998). Messy time series: A unified approach. In T. B. Fomby & R. Carter Hill (Eds.), *Advances in econometrics* (Vol. 13, pp. 103–143). New York, NY, USA: JAI Press.
13. Harvey, A. C., & Trimbur, T. M. (2003). General model-based filters for extracting trends and cycles in economic time series. *Review of Economics and Statistics, 85*, 244–255.
14. Jones, R. H. (1980). Maximum likelihood fitting of ARIMA models to time series with missing observations. *Technometrics, 22*, 389–395.
15. Kalman, R. E. (1960). A new approach to linear filtering and prediction problems. *Journal of Basic Engineering, Transactions ASMA, Series D, 82*, 35–45.
16. Kim, C. J., & Nelson, C. R. (1999). *State space models with regime switching*. Cambridge, MA: MIT Press.
17. Kitagawa, G., & Gersch, W. (1996). *Smoothness priors analysis of time series*. New York: Springer.
18. Koopman, S. J. (1997). Exact initial Kalman filtering and smoothing for non-stationary time series models. *Journal of the American Statistical Association, 92*, 1630–1638.
19. Koopman, S. J, Harvey, A. C., Doornik, J. A., & Shepard, N. (2000). *STAMP 8.2: Structural time series analyser, modeller and predictor*. London: Timberlake Consultants Ltd.
20. Koopman, S. J., & Ooms, M. (2011). Forecasting economic time series using unobserved components time series models. In M. Clements & D. F. Hendry (Eds.), *Oxford handbook of economic forecasting*. New York: Oxford University Press.
21. Koopman, S. J., & Shephard, N. (1992). Exact score for time series models in state space form. *Biometrika, 79*, 823–826.
22. Koopman, S. J., Shephard, N., & Doornik, J. A. (1999). Statistical algorithms for models in state space form using SsfPack 2.2. *Econometrics Journal, 2*, 113–166.
23. Ljung, G. M., & Box, G. E. P. (1978). On a measure of lack of fit in time series models. *Biometrika, 66*, 67–72.
24. Nocedal, J., & Wright, S. J. (1999). *Numerical optimization*. New York: Springer.
25. Proietti, T. (2000). Comparing seasonal components for structural time series models. *International Journal of Forecasting, 16*, 247–260.
26. Schweppe, F. (1965). Evaluation of likelihood functions for Gaussian signals. *IEEE Transactions on Information Theory, 11*, 61–70.
27. Trimbur, T. M. (2006). Properties of higher-order stochastic cycles. *Journal of Time Series Analysis, 27*, 1–17.

Part II
Trend-Cycle Estimation

Chapter 7
Trend-Cycle Estimation

Abstract Economists and statisticians are often interested in the short-term trend of socioeconomic time series. The short-term trend generally includes cyclical fluctuations and is referred to as trend-cycle. In recent years, there has been an increased interest to use trend-cycle estimates. Among other reasons, this interest originated from major economic and financial changes of global nature which have introduced more variability in the data, and, consequently, in the seasonally adjusted numbers. This makes very difficult to determine the direction of the short-term trend, particularly to assess the presence or the upcoming of a turning point. This chapter discusses in detail stochastic and deterministic trend-cycle models formulated for global and local estimation with special reference to the TRAMO-SEATS and STAMP software as well as the Henderson filters used in the Census II-X11 and its variants X11/X12ARIMA. It includes also trend-cycle models obtained with locally weighted averages as the Gaussian kernel, locally weighted regression smoother (LOESS), and cubic smoothing spline. Many illustrative examples are exhibited.

The trend corresponds to sustained and systematic variations over a long period of time. It is associated with the structural causes of the phenomenon in question, for example, population growth, technological progress, new ways of organization, or capital accumulation. For the majority of socioeconomic time series, the trend is very important because it dominates the total variation of the series. The identification of the trend component has always posed a serious statistical problem. The problem is not of mathematical or analytical complexity, but of conceptual complexity. This problem exists because the trend and the remaining components of a time series are latent (not directly observable) variables and, therefore, assumptions must be made on their behavioral pattern. The trend is generally thought of as a smooth and slow movement over a long term. The concept of "long" in this connection is relative and what is identified as trend for a given series span might well be part of a long cycle once the series is considerably augmented, such as the Kondratieff economic cycle. Kondratieff [35] estimated the length of this cycle to be between 47 and 60 years. Often, a long cycle is treated as a trend because the length of the observed time series is shorter than one complete cycle.

To avoid the complexity of the problem posed by a statistically vague definition, statisticians have resorted to two simple solutions. One consists of estimating trend

© Springer International Publishing Switzerland 2016   167
E. Bee Dagum, S. Bianconcini, *Seasonal Adjustment Methods and Real Time*
Trend-Cycle Estimation, Statistics for Social and Behavioral Sciences,
DOI 10.1007/978-3-319-31822-6_7

and cyclical fluctuations together calling this combined movement *trend-cycle*. The other consists of defining the trend in terms of the series length, denoting it as the longest no periodic movement. The estimation of the trend-cycle can be done via a specified model applied to the whole data, called the *global* trend-cycle, or by fitting a *local* polynomial function in such a way that, at any time point, its estimates only depend on the observations at that point and some specified neighboring observations.

Local polynomial fitting has a long history in the smoothing of noisy data. Henderson [30], Whittaker and Robinson[49], and Macauley [37] are some of the earliest classical references. These authors were very much concerned with the smoothing properties of linear estimators, being Henderson [30] the first to show that the smoothing power of a linear filter depends on the shape and values of its weighting system.

On the other hand, more recent contributions (among others, [8, 18–21, 25, 26, 44, 48]) concentrated on the asymptotic statistical properties of optimally estimated smoothing parameters. Optimality is being defined in terms of minimizing a given loss function, usually the mean square error of the estimator or the prediction risk.

In this chapter, we will discuss stochastic and deterministic trend-cycle models formulated for global and local estimation with special reference to those available in TRAMO-SEATS and STAMP, as well as the Henderson filters used in the Census II-X11 and its variants X11/X12ARIMA.

7.1   Deterministic Global Trend Models

Deterministic global trend models are based on the assumption that the trend or nonstationary mean of a time series can be approximated closely by simple functions of time over the entire span of the series.

The most common representation of a deterministic trend is by means of polynomial functions. The time series from which the trend has to be identified is assumed to be generated by a nonstationary process where the nonstationarity results from a deterministic trend. A classical model is the *regression* or *error model* [3] where the observed series is treated as the sum of a systematic part or trend and a random part. This model can be written as

$$y_t = T_t + e_t \tag{7.1}$$

where $\{e_t\}$ is a purely random process, that is, $\{e_t\} \sim i.i.d.(0, \sigma_e^2)$ (independent and identically distributed with expected value 0 and variance, σ_e^2).

In the case of a polynomial trend-cycle,

$$T_t = a_0 + a_1 t + a_2 t^2 + \cdots + a_p t^p \tag{7.2}$$

where generally $n \leq 3$. The trend is said to be of a deterministic character because it is not affected by random shocks which are assumed to be uncorrelated with the systematic part.

Model (7.1) can be generalized by assuming that $\{e_t\}$ is a second order linear stationary stochastic process, that is, its mean and variance are constant and its autocovariance is finite and depends only on the time lag.

Besides polynomials in time, other suitable mathematical functions are used to represent deterministic trends. Three of the most widely applied functions, known as *growth curves*, are the modified exponential, the Gompertz, and the logistic.

The modified exponential trend can be written as

$$T_t = a + bc^t, \quad a \in \mathbb{R}, b \neq 0, c > 0, c \neq 1. \tag{7.3}$$

For $a = 0$, model (7.3) reduces to the unmodified exponential trend

$$T_t = bc^t = T_0 e^{\alpha t}, \quad b = T_0, \alpha = \log c. \tag{7.4}$$

When $b > 0$ and $c > 1$, and so $\alpha > 0$, model (7.4) represents a trend that increases at a constant relative rate α. For $0 < c < 1$, the trend decreases at the rate α.

Models (7.3) and (7.4) are solutions of the differential equation

$$\frac{dT_t}{dt} = \alpha(T_t - a), \quad \alpha = \log c \tag{7.5}$$

which specifies the simple assumption of no inhibited growth.

Several economic variables during periods of sustained growth or of rapid inflation, as well as population growths measured in relative short periods of time, can be well approximated by trend models (7.3) and (7.4). But in the long run, socioeconomic and demographic time series are often subject to obstacles that slow their time path, and if there are no structural changes, their growth tends to a stationary state. Quetelet [40] made this observation with respect to population growth and Verhulst [46] seems to have been the first to formalize it by deducing the logistic model. Adding to Eq. (7.5) an inhibit factor proportional to $-T_t^2$, the result is

$$\frac{dT_t}{dt} = \alpha T_t - \beta T_t^2 = \alpha T_t(1 - T_t/k) \tag{7.6}$$

$$k = \alpha/\beta, \quad \alpha, \beta > 0$$

which is a simple null form of the Riccati differential equation.

Solving Eq. (7.6), we obtain the logistic model,

$$T_t = k(1 + ae^{-\alpha t})^{-1} \tag{7.7}$$

where $a > 0$ is a constant of integration.

Model (7.7) belongs to a family of *S*-shaped curves generated from the differential equation (see Dagum [10]):

$$\frac{dT_t}{dt} = T_t \Psi(t) \Phi(T_t/k), \qquad \Phi(1) = 0. \tag{7.8}$$

Solving Eq. (7.8) for $\Psi = \log c$ and $\Phi = \log(T_t/k)$, we obtain the Gompertz curve used to fit mortality table data. That is,

$$T_t = kb^{c^t}, \qquad b > 0, b \neq 1, 0 < c < 1, \tag{7.9}$$

where b is a constant of integration. It should be noticed that differencing will remove polynomial trends and suitable mathematical transformations plus differencing will remove trends from nonlinear processes. For example, for (7.7) using

$$Z_t = \log \left[\frac{T_t}{k - T_t} \right]$$

and then taking differences give $(1 - B)Z_t = \alpha$.

7.2 Stochastic Global Trend Models

The second major class of global trend models is the one that assumes the trend to be a stochastic process, most commonly that the series from which the trend will be identified has been generated by a homogeneous nonstationary stochastic process [51]. Processes of this kind are nonstationary, but applying a homogeneous filter, usually the difference filter, we obtain a stationary process in the differences of a finite order. In empirical applications, the nonstationarity is often present in the level and/or slope of the series. Hence, the order of the difference is low.

An important class of homogeneous linear nonstationary processes are the Autoregressive Integrated Moving Average (ARIMA) processes which can be written as [4]

$$\phi_p(B)(1 - B)^d T_t = \theta_q(B)a_{T_t}, \qquad a_{T_t} \sim i.i.d.(0, \sigma_T^2), \tag{7.10}$$

where B is the backshift operator, such that $B^n T_t = T_{t-n}$, $\phi_p(B)$ and $\theta_q(B)$ are polynomials in B of order p and q, respectively, satisfying the conditions of stationarity and invertibility. $(1 - B)^d$ is the difference operator of order d, and $\{a_{T_t}\}$ is a purely random process. Model (7.10) is also known as an ARIMA process of order (p, d, q). If $p = 0$, the process follows an $IMA(0, q)$ model.

Three common stochastic trend models are the $IMA(1,1)$, $IMA(2,2)$, and $ARIMA(2,1,2)$ which take the following forms

- $IMA(1,1)$

$$(1 - B)T_t = (1 - \theta B)a_{T_t},$$ (7.11)

$$|\theta| < 1, \quad a_{T_t} \sim i.i.d.(0, \sigma_T^2)$$

or, equivalently,

$$T_t = T_{t-1} + a_{T_t} - \theta a_{T_{t-1}}.$$ (7.12)

- $IMA(2,2)$

$$(1 - B)^2 T_t = (1 - \theta_1 B - \theta_2 B^2)a_{T_t},$$ (7.13)

$$\theta_2 + \theta_1 < 1 , \theta_2 - \theta_1 < 1, |\theta_2| < 1,$$

$$a_{T_t} \sim i.i.d.(0, \sigma_T^2),$$

or, equivalently,

$$T_t = 2T_{t-1} - T_{t-2} + a_{T_t} - \theta_1 a_{T_{t-1}} - \theta_2 a_{T_{t-2}}.$$

- $ARIMA(2,1,2)$

$$(1 - \phi_1 B - \phi_2 B^2)(1 - B)T_t = (1 - \theta_1 B - \theta_2 B^2)a_{T_t},$$ (7.14)

$$\phi_2 + \phi_1 < 1, \phi_2 - \phi_1 < 1, -1 < \phi_2 < 1$$

$$\theta_2 + \theta_1 < 1, \theta_2 - \theta_1 < 1, -1 < \theta_2 < 1$$

$$a_{T_t} \sim i.i.d.(0, \sigma_T^2)$$

or equivalently

$$T_t = (1 + \phi_1)T_{t-1} + (\phi_2 - \phi_1)T_{t-2} - \phi_2 T_{t-3} + a_{T_t} - \theta_1 a_{T_{t-1}} - \theta_2 a_{T_{t-2}}. \quad (7.15)$$

The a_{T_t}'s may be regarded as a series of random shocks that drive the trend and θ can be interpreted as measuring the extent to which the random shocks or innovations incorporate themselves into the subsequent history of the trend. For example, in model (7.11), the smaller the value of θ, the more flexible the trend; the higher the value of θ, the more rigid the trend (less sensitive to new innovations). For $\theta = 1$, model (7.11) reduces to one type of random walk model which has been used mainly for economic time series, such as stock market price data [23]. In such models, as time increases the random variables tend to oscillate about their mean value with an ever increasing amplitude.

Stochastic models for the trend have become widespread in economics (see Stock and Watson [45]), and several modeling approaches have been suggested. Within the context of linear stochastic processes, there are two general ones. First, the so-called ARIMA Model-Based (AMB) approach, adopted by TRAMO-SEATS, in which the model for the trend is derived from the ARIMA model identified for the observed series (see Burman [5]). The other approach starts by directly specifying a model for the trend, as implemented in STAMP. Basic references are Harvey [27], and Harvey and Todd [29].

7.2.1 TRAMO-SEATS Trend Models

Let y_t be a series which is the sum of a trend T_t and a nontrend component e_t,

$$y_t = T_t + e_t, \qquad (7.16)$$

where the two components are uncorrelated and each generated by an ARIMA model as follows

$$\varphi_T(B)T_t = \theta_T(B)a_{T_t}, \qquad (7.17)$$

$$\varphi_e(B)e_t = \theta_e(B)a_{e_t}, \qquad (7.18)$$

where $\varphi_T(B), \varphi_e(B), \theta_T(B)$ and $\theta_e(B)$ are finite polynomials in B. In particular, $\varphi_T(B) = \phi_T(B)\delta_T(B)$ and $\varphi_e(B) = \phi_e(B)\delta_e(B)$ where both $\delta_T(B)$ and $\delta_e(B)$ contain unit autoregressive roots. a_{T_t} and a_{e_t} are orthogonal white noise variables, with variances σ_T^2 and σ_e^2, respectively. It is generally assumed that the roots of the AR polynomials $\phi_T(B)$ and $\phi_e(B)$ are different, since AR roots for the same frequency should be associated with the same component. The two polynomials have also to be prime.

Combining Eqs. (7.16)–(7.18), y_t is assumed to follow an ARIMA model of the type

$$\varphi(B)y_t = \theta(B)a_t, \qquad (7.19)$$

where $\varphi(B) = \phi(B)\delta(B)$, being $\phi(B) = \phi_T(B)\phi_e(B)$, and $\delta(B) = (1-B)^d(1-B^s)^D$ where s is the number of observations for year. It can be easily derived that the polynomial $(1 - B^s)$ can be factorized as $(1 - B^s) = (1 - B)(1 + B + \cdots + B^{s-1})$, where $(1 + B + \cdots + B^{s-1})$ denotes the annual aggregation operator. $\theta(B)a_t$ is a Moving Average (MA) process such that

$$\theta(B)a_t = \varphi_e(B)\theta_T(B)a_{T_t} + \varphi_T(B)\theta_e(B)a_{e_t}, \qquad (7.20)$$

with $a_t \sim WN(0, \sigma_a^2)$. It has to be noticed that in order to allow model (7.19) to be invertible it is assumed that $\theta_T(B)$ and $\theta_e(B)$ do not have common unit root.

In the AMB approach $\varphi(B)$, $\theta(B)$ and σ_a^2 are assumed known, and the parameters of $\varphi_T(B)$, $\varphi_e(B)$, $\theta_T(B)$ and $\theta_e(B)$ as well as the variances σ_T^2 and σ_e^2 have to be derived from them. This is done in the frequency domain. Based on Eq. (7.16), the pseudospectrum of y_t is partitioned into additive spectra associated with the terms in the RHS of Eq. (7.16). Identification of a unique decomposition is achieved by imposing the canonical condition that the minimum of the trend component pseudospectrum be zero (see Burman [5]).

The Wiener–Kolmogorov (WK) filter for estimating T_t is given by the ratio of the T_t and y_t pseudospectra, and yields the Minimum Mean Square Error (MMSE) estimator (also conditional mean) of T_t given y_t. The filter is centered, symmetric, and convergent. Its derivation requires an infinite realization of y_t in the direction of the past and of the future. Hence, to apply the filter to a finite realization, model (7.19) is used to extend y_t with forecasts and backcasts. As new observations become available, forecasts will be updated and eventually replaced by observations. As a consequence, the estimator of T_t near the end of the series is preliminary and will be revised. The duration of the revision process of a preliminary estimator depends on the ARIMA model identified for the series. The spectral factorization provides the time domain expression of the components given in Eqs. (7.17) and (7.18).

Letting F denote the forward shift operator ($F = B^{-1}$) and replacing the ratio of pseudospectra by the ratio of Autocovariance Generating Functions (ACGF), the time domain expression of the WK filter is

$$v_T(F, B) = k_T \frac{\theta_T(B)\theta_T(F)\varphi_e(B)\varphi_e(F)}{\theta(B)\theta(F)},$$

where $k_T = \sigma_T^2/\sigma_e^2$, so that the final trend estimator is given by

$$\hat{T}_t = v_T(B, F)y_t. \tag{7.21}$$

As an example, we consider a general class of models widely applied to observed series, that is,

$$(1 - B)(1 - B^s)y_t = \theta(B)a_t \qquad a_t \sim WN(0, \sigma_a^2). \tag{7.22}$$

In this case, the decomposition (7.20) results

$$\theta(B)a_t = (1 + B + \cdots + B^{s-1})\theta_T(B)a_{T_t} + (1 - B)^2\theta_e(B)a_{e_t}$$

where the order q of the polynomial $\theta(B)$ is equal to $\max(q_T + s - 1, q_e + 2)$, being q_T and q_e the orders of the polynomials $\theta_T(B)$ and $\theta_e(B)$, respectively.

Following Maravall [38], necessary and sufficient conditions for the decomposition (7.20) to be underidentified are that $q_T > 1$ and $q_e > s - 2$.

The AMB decomposition restricts the order of the AR and MA polynomial in model (7.17) for the trend to be of the same order. Hence, the trend model becomes an $IMA(2, 2)$ defined in Eq. (7.13), that is,

$$(1 - B)^2 T_t = (1 - \theta_1 B - \theta_2 B^2)a_{T_t}. \tag{7.23}$$

Based on this trend specification, the decomposition (7.19) is still underidentified and a further condition is imposed. Due to the fact that the randomness of T_t is determined by a_{T_t}, all the models of type (7.23) that are compatible with the stochastic structure of the observed series, that is, with model (7.19), the one with smallest σ_T^2 is chosen. This yields the most stable trend, given the observed ARIMA model. As shown by Hillmer and Tiao [31], minimizing σ_T^2 is equivalent to the requirement that it should not be possible to further decompose T_t as the sum of a trend plus a noise component, with the latter orthogonal to the former. When a component satisfies this "noise-free" requirement, it is called "canonical." The canonical trend is uncontaminated by noise, such that its spectrum should contain a zero, that means there is a unit root in the MA polynomial of the trend. Hence, model (7.23) can be rewritten as

$$(1 - B)^2 T_t = (1 - \theta_1 B)(1 + B)a_{T_t} \tag{7.24}$$

which contains two parameters θ_1 and σ_T^2. The number of the unknowns in the system of covariance equations is $(q_e + 3)$ and the number of equations is greater than $(q_e + 3)$. The decomposition becomes identifiable and there will be a unique model (7.24) which will represent the trend component contained in model (7.22).

As an illustrative example, Fig. 7.1 shows the trend estimates, derived using the TRAMO-SEATS software, of the series of the US New Orders for Durable Goods (NODG) observed for the period February 1992–December 2013. The trend estimates have been obtained through canonical decomposition of the $ARIMA(0, 1, 1)(0, 1, 1)_{12}$ identified by TRAMO-SEATS to be the stochastic generating process of the original series.

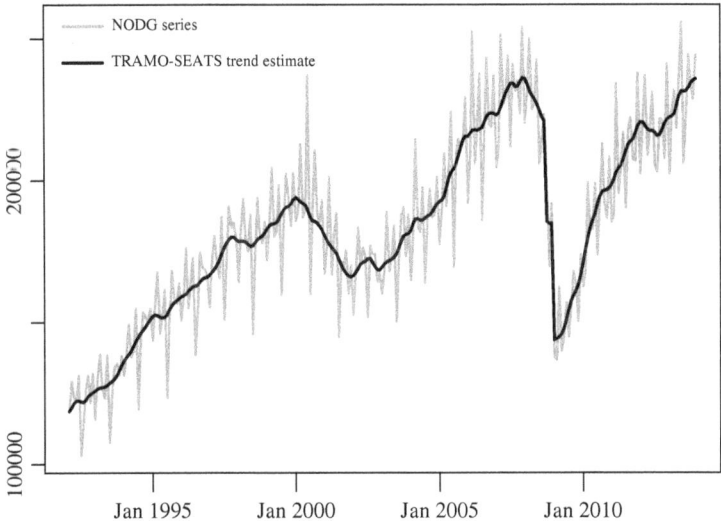

Fig. 7.1 The US New Orders for Durable Goods trend-cycle estimates obtained using the TRAMO-SEATS software

7.2.2 STAMP Trend Models

Structural models can be seen as the sum of unobservable ARIMA models specified to describe the behavior of each time series component separately, and random walk processes are basic elements in many of these models.

The simplest structural model consists of a trend component T_t plus a random disturbance term I_t, where the level T_t, representing the long-term movements in the series, is allowed to fluctuate up and down according to a random walk process. In particular, if the irregular component is assumed to be normally distributed, this *local level model* is specified as

$$y_t = T_t + I_t, \qquad I_t \sim \text{NID}(0, \sigma_I^2), \tag{7.25}$$

$$T_t = T_{t-1} + a_{T_t}, \qquad a_{T_t} \sim \text{NID}(0, \sigma_T^2),$$

for $t = 1, 2, \ldots, n$, where the I_t's and a_{T_t}'s are mutually independent. The model is incomplete without a specification of the trend at T_1, which generally is assumed to be normally distributed $T_1 \sim N(\mu_1, P_1)$, where μ_1 and P_1 are known, and independent of I_t and a_{T_t}.

When σ_I^2 is equal to 0, the observations follow a random walk and the forecast of future observations is the last observation y_T. If σ_T^2 is equal to 0, then the level is constant and the best forecast of future observation is the sample mean. The effect of a_{T_t} is to allow the trend to shift up and down. The larger the variances, the greater the stochastic movements in the trend. Although it is simple, the local level model provides the basis for the treatment of important series in practice. The properties of series generated by this model are studied in detail by Harvey [28].

An extension of the local level model is represented by the *local linear trend model* which is characterized by a stochastic linear trend whose level and slope vary over time according to random walk processes as follows

$$y_t = T_t + I_t, \tag{7.26}$$

$$T_t = T_{t-1} + \beta_{t-1} + a_{T_t},$$

$$\beta_t = \beta_{t-1} + \zeta_t, \qquad \zeta \sim \text{NID}(0, \sigma_\zeta^2),$$

where a_{T_t} and ζ_t are uncorrelated white noise disturbances. The trend is equivalent to an ARIMA(0,2,2) process [41]. However, if both σ_T^2 and σ_ζ^2 are equal to zero, then T_t reduces to a deterministic trend, that is, $T_t = T_1 + \beta t$. On the other hand, if $\sigma_T^2 = 0$, but $\sigma_\zeta^2 > 0$, the trend is an Integrated process of order 2, $I(2)$, that is stationary in the second order differences. A trend component with this feature tends to be relatively smooth. An important issue is therefore whether or not the constraint $\sigma_T^2 = 0$ should be imposed at the outset. There are series where it is unreasonable to assume a smooth trend a priori, and therefore the question whether or not σ_T^2 is set to zero is an empirical one.

In this model, β_t shows the contribution of the slope, and ζ_t allows the slope to change over time. If σ_ζ^2 is equal to zero, the slope is constant and the local linear trend model reduces to a *random walk plus drift model*. On the other hand, when the estimated value of the slope and its variance are very small, the trend follows a local level model without drift described in Eq. (7.25).

For the US NODG series introduced in the previous section, Fig. 7.2 illustrates the trend component estimated by fitting the basic structural model which includes a local linear trend model as specified in Eq. (7.26) as well as a stochastic seasonal component. The estimated component has been obtained using the STAMP (Structural Time series Analyser, Modeller and Predictor) software.

Both the local level model and the local linear trend model can be expressed in matrix form as special cases of the *linear Gaussian state space model* defined as follows

$$y_t = \mathbf{z}_t^T \boldsymbol{\alpha}_t + I_t, \qquad I_t \sim N(0, \sigma_I^2), \tag{7.27}$$

$$\boldsymbol{\alpha}_{t+1} = \boldsymbol{\Gamma}_t \boldsymbol{\alpha}_t + \mathbf{R}_t \boldsymbol{\varepsilon}_t, \qquad \boldsymbol{\varepsilon}_t \sim N(\mathbf{0}, \boldsymbol{\Sigma}_\varepsilon), \tag{7.28}$$

$$\boldsymbol{\alpha}_1 \sim N(\mathbf{a}_1, \mathbf{P}_1).$$

Equation (7.27) is called the *observation equation* and Eq. (7.28) is called the *state equation*, being the unobserved vector $\boldsymbol{\alpha}_t$ defined as the *state*. The vector $\mathbf{z}_t$ and the matrices $\boldsymbol{\Gamma}_t$, $\mathbf{R}_t$, and $\boldsymbol{\Gamma}_\varepsilon$ are assumed known as well as σ_I^2, whereas the disturbances I_t and $\boldsymbol{\varepsilon}_t$ are independent sequences of normal random variables. The matrix $\mathbf{R}_t$, when it is not the identity, is usually a selection matrix whose columns are a subset of the columns of the identity matrix. It is needed when the dimensionality of $\boldsymbol{\alpha}_t$ is

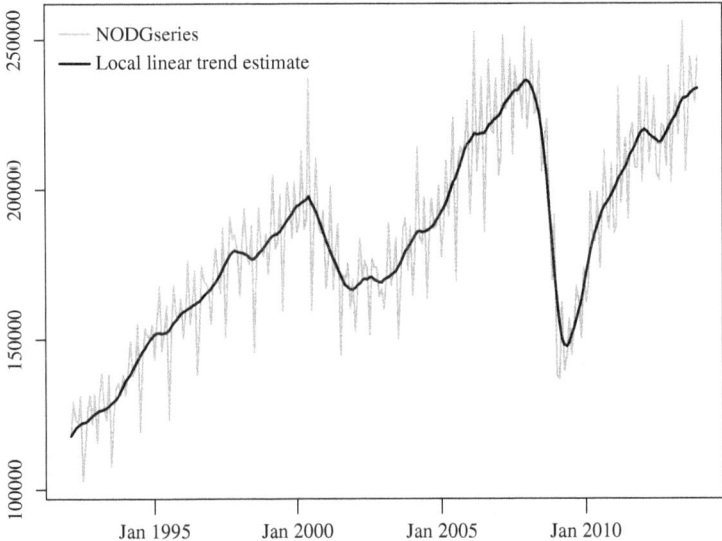

Fig. 7.2 The US New Orders for Durable Goods trend estimates obtained using the STAMP software

greater than that of the disturbance vector $\boldsymbol{\varepsilon}_t$. The structure of model defined through Eqs. (7.27) and (7.28) is a natural one for representing the behavior of many time series as a first approximation. The first equation is a standard multivariate linear regression model whose coefficient vector $\boldsymbol{\alpha}_t$ varies over time. The development over time of $\boldsymbol{\alpha}_t$ is determined by the first order autoregressive vector given in the second equation. The Markovian nature of the model has a remarkably wide range of applications to problems in practical time series analysis where the local level and local linear trend models are special cases. The local level model admits a state space representation by setting in Eq. (7.27) $\boldsymbol{\alpha}_t = T_t$, $\mathbf{z}_t = 1$ and in Eq. (7.28) $\boldsymbol{\Gamma}_t = \mathbf{R}_t = 1$, and $\boldsymbol{\Sigma}_\varepsilon = \sigma_T^2$.

On the other hand, the local linear trend model defined in Eq. (7.26) can be rewritten in matrix form as follows

$$y_t = \begin{pmatrix} 1 & 0 \end{pmatrix} \begin{pmatrix} T_t \\ \beta_t \end{pmatrix} + I_t,$$

$$\begin{pmatrix} T_{t+1} \\ \beta_{t+1} \end{pmatrix} = \begin{pmatrix} 1 & 1 \\ 0 & 1 \end{pmatrix} \begin{pmatrix} T_t \\ \beta_t \end{pmatrix} + \begin{pmatrix} a_{T_t} \\ \zeta_t \end{pmatrix},$$

$$T_1 \sim N(\mu_1, \sigma_{T_1}^2).$$

For the measurement of the short-term trend, the business cycle component has to be included in the model. Hence, the *trend plus cycle model* is specified as follows

$$y_t = T_t + C_t + I_t, \tag{7.29}$$

$$T_t = T_{t-1} + \beta_{t-1} + a_{T_t},$$

$$\beta_t = \beta_{t-1} + \zeta_t,$$

where C_t is a cycle with frequency (in radians) λ_c, being $0 \le \lambda_c \le \pi$. It is generated by the stochastic process

$$C_t = \rho \cos \lambda_c C_{t-1} + \rho \sin \lambda_c C_{t-1}^* + a_{C_t}, \tag{7.30}$$

$$C_t^* = -\rho \sin \lambda_c C_{t-1} + \rho \cos \lambda_c C_{t-1}^* + a_{C_t^*},$$

being ρ a damping factor on the amplitude such that $0 \le \rho \le 1$, and both a_{C_t} and $a_{C_t^*}$ are NID$(0, \sigma_C^2)$. The irregular component I_t is assumed to follow an NID$(0, \sigma_I^2)$ process, and the disturbances in all the three components are taken to be independent of each other. It can be seen that cycle and trend are simply added to each other. When ρ is zero, then this model is not identifiable and, since C_t is assumed to be a stationary process, ρ is strictly less than one. In particular, C_t follows an ARMA(2,1) process in which both the MA and AR parts are subject to restrictions [27]. The period of the cycle, which is the time taken to go through its complete sequence of values, is $2\pi/\lambda_c$. When σ_T^2 is equal to zero and σ_C^2 is small, the trend plus cycle model allocates most of the short run variation in the series to the nonstationary trend component. When both σ_ζ^2 and σ_T^2 are small, then the trend becomes much smoother and most of the variation in the series is allocated to the stationary cyclical component. Finally, the trend plus cycle model reduces to an AR(1) process when λ_c is equal to zero or π.

Estimation of the parameters σ_I^2, σ_T^2, σ_ζ^2, ρ, λ_c, and σ_C^2 can be carried out by maximum likelihood either in the time domain or in the frequency domain. Once this has been done, estimates of the trend, cyclical, and irregular components are obtained from a smoothing algorithm. These calculations may be carried out using the STAMP software.

7.3   Stochastic Local Trend-Cycle Models

Economists and statisticians are often interested in the short-term trend of socioeconomic time series. The short-term trend generally includes cyclical fluctuations and is referred to as *trend-cycle*. In recent years, there has been an increased interest to use trend-cycle estimates or smoothed seasonally adjusted data to facilitate recession and recovery analysis. Among other reasons, this interest originated from major economic and financial changes of global nature which have introduced more variability in the data, and, consequently, in the seasonally adjusted numbers. This makes very difficult to determine the direction of the short-term trend, particularly to assess the presence or the upcoming of a turning point. The local polynomial regression predictors developed by Henderson [30] and LOESS due to Cleveland [7] are widely applied to estimate the short-term trend of seasonally adjusted economic indicators. Particularly, the former is available in nonparametric seasonal adjustment software such as the US Bureau of the Census II-X11 method [43] and its variants the X11ARIMA ([11] and [12]) and X12ARIMA [22], the latter, in the STL (Seasonal-Trend decomposition procedure based on LOESS) [6].

The basic assumption is that the input series $\{y_t, t = 1, \ldots, n\}$, that is seasonally adjusted or without seasonality, can be decomposed into the sum of a systematic component called the *signal* (or nonstationary mean) TC_t, that represents the trend and cyclical components, usually referred to as trend-cycle for they are estimated jointly, plus an erratic component, called the noise I_t, such that

$$y_t = TC_t + I_t. \tag{7.31}$$

The irregular component I_t is assumed to be either a white noise, $WN(0, \sigma_I^2)$, or, more generally, to follow a stationary and invertible ARMA process. The trend-cycle can be represented locally by a polynomial of degree of the time distance j between y_t and the neighboring observations y_{t+j}. Hence, it is possible to find a local polynomial trend-cycle estimator

$$TC_t(j) = b_0 + b_1 j + \cdots + b_p j^p + a_{TC_t}(j), \tag{7.32}$$

where $b_0, b_1, \ldots, b_p$ are real and a_{TC_t} is assumed to be purely random and mutually uncorrelated with I_t. The coefficients, $b_0, b_1, \ldots, b_p$, can be estimated by ordinary or weighted least squares or by summation formulae. The solution for $\widehat{TC_t}(0)$ provides the trend-cycle estimate, which equivalently is a weighted average applied in a moving manner [32], such that

$$\widehat{TC_t}(0) = \widehat{TC_t} = \sum_{j=-m}^{m} w_j y_{t-j}, \tag{7.33}$$

where $w_j, j = -m, \ldots, m$, denotes the weights to be applied to the observations y_{t-j} to get the estimate $\widehat{TC_t}$ for each point in time $t = 1, 2, \ldots, n$.

The weights depend on:

(1) the degree p of the fitted polynomial,
(2) the amplitude $(2m + 1)$ of the neighborhood, and
(3) the shape of the function used to average the observations in each neighborhood.

Once a (symmetric) span $2m + 1$ of the neighborhood has been selected, the weights associated with the observations corresponding to points falling out of the neighborhood of any target point are null or approximately null, such that the estimates of the $n - 2m$ central observations are obtained by applying $2m + 1$ symmetric weights to the observations neighboring the target point. The missing estimates for the first and last m observations can be obtained by applying asymmetric moving averages of variable length to the first and last m observations, respectively. The length of the moving average or time invariant symmetric linear filter is $2m + 1$, whereas the asymmetric linear filters length is time-varying.

Using the backshift operator B, such that $B^n y_t = y_{t-n}$, Eq. (7.33) can be written as

$$\widehat{TC}_t = \sum_{j=-m}^{m} w_j B^j y_t = W(B)y_t, \qquad t = 1, 2, \ldots, n, \qquad (7.34)$$

where $W(B)$ is a linear nonparametric estimator. The nonparametric estimator $W(B)$ is said to be a second order estimator if it satisfies the conditions

$$\sum_{j=-m}^{m} w_j = 1, \qquad \sum_{j=-m}^{m} j w_j = 0, \qquad (7.35)$$

such that it preserves a constant and a linear trend. On the other hand, $W(B)$ is a higher order estimator if

$$\sum_{j=-m}^{m} w_j = 1, \qquad \sum_{j=-m}^{m} j^i w_j = 0 \qquad i = 1, 2, \ldots, p. \qquad (7.36)$$

In other words, it will reproduce a polynomial trend of degree $(p - 1)$ without distortion.

Nonparametric estimators are based on different assumptions of smoothing. We shall next discuss: (1) the local polynomial regression predictor LOESS due to Cleveland [7], (2) the Henderson linear filters, (3) the Gaussian kernel, and (4) the cubic smoothing spline. The first three estimators are often applied to estimate the short-term trend of seasonally adjusted economic indicators.

7.3.1 Locally Weighted Regression Smoother (LOESS)

The locally weighted regression smoother (LOESS) developed by Cleveland [7] fits local polynomials of a degree p where the parameters are estimated either by ordinary or weighted least squares. Given a series of equally spaced observations and corresponding target points $\{(y_{t_i}, t_i), i = 1, \ldots, n\}, t_1 < \cdots < t_n$, where t_i denotes the time the observation y_{t_i} is taken, LOESS produces a smoothed estimate as follows

$$\widehat{TC}_{t_i} = \mathbf{t}_i' \mathbf{b}_i \qquad (7.37)$$

where $\mathbf{t}_i$ is a $(p + 1)$-dimensional vector of generic component t_i^d, $d = 0, \ldots, p$, $p = 0, 1, 2, \ldots$ denotes the degree of the fitted polynomial, and $\mathbf{b}_i$ is the $(p + 1)$-dimensional least squares estimate of a weighted regression computed over a neighborhood of t_i constituting a subset of the full span of the series.

The weights of the regression depend on the distance between the target point t_i and any other point belonging to its neighborhood, through a weighting function $W(t)$. The weighting function most often used is the tricube proposed by Cleveland et al. [6], that is,

$$W(t) = \left(1 - |t|^3\right)^3 I_{[-1,1]}(t). \tag{7.38}$$

In particular, at each point t_k in the neighborhood of the target point t_i, denoted by $N(t_i)$, has assigned a weight

$$w(t_k) = W\left(\frac{|t_i - t_k|}{\nabla(t_i)}\right) \qquad \forall t_k \in N(t_i) \tag{7.39}$$

with $\nabla(t_i)$ representing the distance of the furthest near-neighbor from t_i.

Each neighborhood is made of the same number of points chosen to be the nearest to t_i, and the ratio between the amplitude of the neighborhood and the full span of the series defines the bandwidth or smoothing parameter. For the first and last observations, Cleveland [7] derived the filters by weighting the data belonging to an asymmetric neighborhood which contains the same number of data points of the symmetric one.

7.3.2   Henderson Smoothing Filter

The Henderson smoothing filters are derived from the graduation theory and used in Census II-X11 method and its variants X11/X12ARIMA. The basic principle of the graduation theory is the combination of operations of differencing and summation in such a manner that, when differencing above a certain order is ignored, they will reproduce the functions operated on. The merit of these procedures is that the smoothed values thus obtained are functions of a large number of observations whose errors, to a considerable extent, cancel out. These smoothers have the properties that when fitted to second or third degree parabolas, their output will fall exactly on those parabolas and when fitted to stochastic data, they will give smoother results than can be obtained from weights that give the middle point to a second degree parabola fitted by ordinary least squares.

Recognition of the fact that the smoothness of the resulting graduation depends directly on the smoothness of the weight diagram led Henderson [30] to develop a formula which makes the sum of squares of the third differences of the smoothed series a minimum for any number of terms.

Henderson's starting point was the requirement that the filter should reproduce a cubic polynomial trend without distortion. Henderson showed that three alternative smoothing criteria lead to the same formula:

1. minimization of the variance of the third differences of the series defined by the application of the moving average;
2. minimization of the sum of squares of the third differences of the coefficients of the moving average formula; and
3. fitting a cubic polynomial by weighted least squares, where the weights are chosen as to minimize the sum of squares of their third differences.

Kenny and Durbin [33] and Gray and Thomson [24] showed the equivalence of these three criteria. The problem is one of locally fitting a cubic trend by weighted least squares to the observations where the weights are chosen to minimize the sum of squares of their third differences (*smoothing criterion*). The objective function to be minimized is

$$\sum_{j=-m}^{m} W_j \left[y_{t+j}^a - a_0 - a_1 j - a_2 j^2 - a_3 j^3 \right]^2, \tag{7.40}$$

where the solution for the constant term $\hat{a}_0$ is the smoothed observation $\widehat{TC}_t$, $W_j = W_{-j}$, and the filter length is $2m + 1$. The solution is a local cubic smoother with weights

$$W_j \propto \left[(m+1)^2 - j^2 \right] \left[(m+2)^2 - j^2 \right] \left[(m+3)^2 - j^2 \right], \tag{7.41}$$

and the weight diagram known as Henderson's ideal formula is obtained, for a filter length equal to $2m - 3$,

$$w_j = \frac{315 \times \left[(m-1)^2 - j^2 \right] (m^2 - j^2) \left[(m+1)^2 - j^2 \right] (3m^2 - 16 - 11j^2)}{8m(m^2 - 1)(4m^2 - 1)(4m^2 - 9)(4m^2 - 25)}. \tag{7.42}$$

By making $m = 8$ the w_j values are obtained for each j of the 13-term filter ($m = 6$ for the 9-term Henderson filter, and $m = 13$ for the 23-term Henderson).

On the contrary, the weights of the usually known as asymmetric Henderson filters developed by Musgrave [39] are based on the minimization of the mean squared revision between the final estimates and the preliminary estimates subject to the constraint that the sum of the weights is equal to one. The assumption made is that the most recent values of the series (where seasonality has been removed if present in the original observations) follow a constant linear trend plus an erratic component (see Laniel [36] and Doherty [17]). The equation used is

$$E\left[r_t^{(i,m)}\right]^2 = c_1^2 \left(t - \sum_{j=-a}^{m} w_{a,j}(t-j)\right)^2 + \sigma_I^2 \sum_{j=-m}^{m} (w_j - w_{a,j})^2, \qquad (7.43)$$

where w and w_a are the weights of the symmetric (central) filter and the asymmetric filters, respectively; $w_{a,j} = 0$ for $j = -m, \ldots, -(a+1)$, c_1 is the slope of the line and σ_I^2 denotes the noise variance. There is a relation between c_1 and σ_I such that the noise to signal ratio, I/C, is given by

$$I/C = \frac{(4\sigma_I^2/\pi)^{1/2}}{|c_1|} \quad \text{or} \quad \frac{c_1^2}{\sigma_I^2} = \frac{4}{\pi(I/C)^2}. \qquad (7.44)$$

The noise to signal ratio I/C (7.44) determines the length of the Henderson trend-cycle filter to be applied. Thus, setting $t = 0$ and $m = 6$ for the end weights of the 13-term Henderson filter, we have

$$\frac{E[r_0^{(i,6)}]^2}{\sigma_I^2} = \frac{4}{\pi(I/C)^2} \left(\sum_{j=-a}^{6} w_{a,j}\right)^2 + \sum_{j=-6}^{6} (w_j - w_{a,j})^2. \qquad (7.45)$$

Making $I/C = 3.5$ (the most noisy situation where the 13-term Henderson filter is applied), Eq. (7.45) gives the same set of end weights of Census II-X11 variant [43]. The end weights for the remaining monthly Henderson filters are calculated using $I/C = 0.99$ for the 9-term filter and $I/C = 4.5$ for the 23-term filter.

The estimated final trend-cycle is obtained by cascade filtering that results from the convolution of various linear trend and seasonal filters. In fact, if the output from the filtering operation H is the input to the filtering operation Q, the coefficients of the cascade filter C result from the convolution of $H * G$. For symmetric filters, $H * G = G * H$, but this is not valid for asymmetric filters. Assuming an input series $y_t, t = 1, 2, \ldots, n$, we can define a matrix $W = [w_{a,j}], a = 1, 2, \ldots, m; j = 1, 2, \ldots, 2m+1$, where each row is a filter and m is the half length of the symmetric filter. $w_{0,\cdot}$ denotes an asymmetric filter where the first m coefficients are zeroes, whereas $w_{m,\cdot}$ denotes the symmetric filter.

Given data up to time n, the $m+1$ most recent values of the output (*filtered series*) are given by

$$\widehat{TC}_{n+1-a} = \sum_{j=m-a+2}^{2m+1} w_{a,j} y_{n-2m-1+j} \qquad a = 1, 2, \ldots, m+1. \qquad (7.46)$$

The 13-term Henderson filter, which is the most often applied, can then be put in matrix form as follows:

$$
\begin{bmatrix}
-0.019 & -0.028 & 0.000 & 0.066 & 0.147 & 0.214 & 0.240 & 0.214\ 0.147\ 0.066\ 0.000 & -0.028 & -0.019 \\
0 & -0.017 & -0.025 & 0.001 & 0.066 & 0.147 & 0.213 & 0.238\ 0.212\ 0.144\ 0.061 & -0.006 & -0.034 \\
0 & 0 & -0.011 & -0.022 & 0.003 & 0.067 & 0.145 & 0.210\ 0.235\ 0.205\ 0.136 & 0.050 & -0.018 \\
0 & 0 & 0 & -0.009 & -0.022 & 0.004 & 0.066 & 0.145\ 0.208\ 0.230\ 0.201 & 0.131 & 0.046 \\
0 & 0 & 0 & 0 & -0.016 & -0.025 & 0.003 & 0.068\ 0.149\ 0.216\ 0.241 & 0.216 & 0.148 \\
0 & 0 & 0 & 0 & 0 & -0.043 & -0.038 & -0.002\ 0.080\ 0.174\ 0.254 & 0.292 & 0.279 \\
0 & 0 & 0 & 0 & 0 & 0 & -0.092 & -0.058\ 0.012\ 0.120\ 0.244 & 0.353 & 0.421
\end{bmatrix}.
$$

$$(7.47)$$

7.3.3 Gaussian Kernel Smoother

Kernel type smoothers are locally weighted averages. Given a series of seasonally adjusted observations and corresponding target points $(y_{t_j}, t_j), j = 1, \ldots, n$, a kernel smoothing gives, at time $t_h, 1 \leq h \leq n$, the smoothed estimate

$$
\widehat{\text{TC}}_{t_h} = \sum_{j=1}^{n} w_{hj} y_{t_j}, \tag{7.48}
$$

where

$$
w_{hj} = \frac{K\left(\frac{t_h - t_j}{b}\right)}{\sum_{j=1}^{n} K\left(\frac{t_h - t_j}{b}\right)} \tag{7.49}
$$

are the weights from a parametric kernel which is a nonnegative symmetric function that integrated over its domain gives unity, and $b \neq 0$ is the smoothing parameter.

Kernel smoothers are local functions since their weighting systems are local. In fact, for any observed value y_{t_h}, a weighted average is computed. Each weight is obtained as a function of the distance between the target point t_h and the t_j's close to t_h that belong to an interval whose amplitude is established by the smoothing parameter b, otherwise said *bandwidth parameter*. Increasing the distance between t_h and $t_j, j = 1, \ldots, n$, the weights assigned to the corresponding observations decrease till a certain point when they become zero. Such a point depends on the bandwidth parameter that, in practice, determines the lengths of the smoother, i.e., the number of observations near y_{t_h} that have non-null weight.

In matrix form the action of kernel smoother can be represented by the relation

$$
\widehat{\text{TC}} = \mathbf{W}_b \mathbf{y}^a \tag{7.50}
$$

where $\mathbf{W}_b$ is the smoothing matrix of generic element w_{hj} as defined in (7.49) for $h, j = 1, \ldots, n$.

Once the smoothing parameter has been selected, the w_{hj}'s corresponding to the target points falling out of the bandwidth of any t_h turn out to be null, where the number of no null weights depends on both the value of the bandwidth parameter and on the number of decimals chosen for each weight. Notice that as long as the kernel is a symmetric function, the number of no null weights turns out to be odd. For instance, for a standard Gaussian kernel function

$$K\left(\frac{t_h - t_j}{b}\right) = \frac{1}{\sqrt{2\pi}} \exp\left\{-\frac{1}{2}\left(\frac{t_h - t_j}{b}\right)^2\right\} \tag{7.51}$$

with $b = 5$ and weights w_{hj} taken with three decimals, the smoothing matrix $\mathbf{W}_5$ of no null weights is of size 31×31. For further details, we refer the reader to Dagum and Luati [15, 16].

In kernel smoothing, the bandwidth parameter is of great importance. Similar to LOESS, increasing the smoothing parameter from the extreme case of interpolation ($b \to 0$) to that of oversmoothing ($b \to \infty$) produces an increase in bias and a decrease in variance.

7.3.4 Cubic Smoothing Spline

The current literature on spline functions, particularly on smoothing splines, is very large and we refer the reader to Wahba [47] for an excellent summary of the most important contributions on this topic. The name is due to the resemblance with the curves obtained by draftsmen using a mechanical spline, that is, a thin and flexible rod with weights or "ducks" used to position the rod at points through which it was desired to draw a smooth interpolating curve.

The problem of smoothing via spline functions is closely related to that of smoothing priors and signal extraction in time series, where these latter are approached from a parametric point of view (see, among others, Akaike [1, 2], and Kitagawa and Gersch [34]).

Similar to the Henderson filters, the original work on smoothing spline functions was based on the theory of graduation. The first two seminal works are due to Whittaker [50] and Whittaker and Robinson [49] who proposed a new graduation method that basically consisted of a trade-off between fidelity and smoothing. The problem was that of estimating an unknown "smooth" function, in our case the trend-cycle, TC, observed with errors assumed to be white noise. That is, given a set of seasonally adjusted observations $y_t, t = 1, 2, \ldots, n$ such that,

$$y_t = \mathrm{TC}_t + I_t, \qquad I_t \sim i.i.d.(0, \sigma_I^2), \tag{7.52}$$

we want to minimize

$$\sum_{t=1}^{n} (y_t - TC_t)^2 + \lambda^2 \sum_{t=k+1}^{n} \left((1-B)^k TC_t\right)^2, \tag{7.53}$$

where $(1-B)^k$ denotes the k-th order difference of TC_t, e.g., $(1-B)TC_t = TC_t - TC_{t-1}$, $(1-B)^2 TC_t = (1-B)((1-B)TC_t)$, and so on. The smoothing trade-off parameter λ must be appropriately chosen.

Following this direction, Schoenberg [42] extended Whittaker smoothing method to the fitting of a continuous function to observed data, not necessarily evenly spaced. In this case, the model is

$$y_t = TC(t) + I_t, \qquad I_t \sim i.i.d.(0, \sigma_I^2), \tag{7.54}$$

where the unobserved function $TC(t), t \in [a, b]$, is assumed to be smooth on the interval $[a, b]$, and the observations are at the n points, $t_1, t_2, \ldots, t_n$. The problem is to find

$$\min_{TC \in C^m} \frac{1}{n} \sum_{i=1}^{n} (y_{t_i} - TC(t_i))^2 + \lambda \int_a^b \left[TC^{(m)}(t)\right]^2, \tag{7.55}$$

where C^m is the class of functions with m continuous derivatives and $\lambda > 0$.

The solution to (7.55) known as a *smoothing spline* is unique and given by a univariate natural polynomial or piecewise polynomial spline of degree $2m - 1$ with knots at the data points $t_1, t_2, \ldots, t_T$. The smoothing trade-off parameter λ controls the balance between the *fit to the data* as measured by the residual sum of squares, and the *smoothness* as measured by the integrated squared m-th derivative of the function. When $m = 2$, which is the case of a cubic smoothing spline, the integral of the squared second order derivative $TC^{(2)}$ is the curvature, and a small value for the integral corresponds visually to a smooth curve. As $\lambda \to 0$, the solution $\widehat{TC}_\lambda$ tends to the univariate natural polynomial spline which interpolates the data, and as $\lambda \to \infty$, the solution tends to the polynomial of degree m that represents the best fitting of the data in the least squares sense. The smoothing trade-off parameter λ is known as *hyperparameter* in the Bayesian terminology, and it has the interpretation of a noise to signal ratio: the larger the λ, the smoother the trend-cycle.

The estimation of λ was first done using Ordinary Cross Validation (OCV). OCV consisted of deleting one observation and solving the optimization problem with a trial value of λ, computing the difference between the predicted value and the deleted observation, accumulating the sums of squares of these differences as one runs through each of the data points in turn, and finally choosing the λ for which the accumulated sum is the smallest. This procedure was improved by Craven and Wahba [9] who developed the Generalized Cross Validation (GCV) method currently available in most computer packages. GCV can be obtained from OCV by rotating the system to a standard coordinate system, doing OCV, and rotating back. The GCV estimate of λ is obtained by minimizing

$$V(\lambda) = \frac{(1/T) \left[(\mathbf{I}_n - \mathbf{W}(\lambda)) \mathbf{y} \right]^2}{[(1/n) tr(\mathbf{I}_n - \mathbf{W}(\lambda))]^2}, \tag{7.56}$$

where $\mathbf{W}(\lambda)$ is the influential matrix associated with $\widehat{TC}$, that is,

$$\begin{bmatrix} \widehat{TC}_{t_1} \\ \widehat{TC}_{t_2} \\ \vdots \\ \widehat{TC}_{t_n} \end{bmatrix} = \mathbf{W}(\lambda) \begin{bmatrix} y_{t_1} \\ y_{t_2} \\ \vdots \\ y_{t_n} \end{bmatrix} \tag{7.57}$$

or in compact form

$$\widehat{TC} = \mathbf{W}(\lambda)\mathbf{y}.$$

The trace of $\mathbf{W}(\lambda)$ can be viewed as the degrees of freedom for the signal and so, (7.56) can be interpreted as minimizing the standardized sum of squares of the residuals. Since λ is usually estimated by cross validation, cubic smoothing splines are considered nonlinear smoothers. However, Dagum and Capitanio [13] showed that if λ is fixed, i.e., $\lambda = \lambda_0$, then $\mathbf{W}(\lambda_0) = \mathbf{W}_{\lambda_0}$ has the form

$$\mathbf{W}_{\lambda_0} = \left[\mathbf{I}_n - \mathbf{D}^T \left(\frac{1}{\lambda_0} \mathbf{B} + \mathbf{D} \mathbf{D}^T \right)^{-1} \mathbf{D} \right], \tag{7.58}$$

where $\mathbf{B} \in \mathbb{R}^{(n-2) \times (n-2)}$ and $\mathbf{D} \in \mathbb{R}^{(n-2) \times T}$ are as follows:

$$\mathbf{B} = \begin{bmatrix} \frac{1}{3}(t_3 - t_1) & \frac{1}{6}(t_3 - t_2) & 0 & \cdots & & 0 \\ \frac{1}{6}(t_3 - t_2) & \frac{1}{3}(t_4 - t_2) & \frac{1}{6}(t_4 - t_3) & \cdots & & 0 \\ 0 & \vdots & \vdots & \cdots & & \cdots \\ \vdots & \vdots & \vdots & \cdots & & \cdots \\ 0 & \cdots & & \cdots & \frac{1}{6}(t_n - t_{n-1}) & \frac{1}{3}(t_n - t_{n-2}) \end{bmatrix} \tag{7.59}$$

and

$$
D = \begin{bmatrix}
\frac{1}{t_2-t_1} & -\frac{1}{t_2-t_1}-\frac{1}{t_3-t_2} & \frac{1}{t_3-t_2} & \cdots & 0 \\
0 & \frac{1}{t_3-t_2} & -\frac{1}{t_3-t_2}-\frac{1}{t_4-t_3} & \vdots & 0 \\
\vdots & \vdots & \vdots & \cdots & \cdots \\
\vdots & \vdots & \vdots & \cdots & \cdots \\
0 & \cdots & \frac{1}{t_{n-1}-t_{n-2}} & -\frac{1}{t_{n-1}-t_{n-2}}-\frac{1}{t_n-t_{n-1}} & \frac{1}{t_n-t_{n-2}}
\end{bmatrix}
\tag{7.60}
$$

Making $t_j = j, j = 1, \ldots, n$, (7.59) and (7.60) become, respectively,

$$
B = \begin{bmatrix}
\frac{2}{3} & \frac{1}{6} & 0 & \cdots & 0 \\
\frac{1}{6} & \frac{2}{3} & \frac{1}{6} & \cdots & 0 \\
0 & \vdots & \vdots & \cdots & \\
\vdots & \vdots & \vdots & \cdots & \\
0 & \cdots & & \frac{1}{6} & \frac{2}{3}
\end{bmatrix}
\quad \text{and} \quad
D = \begin{bmatrix}
1 & -2 & 1 & \cdots & 0 \\
0 & 1 & -2 & \vdots & 0 \\
\vdots & \vdots & \vdots & \cdots & \\
\vdots & \vdots & \vdots & \cdots & \\
0 & \cdots & 0 & \cdots & 1
\end{bmatrix}.
\tag{7.61}
$$

7.3.5 Theoretical Properties of Symmetric and Asymmetric Linear Trend-Cycle Filters

The theoretical properties of the symmetric and asymmetric linear filters can be studied by analyzing their frequency response functions. It is defined by

$$
H(\omega) = \sum_{j=-m}^{m} w_j e^{i\omega j}, \qquad 0 \le \omega \le 1/2,
\tag{7.62}
$$

where w_j are the weights of the filter and ω is the frequency in cycles per unit of time. In general, the frequency response functions can be expressed in polar form as follows:

$$
H(\omega) = A(\omega) + iB(\omega) = G(\omega)e^{i\phi(\omega)},
\tag{7.63}
$$

where $G(\omega) = \left[A^2(\omega) + B^2(\omega)\right]^{1/2}$ is called the *gain of the filter* and $\phi(\omega) = \tan^{-1}(-B(\omega)/A(\omega))$ is called the *phase shift of the filter* and is usually expressed in radians. The expression (7.63) shows that if the input function is a sinusoidal variation of unit amplitude and constant phase shift $\psi(\omega)$, the output function will also be sinusoidal but of amplitude $G(\omega)$ and phase shift $\psi(\omega) + \phi(\omega)$. The gain and phase shift vary with ω. For symmetric filters the phase shift is 0 or $\pm\pi$, and for asymmetric filters take values between $\pm\pi$ at those frequencies where the gain function is zero. For a better interpretation the phase shifts will be here given in months instead of radians, that is, given by $\phi(\omega)/2\pi\omega$ for $\omega \neq 0$.

The gain function should be interpreted as relating the spectrum of the original series to the spectrum of the output obtained with a linear time invariant filter. For example, let $\hat{y}_t$ be the estimated seasonally adjusted observation for the current period based on data $y_1, y_2, \ldots, y_n$ by application of the concurrent linear time-invariant filter $w_{0,j}$. Thus, the gain function shown in the below figures relates the spectrum of $\{y_t, t = 1, \ldots, n\}$ to the spectrum of $\hat{y}_t$, and not to the spectrum of the complete seasonally adjusted series produced at time n (which includes y_t, a first revision of time $t - 1$, a second revision at time $t - 2$, and so on). Dagum et al. [14] derived the gain functions of standard, short, and long convolutions corresponding to the 13-term (H13), 9-term (H9), and 23-term (H23) symmetric Henderson filters, respectively. These authors showed how cycles of 9 and 10 months periodicity (in the 0.08–0.18 frequency band) are not suppressed by any of the cascade filters, particularly, those using H13 and H9 filters. In fact, about 90 %, 72 %, and 21 % of the power of these short cycles are left in the output by the 9-, 13-, and 23-term Henderson filters, respectively. Figure 7.3 shows the gain functions of four symmetric trend-cycle filters of 13-term each. In the context of trend-cycle estimation, it is useful to divide the total range of $\omega \in [0, 0.50]$ into two major intervals, one for the signal, and another for the noise. There are no-fixed rules on defining the cut-off frequency, but for monthly data the intervals are

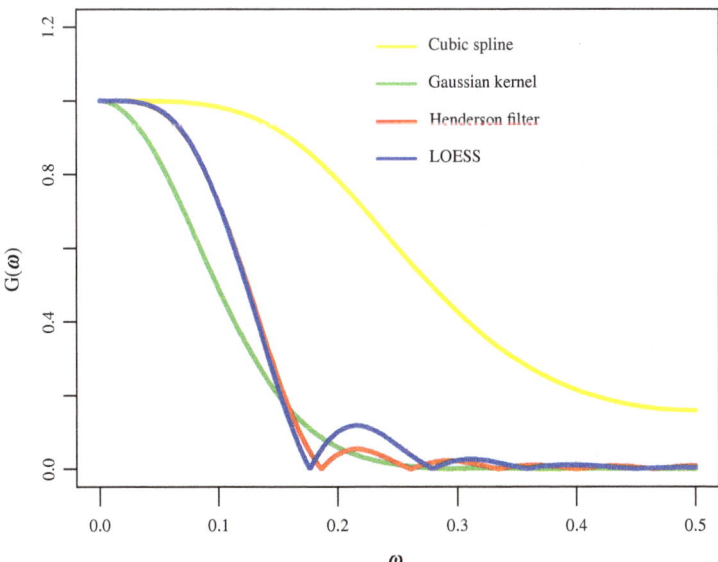

Fig. 7.3 Gain functions $G(\omega), 0 \leq \omega \leq 0.5$, of several symmetric 13-term filters

usually given by:

1. $0 \leq \omega \leq 0.06$ associated with cycles of 16 months and longer attributed to the
   trend-cycle of the series, and
2. $0.06 \leq \omega \leq 0.50$ corresponding to seasonality, if present, and noise.

It is apparent that the 13-term Henderson and LOESS are very close to each other
and pass well the power in the frequency band $0 \leq \omega \leq 0.06$ reproducing well the
cycles associated with the trend-cycle and suppressing a large amount of noise. But
they have the limitation of passing too much power at $\omega = 0.10$, that will produce
a large number of 10-month cycles in the output, also known as unwanted ripples.

On the other hand, the Gaussian kernel will not pass too much power at $\omega =$
0.10, but will suppress also the power attributed to the trend-cycle. In other words,
it will produce a very smooth trend-cycle whereas the 13-term cubic spline will
do the opposite that is why in empirical applications the length of this filter is
selected according to other criteria. Figures 7.4 and 7.5 exhibit the gain and phase
shifts of the last point of the corresponding asymmetric filters. The gains of the
asymmetric filters suppress less noise and pass more power related to the trend-
cycle (particularly, H13 and LOESS amplify the power) relative to the symmetric
filters. All the filters introduce a time lag to detect a true turning point as shown by
their phase shift functions.

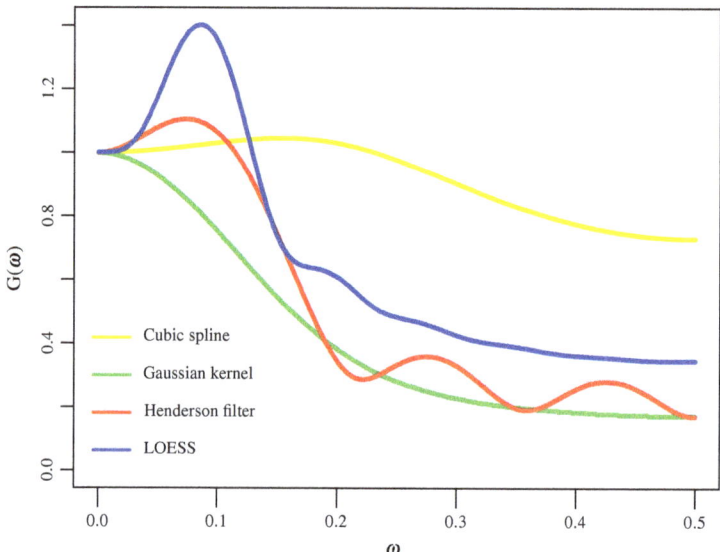

Fig. 7.4 Gain functions $G(\omega), 0 \leq \omega \leq 0.5$, of last point asymmetric filters

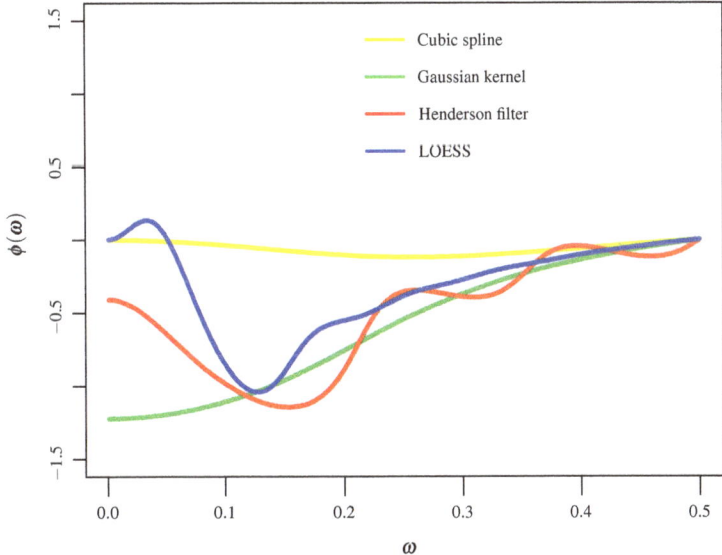

Fig. 7.5 Phase shift functions $\phi(\omega), 0 < \omega \le 0.5$, of last point asymmetric filters

7.4   Illustrative Results

We now illustrate with a real series the various trend-cycle values obtained with the different filters discussed before. To select the appropriate lengths of the Henderson, LOESS, and Gaussian kernel filters we use the irregular/trend-cycle (I/C) ratio, whereas for the cubic spline we look at the smoothing parameter estimated by means of the generalized cross validation criterion. In particular, we consider the US New Orders for Durable Goods (NODG) observed from February 1992 till December 2013.

Figures 7.6, 7.7, and 7.8 exhibit the various trend-cycle estimates for the US New Orders for Durable Goods series, where 13-term filters have been chosen for the Henderson, the LOESS, and the Gaussian kernel and the smoothing parameter for the cubic spline is 16.438 and the equivalent degrees of freedom is 47.289. Figure 7.6 shows that the LOESS and Henderson filter give values close to each other whereas the Gaussian kernel (Fig. 7.7) produces the smoothest trend, and the cubic spline (Fig. 7.8) gives the most volatile. The difference is more noticeable at cyclical turning points where the Gaussian kernel cuts them more whereas the cubic spline cuts them the least.

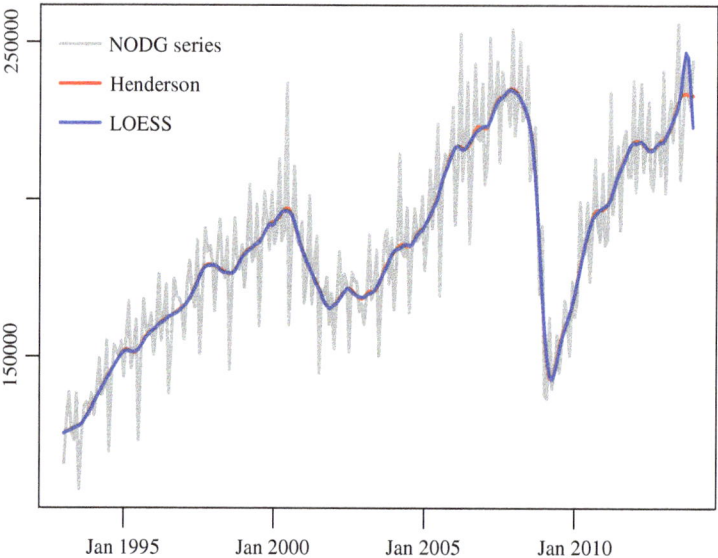

Fig. 7.6 US New Orders for Durable Goods trend-cycle estimates based on LOESS and Henderson 13-term filters

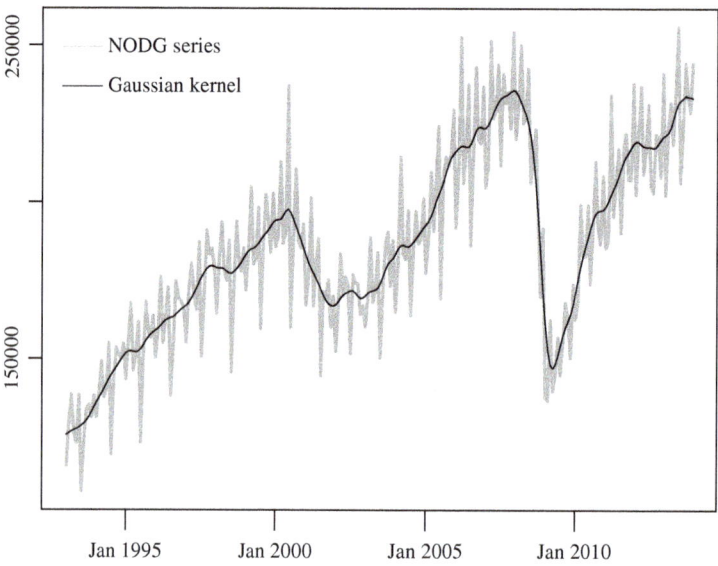

Fig. 7.7 The US New Orders for Durable Goods trend-cycle estimates based on the 13-term Gaussian kernel

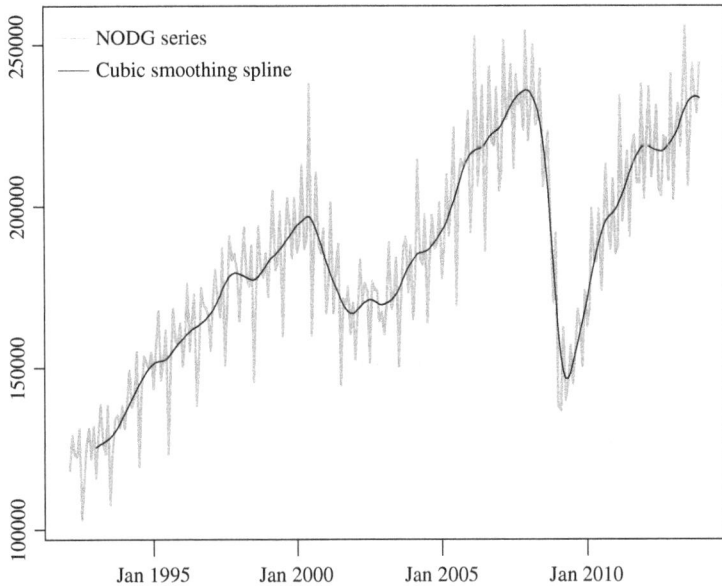

Fig. 7.8 The US New Orders for Durable Goods trend-cycle estimates based on cubic smoothing spline

References

1. Akaike, H. (1980). Likelihood and the Bayes procedure (with discussion). In J. M. Bernardo, M. H. De Groot, D. V. Lindley, & A. F. M. Smith (Eds.), *Bayesian statistics* (pp. 143–203). Valencia, Spain: University Press.
2. Akaike, H. (1981). Likelihood of a model and information criteria. *Journal of Econometrics, 16*, 3–14.
3. Anderson, T. W. (1971). *The statistical analysis of time series*. New York: Wiley.
4. Box, G. E. P., & Jenkins, G. M. (1970). *Time series analysis: Forecasting and control*. San Francisco, CA, USA: Holden-Day (Second Edition 1976).
5. Burman, J. P. (1980). Seasonal adjustment by signal extraction. *Journal of the Royal Statistical Society, Series A, 143*, 321–337.
6. Cleveland, R. B., Cleveland, W. S., McRae, J. E., & Terpenning, I. (1990). STL: A seasonal trend decomposition procedure based on LOESS. *Journal of Official Statistics, 6*(1), 3–33.
7. Cleveland, W. S. (1979). Robust locally regression and smoothing scatterplots. *Journal of the American Statistical Association, 74*, 829–836.
8. Cleveland, W. S., & Devlin, S. J. (1988). Locally weighted regression: An approach to regression analysis by local fitting. *Journal of the American Statistical Association, 83*(403), 596–610.
9. Craven, P., & Wahba, G. (1979). Smoothing noisy data with spline functions. *Numerical Mathematics, 31*, 377–403.
10. Dagum, C. (1985). Analyses of income distribution and inequality by education and sex in Canada. In R. L. Basmann & G. F. Rhodes Jr. (Eds.), *Advances in econometrics* (Vol. IV, pp. 167–227). Greenwich, CN: JAI Press.
11. Dagum, E. B. (1980). *The X11ARIMA seasonal adjustment method*. Ottawa, Canada: Statistics Canada. Catalogue No. 12-564.

12. Dagum, E. B. (1988). *The X11ARIMA/88 seasonal adjustment method-foundations and user's manual*. Ottawa, Canada: Time Series Research and Analysis Centre, Statistics Canada.
13. Dagum, E. B., & Capitanio, A. (1998). Smoothing methods for short-term trend analysis: Cubic splines and Henderson filters. *Statistica, LVIII*(1), 5–24.
14. Dagum, E. B., Chhab, N., & Chiu, K. (1996). Derivation and analysis of the X11ARIMA and Census X11 linear filters. *Journal of Official Statistics, 12*(4), 329–347.
15. Dagum, E. B., & Luati, A. (2000). Predictive performance of some nonparametric linear and nonlinear smoothers for noisy data. *Statistica, LX*(4), 635–654.
16. Dagum, E. B., & Luati, A. (2001). A study of asymmetric and symmetric weights of kernel smoothers and their spectral properties. *Estadistica, Journal of the Inter-American Statistical Institute, 53*(160–161), 215–258.
17. Doherty, M. (2001). The surrogate Henderson filters in X-11. *Australian and New Zealand Journal of Statistics, 43*(4), 385–392.
18. Eubank, R. L. (1999). *Spline smoothing and nonparametric regression*. New York: Marcel Dekker.
19. Fan, J. (1992). Design-adaptive nonparametric regression. *Journal of the American Statistical Association, 87*, 998–1004.
20. Fan, J. (1993). Local linear regression smoothers and their minimax efficiencies. *Annals of Statistics, 21*, 196–216.
21. Fan, J., & Gijbels, I. (1996). *Local polynomial modelling and its applications*. New York: Chapman and Hall.
22. Findley, D. F., Monsell, B. C., Bell, W. R., Otto, M. C., & Chen, B. C. (1998). New capabilities and methods of the X12ARIMA seasonal adjustment program. *Journal of Business and Economic Statistics, 16*(2), 127–152.
23. Granger, C. W., & Morgenstern, O. (1970). *Predictability of stock market prices*. Lexington, MA: D.C. Heath.
24. Gray, A., & Thomson, P. (1996). Design of moving-average trend filters using fidelity and smoothness criteria. In P. M. Robinson & M. Rosenblatt (Eds.), *Time series analysis* (in memory of E.J. Hannan), Vol. II. Springer Lecture notes in statistics (Vol. 115, pp. 205–219). New York: Springer.
25. Green, P. J., & Silverman, B. W. (1994). *Nonparametric regression and generalized linear models*. London: Chapman and Hall.
26. Hardle, W. (1990). *Applied nonparametric regression*. Cambridge: Cambridge University Press.
27. Harvey, A. C. (1985). Trends and cycles in macroeconomic time series. *Journal of Business and Economic Statistics, 3*, 216–227.
28. Harvey, A. C. (1989). *Forecasting, structural time series models and the Kalman filter*. Cambridge: Cambridge University Press.
29. Harvey, A. C., & Todd, P. H. J. (1983). Forecasting economic time series with structural and Box-Jenkins models: A case study. *Journal of Business and Economic Statistics, 1*, 299–306.
30. Henderson, R. (1916). Note on graduation by adjusted average. *Transaction of Actuarial Society of America, 17*, 43–48.
31. Hillmer, S. C., & Tiao, G. C. (1982). An ARIMA-model-based approach to seasonal adjustment. *Journal of the American Statistical Association, 77*, 63–70.
32. Kendall, M. G., Stuart, A., & Ord, J. (1983). *The advanced theory of statistics* (Vol. 3). New York: Wiley.
33. Kenny, P., & Durbin, J. (1982). Local trend estimation and seasonal adjustment of economic and social time series. *Journal of the Royal Statistical Society A, 145*, 1–41.
34. Kitagawa, G., & Gersch, W. (1996). *Smoothness priors analysis of time series*. Lecture notes in statistics (Vol. 116). New York: Springer.
35. Kondratieff, N. (1925). *Long economic cycles*. Voprosy Konyuktury (Vol. 1, No. (1)) (English translation: *The long wave cycle*. Richardson and Snyder: New York, 1984).
36. Laniel, N. (1985). Design criteria for the 13-term Henderson end weights. Working paper. Methodology Branch. Statistics Canada, Ottawa.

37. Macauley, F. (1931). *The smoothing of time series*. New York: National Bureau of Economic Research.
38. Maravall, A. (1993). Stochastic linear trends: Models and estimators. *Journal of Econometrics, 56*, 5–37.
39. Musgrave, J. (1964). A set of end weights to end all end weights. Working paper. U.S. Bureau of Census, Washington, DC.
40. Quetelet, A. (1835). *Sur l'homme et le développement de ses facultés*. Paris: Bachelier.
41. Sbrana, G. (2011). Structural time series models and aggregation: Some analytical results. *Journal of Time Series Analysis, 32*, 315–316.
42. Schoenberg, I. (1964). Monosplines and quadrature formulae. In T. Greville (Ed.), *Theory and applications of spline functions*. Madison, WI: University of Wisconsin Press.
43. Shiskin, J., Young, A. H., & Musgrave, J. C. (1967). The X-11 variant of the Census Method II seasonal adjustment program. Technical Paper 15 (revised). US Department of Commerce, Bureau of the Census, Washington, DC.
44. Simonoff, J. S. (1995). *Smoothing methods in statistics*. New York: Springer.
45. Stock, J. H., & Watson, M. W. (1988). Variable trends in economic time series. *Journal of Economic Perspectives, 2*, 147–174.
46. Verhulst, P. F. (1838). Notice sur la loi que la population suit dans son accroissement. In A. Quetelet (Ed.), *Correspondance Mathematique et Physique* (Tome X, pp. 113–121).
47. Wahba, G. (1990). *Spline models for observational data*. Philadelphia: SIAM.
48. Wand, M., & Jones, M. (1995). *Kernel smoothing*. Monographs on statistics and applied probability (Vol. 60). London: Chapman and Hall.
49. Whittaker, E., & Robinson, G. (1924). *Calculus of observations: A treasure on numerical calculations*. London: Blackie and Son.
50. Whittaker, E. T. (1923). On a new method of graduation. In *Proceedings of the Edinburg Mathematical Association* (Vol. 78, pp. 81–89).
51. Yaglom, A. M. (1962). *An introduction to the theory of stationary random functions*. Englewood Cliffs, NJ: Prentice-Hall.

Chapter 8
Further Developments on the Henderson Trend-Cycle Filter

Abstract The linear filter developed by Henderson is the most widely applied to estimate the trend-cycle component in seasonal adjustment software such as the US Bureau of Census II-X11 and its variants, the X11/X12ARIMA. Major studies have been done on trend-cycle estimation during the last 20 years by making changes to the Henderson filters. The emphasis has been on determining the direction of the short-term trend for an early detection of a true turning point. This chapter introduces in detail three major contributions: (1) a nonlinear trend-cycle estimator also known as Nonlinear Dagum Filter (NLDF), (2) a Cascade Linear Filter (CLF) that closely approximates the NLDF, and (3) an approximation to the Henderson filter via the Reproducing Kernel Hilbert Space (RKHS) methodology.

The linear filter developed by Henderson [22] is the most widely applied to estimate the trend-cycle component in seasonal adjustment software such as the US Bureau of Census II-X11 method [30] and its variants, the X11ARIMA [8] and X12ARIMA [20]. Major studies have been done on trend-cycle estimation during the last 20 years by making changes to the Henderson filters. The emphasis has been made on determining the direction of the short term trend for an early detection of a true turning point. Major changes of global character in the financial and economic sectors have introduced high levels of variability in time series making difficult to detect the direction of the short-term trend by simply looking at seasonally adjusted data and the use of trend-cycle data has been supported.

In 1996, Dagum [9] developed a nonlinear trend-cycle estimator to improve on the classical 13-term Henderson filter [22]. The Nonlinear Dagum Filter (NLDF) results from applying the 13-term symmetric Henderson filter (H13) to seasonally adjusted series where outliers and extreme observations have been replaced and which have been extended with extrapolations from an ARIMA model [3]. Later on, in 2009, Dagum and Luati [16] developed a linear approximation to the NLDF to facilitate a broader application since the new linear approximation does not need an ARIMA model identification.

From another perspective, in 2008 Dagum and Bianconcini [11] developed an approximation to the Henderson filter via reproducing kernel methodology that was shown to produce better results for real time trend-cycle estimation. All three papers make explicit reference to the Henderson filters applied to monthly time series

© Springer International Publishing Switzerland 2016

E. Bee Dagum, S. Bianconcini, *Seasonal Adjustment Methods and Real Time Trend-Cycle Estimation*, Statistics for Social and Behavioral Sciences, DOI 10.1007/978-3-319-31822-6_8

197

but the results can be easily extended to quarterly series. It should be noticed that following the tradition of statistical agencies, in these papers it is called asymmetric Henderson filter the one that corresponds to the asymmetric weights developed by Musgrave [27]. In the next sections, we will discuss the major assumptions and modifications introduced in each of these studies and refer the reader to the published papers for more details.

8.1 The Nonlinear Dagum Filter (NLDF)

The modified Henderson filter developed by Dagum [9] is nonlinear and basically consists of:

(a) extending the seasonally adjusted series with ARIMA extrapolated values, and
(b) applying the 13-term Henderson filter to the extended series where extreme values have been modified using very strict sigma limits.

To facilitate the identification and fitting of simple ARIMA models, Dagum [9] recommends, at step (a), to modify the input series for the presence of extreme values using the standard $\pm 2.5\sigma$ limits of X11/X12ARIMA and X13. These computer programs are those that can be used to implement the Henderson filter. In this way, a simple and very parsimonious ARIMA model, the ARIMA(0,1,1), is often found to fit a large number of series.

Concerning step (b), it is recommended to use very strict sigma limits, such as $\pm 0.7\sigma$ and $\pm 1.0\sigma$. The main purpose of the ARIMA extrapolations is to reduce the size of the revisions of the most recent estimates, whereas that of extreme values replacement is to reduce the number of unwanted ripples produced by H13. An unwanted ripple is a 10-month cycle (identified by the presence of high power at $\omega = 0.10$ in the frequency domain) which, due to its periodicity, often leads to the wrong identification of a true turning point. In fact, it falls in the neighborhood between the fundamental seasonal frequency and its first harmonic. On the other hand, a high frequency cycle is generally assumed to be part of the noise pertaining to the frequency band $0.10 \leq \omega < 0.50$. The problem of the unwanted ripples is specific of H13 when applied to seasonally adjusted series.

The NLDF can be formally described in matrix notation as follows. Let $\mathbf{y} \in \mathbb{R}^n$ be the n-dimensional seasonally adjusted time series to be smoothed, which consists of a nonstationary trend-cycle $\mathbf{TC}$ plus an erratic component $\mathbf{e}$, that is,

$$\mathbf{y} = \mathbf{TC} + \mathbf{e}. \tag{8.1}$$

It is assumed that the trend-cycle is smooth and can be well estimated by means of the 13-term Henderson filter applied to $\mathbf{y}$. Hence,

$$\widehat{TC} = \mathbf{Hy}, \tag{8.2}$$

where $\mathbf{H}$ is the $n \times n$ matrix (canonically) associated with the 13-term Henderson filter. Replacing $\widehat{TC}$ in Eq. (8.1) by Eq. (8.2), we have

$$\mathbf{y} = \mathbf{Hy} + \mathbf{e}, \tag{8.3}$$

or,

$$(\mathbf{I}_n - \mathbf{H})\mathbf{y} = \mathbf{e}, \tag{8.4}$$

where $\mathbf{I}_n$ is the $n \times n$ identity operator on $\mathbb{R}^n$. Assign now a weight to the residuals in such a way that if the observation $y_t, t = 1, \ldots, n$, is recognized to be an extreme value (with respect to $\pm 2.5\sigma$ limits, where σ is a 5-year moving standard deviation), then the corresponding residual e_t is zero weighted (i.e., the extreme value is replaced by $\widehat{TC}_t$ which is a preliminary estimate of the trend). If y_t is not an extreme value, then the weight for e_t is one (i.e., the value y_t is not modified). In symbols,

$$\mathbf{W}_0\mathbf{e} = \mathbf{W}_0(\mathbf{I}_n - \mathbf{H})\mathbf{y}, \tag{8.5}$$

where $\mathbf{W}_0$ is a zero-one diagonal matrix, being the diagonal element w_{tt} equal to zero when the corresponding element y_t of the vector $\mathbf{y}$ is identified as an outlier. For instance, if in the series $\mathbf{y}$ the only extreme value is y_2, then the weight matrix for the residuals will be

$$\mathbf{W}_0 = \begin{bmatrix} 1 & 0 & 0 & \cdots & 0 \\ 0 & 0 & 0 & \cdots & 0 \\ 0 & 0 & 1 & \cdots & 0 \\ & & \cdots & & \\ 0 & 0 & 0 & \cdots & 1 \end{bmatrix}. \tag{8.6}$$

Denoting by

$$\mathbf{e}_0 = \mathbf{W}_0\mathbf{e} \tag{8.7}$$

the vector of the modified residuals, then the series modified by extreme values with zero weights becomes

$$\mathbf{y}_0 = \mathbf{TC} + \mathbf{e}_0, \tag{8.8}$$

which can be written as

$$\mathbf{y}_0 = \mathbf{H}\mathbf{y} + \mathbf{W}_0(\mathbf{I}_n - \mathbf{H})\mathbf{y} = [\mathbf{H} + \mathbf{W}_0(\mathbf{I}_n - \mathbf{H})]\,\mathbf{y}. \tag{8.9}$$

Using (8.9), one year of ARIMA extrapolations is obtained in order to extend the series modified by extreme values. Denoting with $\mathbf{y}_0^E$ the extended series, that is, the $n + 12$ vector whose first n elements are given by $\mathbf{y}_0$ while the last 12 are the extrapolated ones, in block-matrix notation we have

$$\mathbf{y}_0^E = \begin{bmatrix} [\mathbf{H} + \mathbf{W}_0(\mathbf{I}_n - \mathbf{H})]\,\mathbf{y} \\ \mathbf{y}^{12} \end{bmatrix}, \tag{8.10}$$

where $\mathbf{y}^{12}$ is the 12×1 block of extrapolated values. Setting

$$[\mathbf{H} + \mathbf{W}_0(\mathbf{I}_n - \mathbf{H})]^{12} = \begin{bmatrix} [\mathbf{H} + \mathbf{W}_0(\mathbf{I}_n - \mathbf{H})]\ \mathbf{O}_{n\times 12} \\ \mathbf{O}_{12\times n} \qquad \mathbf{I}_{12} \end{bmatrix}, \tag{8.11}$$

and

$$\mathbf{y}^{+,12} = \begin{bmatrix} \mathbf{y} \\ \mathbf{y}^{12} \end{bmatrix}, \tag{8.12}$$

$\mathbf{y}_0^E$ becomes

$$\mathbf{y}_0^E = [\mathbf{H} + \mathbf{W}_0(\mathbf{I}_n - \mathbf{H})]^{12}\,\mathbf{y}^{+,12}. \tag{8.13}$$

This concludes the operations involved in step (a) of the NLDF.

Step (b) follows. The procedure for obtaining $\mathbf{y}_0$ on the series $\mathbf{y}_0^E$ is repeated, but with stricter sigma limits (such as $\pm 0.7\sigma$ and $\pm 1.0\sigma$) and with different weights assigned to the residuals for the replacement of the extreme values. The estimates $\mathbf{y}^E$ computed over the series $\mathbf{y}_0^E$ are

$$\mathbf{y}^E = [\mathbf{H} + \mathbf{W}(\mathbf{I}_n - \mathbf{H})]^E\,\mathbf{y}_0^E. \tag{8.14}$$

The $(n+12)\times(n+12)$ matrix $[\mathbf{H} + \mathbf{W}(\mathbf{I}_n - \mathbf{H})]^E$ is analogue to $[\mathbf{H} + \mathbf{W}_0(\mathbf{I}_n - \mathbf{H})]^{12}$ except for the matrix $\mathbf{W}$ that is also diagonal, but with generic diagonal element w_{tt}, such that $w_{tt} = 0$ if the corresponding value y_t falls out of the upper bound selected limits, say, $\pm 1.0\sigma$, and $w_{tt} = 1$ if the corresponding y_t falls within the lower bound selected limits, say, $\pm 0.7\sigma$ and w_{tt} decreases linearly (angular coefficient equal to -1) from 1 to 0 in the range from $\pm 0.7\sigma$ to $\pm 1.0\sigma$. Under the assumption of normality, these sigma limits imply that 48 % of the values will be modified (replaced by the preliminary smoothed trend), 32 % will be zero weighted while the remaining 16 % will get increasing weights from zero to one.

Notice that $\mathbf{y}^E$ can also be written as

$$\mathbf{y}^E = [\mathbf{H} + \mathbf{W}(\mathbf{I}_n - \mathbf{H})]^E [\mathbf{H} + \mathbf{W}_0(\mathbf{I}_n - \mathbf{H})]^{12} \mathbf{y}^{+,12}. \tag{8.15}$$

Finally, the NLDF estimates are given by applying a 13-term Henderson filter to Eq. (8.15), that is,

$$\mathbf{H}\mathbf{y}^E = \mathbf{H} [\mathbf{H} + \mathbf{W}(\mathbf{I}_n - \mathbf{H})]^E [\mathbf{H} + \mathbf{W}_0(\mathbf{I}_n - \mathbf{H})]^{12} \mathbf{y}^{+,12} \tag{8.16}$$

$$= \begin{bmatrix} \widehat{\mathbf{TC}} \\ \widehat{\mathbf{TC}}^{12} \end{bmatrix},$$

where $\widehat{\mathbf{TC}}$ is the n-dimensional vector of smooth estimates of $\mathbf{y}$.

It is apparent that the NLDF method reduces drastically the effects of extreme values by repeatedly smoothing the input data via downweighting points with large residuals. Furthermore, the ARIMA extension enables the use of the symmetric weights of the 13-term Henderson filter for the last six observations and, thus, reduces the size of the revisions of the last estimates.

Since the values of the matrices $\mathbf{W}_0$ and $\mathbf{W}$ corresponding to extreme values replacement and matrix $[\mathbf{H} + \mathbf{W}_0(\mathbf{I}_n - \mathbf{H})]^{12}$ pertaining to ARIMA extrapolations are data dependent, this filter is nonlinear. Studies by Dagum, Chhab, and Morry [5, 12], and Darnè [18] showed the superior performance of the NLDF with respect to both structural and ARIMA standard parametric trend-cycle models applied to series with different degrees of signal-to-noise ratios. The criteria evaluated were:

1. the number of unwanted ripples,
2. size of revisions, and
3. time delay to detect a turning point.

In another study, the good performance of the NLDF is shown relative to nonparametric smoothers, namely: locally weighted regression (LOESS), Gaussian kernel, cubic smoothing spline, and supersmoother [14]. The NLDF is currently used by many statistical agencies to produce trend-cycle estimates or trend-cycle figures to show the direction of the short-term trend.

8.2   The Cascade Linear Filter

Given the excellent performance of the NLDF according to the three criteria mentioned above, Dagum and Luati [16] developed a cascade linear filter that closely approximates it. The cascading is done via the convolution of several filters chosen for noise suppression, trend estimation, and extrapolation.

A linear filter offers many advantages over a nonlinear one. First, its application is direct and hence, does not require knowledge of ARIMA model identification.

Furthermore, linear filtering preserves the crucial additive constraint by which the trend of an aggregated variable should be equal to the algebraic addition of its component trends, thus avoiding the selection problem of direct versus indirect adjustments. Finally, the properties of a linear filter concerning signal passing and noise suppression can always be compared to those of other linear filters by means of spectral analysis.

The symmetric filter is the one applied to all central observations. In this case, the purpose is to offer a linear solution to the unwanted ripples problem. To avoid the latter, the NLDF largely suppresses the noise in the frequency band between the fundamental seasonal and first harmonic. In this regard, a cascade linear filter is derived by double smoothing the residuals obtained from a sequential application of H13 to the input data. The residuals smoothing is done by the convolution of two short smoothers, a weighted 5-term and a simple 7-term linear filters. The linear approximation for the symmetric part of the NLDF is truncated to 13 terms with weights normalized to add to one.

On the other hand, the asymmetric filter is applied to the last six data points which are crucial for current analysis. It is obtained by means of the convolution between the symmetric filter and the linear extrapolation filters for the last six data points. The extrapolation filters are linearized by fixing both the ARIMA model and its parameter values. The latter are chosen so as to minimize the size of revisions and phase shifts. The model was selected among some parsimonious processes found to fit and extrapolate well a large number of seasonally adjusted series. Such model is the ARIMA(0,1,1) with $\theta = 0.40$. A simple linear transformation [15] allows to apply the asymmetric filter to the first six observations.

The new filter is called the Cascade Linear Filter (CLF) and there is a distinction between the Symmetric (SLF) and the Asymmetric Linear Filter (ALF).

8.2.1   The Symmetric Linear Filter

The smoothing matrix associated with the symmetric linear filter results

$$\mathbf{H}\left[\mathbf{H} + \mathbf{M}_{7(0.14)}\left(\mathbf{I}_n - \mathbf{H}\right)\right]\left[\mathbf{H} + \mathbf{M}_{5(0.25)}\left(\mathbf{I}_n - \mathbf{H}\right)\right], \qquad (8.17)$$

where $M_{5(0.25)}$ is the matrix representative of a 5-term moving average with weights $(0.250, 0.250, 0.000, 0.250, 0.250)$, and $M_{7(0.14)}$ is the matrix representative of a 7-term filter with all weights equal to 0.143.

5- and 7-term filters are chosen following the standard filters length selected in Census II-X11 and X11/X12ARIMA software for the replacement of extreme Seasonal-Irregular (SI) values. In these computer packages, a first iteration is made by means of a short smoother, a 5-term (weighted) moving average, and a second one by a 7-term (weighted) average. In this case, 5- and 7-term filters are applied to the residuals from a first pass of the H13 filter. These two filters have the good property of suppressing large amounts of power at the frequency $\omega = 0.10$,

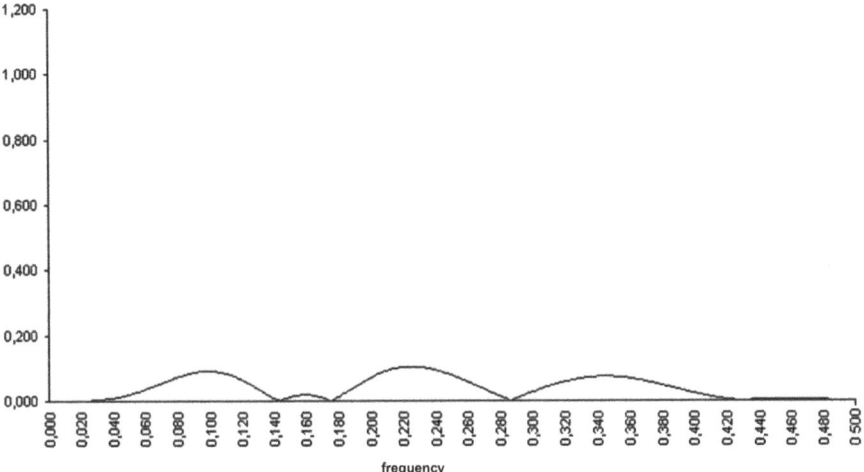

Fig. 8.1   Gain function of the symmetric weights of the modified irregulars filter

corresponding to cycle of 10 months periodicity. Figure 8.1 shows the gain function of the filter convolution $\left[\mathbf{M}_{7(0.14)}\,(\mathbf{I}_n-\mathbf{H})\right]\left[\mathbf{H}+\mathbf{M}_{5(0.25)}\,(\mathbf{I}_n-\mathbf{H})\right]$.

It is apparent that the filter convolution applied to a series consisting of trend plus irregulars suppresses all the trend power and a great deal of the irregular variations. Hence given the input series, the results from the convolution are the modified irregulars needed to produce a new series which will be extended with ARIMA extrapolations, and then smoothed with H13.

Equation (8.17) produces a symmetric filter of 31 terms with very small weights at both ends. This long filter is truncated to 13 terms, and normalized such that its weights add up to unity. Normalization is needed to avoid a biased mean output.

To normalize the filter, the total weight discrepancy (in our case the 13 truncated weights add up to 1.065) is distributed over the 13 weights, w_j, $j = -6,\dots,6$, according to a well-defined pattern. This is a very critical point since adjustment for different distributions produces linear filters with very distinctive properties.

The weights of the symmetric linear filter with mixed distribution normalization, denoted by SLF, are given by $\left(-0.027,\ -0.007,\ 0.031,\ 0.067,\ 0.136,\ 0.188,\ 0.224\right)$.

8.2.1.1   Theoretical Properties of SLF

From a theoretical viewpoint, the properties of SLF were studied by means of classical spectral analysis techniques. It was found that in comparison to H13, SLF suppressed more signal only in the frequency band pertaining to very short cycles, ranging from 15 to 24 months periodicity (of corresponding frequency $0.03 < \omega \leq 0.06$), whereas it passed without modification cycles of 3 year and longer periodicity. Furthermore, it reduced by 14 % the power of the gain corresponding to the unwanted ripples frequency $\omega = 0.10$. We refer the reader to Dagum and Luati [16] for further details.

8.2.1.2 Empirical Properties of SLF

To perform an empirical evaluation of the fitting and smoothing performances of SLF relative to H13, applications were made to a large sample of 100 seasonally adjusted real time series pertaining to various socioeconomic areas, characterized by different degrees of variability. As a measure of fitting the Root Mean Square Error (RMSE) is used, and defined as follows:

$$\text{RMSE} = \sqrt{\frac{1}{n-12} \sum_{t=7}^{n-5} \left(\frac{\widehat{TC}_t - y_t}{y_t} \right)^2},$$

where y_t denotes the original seasonally adjusted series, and $\widehat{TC}_t$ the estimated trend-cycle values. The main goal was to assess the extent to which the fitting properties of the SLF were equivalent to those of H13, which is known to be a good signal estimator but at the expense of producing many unwanted ripples. There was also the interest to reduce the number of unwanted ripples that may lead to the detection of false turning points. In this regard, the accepted definition of a turning point for smoothed data (see, among others, Zellner et al. [32]) is used. A turning point occurs at time t if (*downturn*) $y_{t-k} \leq \cdots \leq y_{t-1} > y_t \geq y_{t+1} \geq \cdots \geq y_{t+m}$ or (*upturn*) $y_{t-k} \geq \cdots \geq y_{t-1} < y_t \leq y_{t+1} \leq \cdots \leq y_{t+m}$ for $k = 3$ and $m = 1$. An unwanted ripple arises whenever two turning points occur within a 10-month period.

Table 8.1 shows the mean values of the RMSE calculated over the 100 series and standardized with respect to H13 to facilitate the comparison. In the same way, the number of false turning points produced in the final estimates from SLF is given relative to that of H13. The empirical results were consistent with those inferred from the theoretical analysis. Furthermore, SLF reduced by 20% the number of false turning points produced by H13.

Table 8.1 Empirical RMSE and the number of false turning points (ftp) for SLF and H13 filters applied to real time series (mean values standardized by those of H13)

Empirical measures of fitting and smoothing	SLF	H13
$\text{RMSE}/\text{RMSE}_{H13}$	1.05	1
$\text{ftp}/\text{ftp}_{H13}$	0.80	1

8.2.2 The Asymmetric Linear Filter

The smoothing matrix associated with the asymmetric linear filter for the last six data points is obtained in two steps:

1. A linear extrapolation filter for six data points is applied to the input series. This filter is represented by a $(n + 6) \times n$ matrix $\mathbf{A}^*$

$$\mathbf{A}^* = \begin{bmatrix} \mathbf{I}_n \\ \mathbf{O}_{6 \times n-12} \ \ \boldsymbol{\Pi}^*_{6 \times 12} \end{bmatrix},$$

where $\boldsymbol{\Pi}^*_{6 \times 12}$ is the submatrix containing the weights for the $n-5, n-4, \ldots, n$ data points. $\boldsymbol{\Pi}^*_{6 \times 12}$ results from the convolution

$$\mathbf{H} \left[\mathbf{H} + \mathbf{M}_{7(0.14)} \left(\mathbf{I}_{n+12} - \mathbf{H} \right) \right]^E \mathbf{A} \left[\mathbf{H} + \mathbf{M}_{5(0.25)} \left(\mathbf{I}_n - \mathbf{H} \right) \right], \tag{8.18}$$

where $\left[\mathbf{H} + \mathbf{M}_{5(0.25)} \left(\mathbf{I}_n - \mathbf{H} \right) \right]$ is the $n \times n$ matrix representative of trend filter plus a first suppression of extreme values, $\left[\mathbf{H} + \mathbf{M}_{7(0.14)} \left(\mathbf{I}_{n+12} - \mathbf{H} \right) \right]^E$ is the $n \times (n + 12)$ matrix for the second suppression of the irregulars applied to the input series plus 12 extrapolated values, generated by

$$\mathbf{A} = \begin{bmatrix} \mathbf{I}_n \\ \boldsymbol{\Pi}_{12 \times n} \end{bmatrix}.$$

This $(n + 12) \times n$ matrix $\mathbf{A}$ is associated with an ARIMA(0,1,1) linear extrapolations filter with parameter value $\theta = 0.40$.

It is well-known that for all ARIMA models that admit a convergent AR(∞) representation of the form $\left(1 - \pi_1 B - \pi_2 B^2 - \cdots \right) y_t = a_t$, the coefficients $\pi_j, j = 1, 2, \ldots$ can be explicitly calculated. For any lead time τ, the extrapolated values $y_t(\tau)$ may be expressed as a linear function of current and past observations y_t with weights $\pi_j^{(\tau)}$ that decrease rapidly as they depart from the current observations. That is,

$$y_t(\tau) = \sum_{j=1}^{\infty} \pi_j^{(\tau)} y_{t-j+1},$$

where

$$\pi_j^{(\tau)} = \pi_{j+\tau-1} + \sum_{h=1}^{\tau-1} \pi_h \pi_j^{(\tau-h)},$$

for $j = 1, 2, \ldots$ and $\pi_j^{(1)} = \pi_j$. Strictly speaking, the y_t's go back to infinite past, but since the power series is convergent, their dependence on y_{t-j} can be ignored after some time has elapsed. For this reason, $\pi_j^{(\tau)} = 0$ for $j > 12$. Furthermore, to generate one year of ARIMA extrapolation $\tau = 12$ is fixed. In terms of signal passing and noise suppression the model chosen is $\theta = 0.40$. For an ARIMA(0,1,1) model, the coefficients are $\pi_j = (1 - \theta)\,\theta^{j-1}$, and it can be shown that $\pi_j^{(\tau)} = \pi_j$, $\forall \tau$. Hence, $\boldsymbol{\Pi}_{12 \times n}$ is the matrix whose generic row is $(0, \ldots, 0, 0.001, 0.002, 0.006, 0.015, 0.038, 0.096, 0.24, 0.6)$.

Because only six observations are needed to be extrapolated, the centered 12-term filter is truncated and the weights are uniformly normalized to obtain the 6×12 matrix $\boldsymbol{\Pi}^*_{6 \times 12}$ given by

−0,023	−0,004	0,004	0,086	0,143	0,187	0,200	0,180	0,130	0,071	0,021	−0,024
0,000	−0,023	−0,004	0,033	0,088	0,148	0,186	0,195	0,160	0,116	0,085	0,017
0,000	0,000	−0,023	−0,005	0,035	0,093	0,146	0,180	0,166	0,155	0,165	0,085
0,000	0,000	0,000	−0,024	−0,003	0,037	0,090	0,141	0,148	0,182	0,255	0,173
0,000	0,000	0,000	0,000	−0,021	−0,006	0,034	0,089	0,114	0,196	0,331	0,264
0,000	0,000	0,000	0,000	0,000	−0,032	−0,009	0,039	0,075	0,200	0,386	0,342

2. The symmetric filter is applied to the series extrapolated by $\mathbf{A}^*$, that is,

$$\widehat{\mathbf{y}} = \mathbf{S}\mathbf{A}^*\mathbf{y}$$

where $\mathbf{S}$ is the $n \times (n + 6)$ matrix given by

$$\mathbf{H}\left[\mathbf{H} + \mathbf{M}_{7(0.14)}\left(\mathbf{I}_n - \mathbf{H}\right)\right]\left[\mathbf{H} + \mathbf{M}_{5(0.25)}\left(\mathbf{I}_n - \mathbf{H}\right)\right].$$

The convolution $\mathbf{S}\mathbf{A}^*$ produces 12-term asymmetric filters for the last six observations, that are truncated and uniformly normalized in order to obtain the following final Asymmetric Linear Filters (ALF) for the last observations:

−0,026	−0,007	0,030	0,065	0,132	0,183	0,219	0,183	0,132	0,066	0,030	−0,006
0,000	−0,026	−0,007	0,030	0,064	0,131	0,182	0,218	0,183	0,132	0,065	0,031
0,000	0,000	−0,025	−0,004	0,034	0,069	0,137	0,187	0,222	0,185	0,131	0,064
0,000	0,000	0,000	−0,020	0,006	0,046	0,083	0,149	0,196	0,226	0,184	0,130
0,000	0,000	0,000	0,000	0,001	0,033	0,075	0,108	0,167	0,205	0,229	0,182
0,000	0,000	0,000	0,000	0,000	0,045	0,076	0,114	0,134	0,182	0,218	0,230

Hence, the asymmetric filters for the last six data points result from the convolution of:

1. the asymmetric weights of an ARIMA(0,1,1) model with $\theta = 0.40$,
2. the weights of $\mathbf{M}_{7(0.14)}$ and $\mathbf{M}_{5(0.25)}$ filters repeatedly used for noise suppression, and
3. the weights of the final linear symmetric filter SLF.

8.2.2.1 Theoretical Properties of ALF

The convergence pattern of the asymmetric filters corresponding to H13 and ALF is shown in Figs. 8.2 and 8.3, respectively. It is evident that the ALF asymmetric filters are very close one another, and converge faster to the SLF relative to H13. The distance of each asymmetric filter with respect to the symmetric one gives an indication of the size of the revisions due to filter changes, when new observations are added to the series. Figure 8.4 shows that the gain function of the last point ALF does not amplify the signal as H13 and suppresses significantly the power at the frequency $\omega < 0.10$. From the viewpoint of the gain function, the ALF is superior to H13 concerning the unwanted ripples problem as well as in terms of faster convergence to the corresponding symmetric filter which implies smaller revisions.

On the other hand, the phase shift of the last point ALF is much greater (near 2 months at very low frequencies) relative to H13 as exhibited in Fig. 8.5. For the

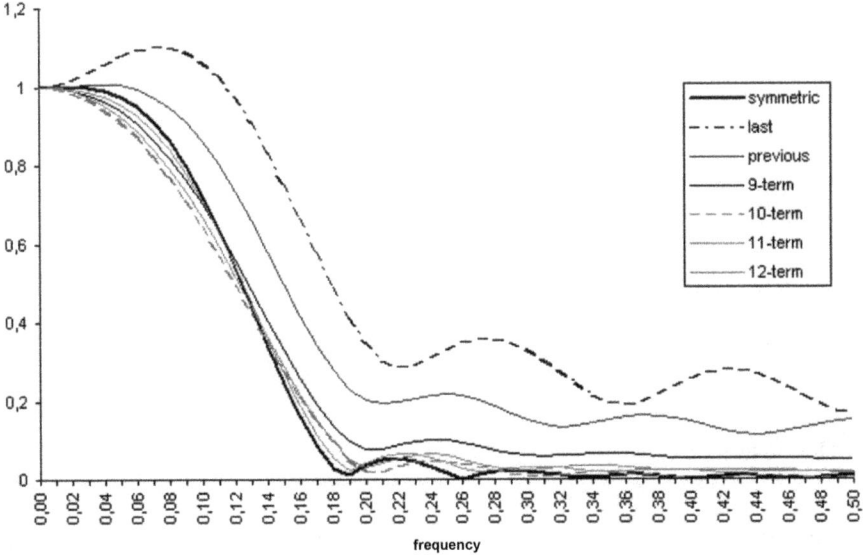

Fig. 8.2 Gain functions showing the convergence pattern of H13 asymmetric weights to the symmetric H13

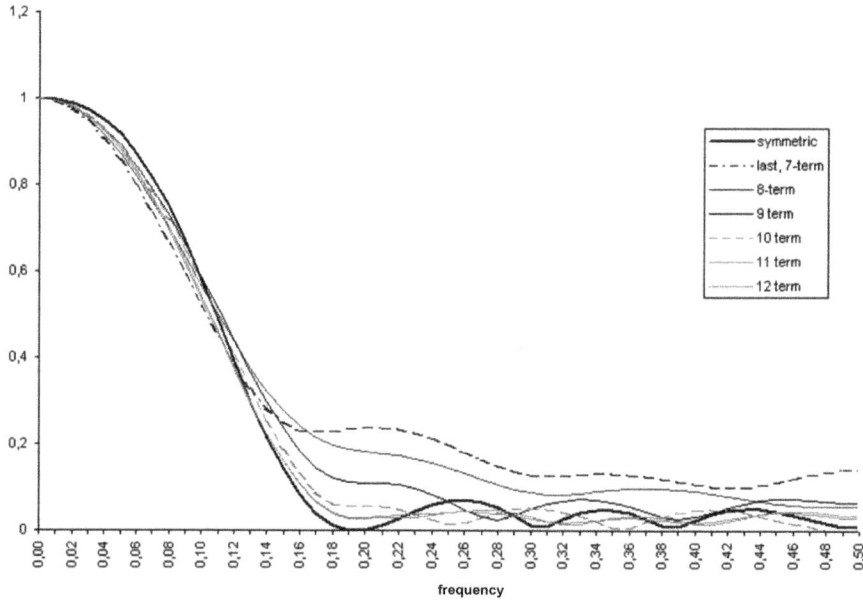

Fig. 8.3 Gain function showing the convergence pattern of the ALF asymmetric weights to the symmetric SLF

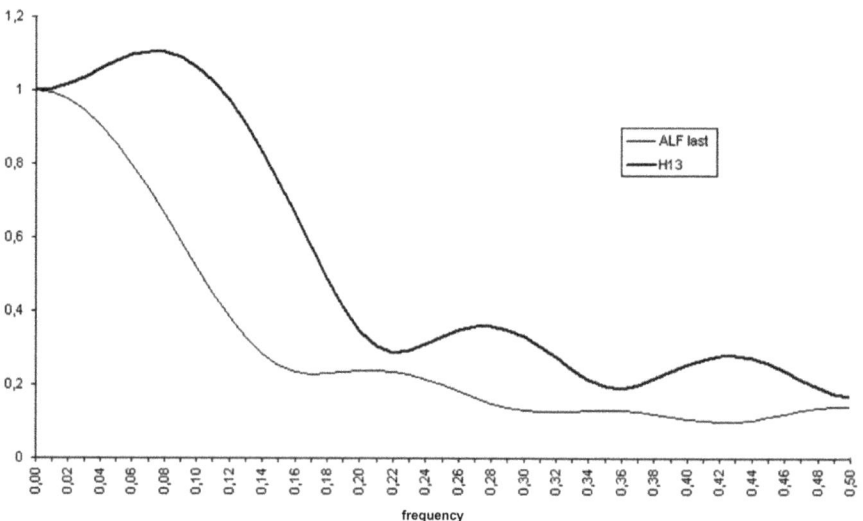

Fig. 8.4 Gain functions of the last point filter of H13 and the last point ALF

remaining asymmetric filters, the differences are much smaller. Nevertheless, the impact of the phase shift in a given input cannot be studied in isolation of the corresponding gain function. It is well-known that a small phase shift associated

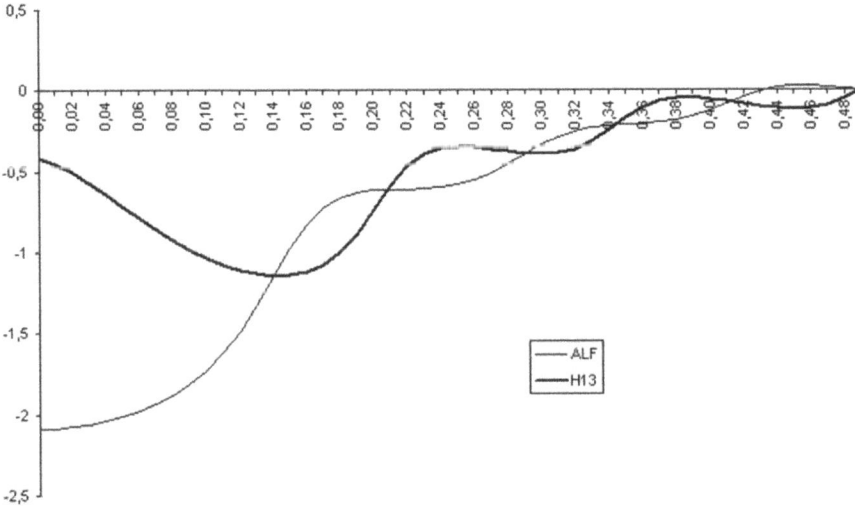

Fig. 8.5 Phase shifts of the last point filter of H13 and last point ALF

Table 8.2 Mean values of revisions of last and previous to the last point asymmetric filters of ALF and H13 applied to a large sample of real time series	Absolute size of revisions	ALF	H13
	$\lvert D_{s,l} \rvert$	0.87	1
	$D_{s,l}^2$	0.42	1
	Mean squared size of revisions		1
	$\lvert D_{s,p} \rvert$	0.60	1
	$D_{s,p}^2$	0.11	1

with frequency gain amplifications may produce as poor results as a much larger phase shift without frequency gain amplifications.

8.2.2.2 Empirical Properties of ALF

The performance of last and previous to the last point ALF was evaluated empirically using a large sample characterized by different degrees of variability. The purpose is to reduce the size of the revision of these ALF with respect to the asymmetric H13 filter. We denote with $\lvert D_{s,l} \rvert$ the mean absolute revision between the final estimates obtained with the symmetric filter, $\hat{y}_k^s$, and the last point estimate obtained with the asymmetric filter for the last point, $\hat{y}_k^l$, calculated over the whole sample. We denote with $\lvert D_{s,p} \rvert$ the mean absolute error between the previous $\hat{y}_k^p$ and the last point $\hat{y}_k^l$. We also calculate the corresponding mean square errors. The results are shown in Table 8.2 standardized by last point H13. It is evident that the size of revisions for the most recent estimates is smaller for ALF relative to last point asymmetric H13. This indicates a faster convergence to the symmetric filter, which was also shown theoretically by means of spectral analysis. Similarly, the distance between the last and previous to the last filters is smaller for ALF relative to H13.

In summary, the symmetric and asymmetric weights of the cascade filter are:

• Symmetric weights

  −0.027 −0.007 0.031 0.067 0.136 0.188 0.224 0.188 0.136 0.067 0.031 −0.007 −0.027

• Asymmetric weights

−0.026	−0.007	0.030	0.065	0.132	0.183	0.219	0.183	0.132	0.065	0.030	−0.006
0.000	−0.026	−0.007	0.030	0.064	0.131	0.182	0.218	0.183	0.132	0.065	0.031
0.000	0.000	−0.025	−0.004	0.034	0.069	0.137	0.187	0.222	0.185	0.131	0.064
0.000	0.000	0.000	−0.020	0.005	0.046	0.083	0.149	0.196	0.226	0.184	0.130
0.000	0.000	0.000	0.000	0.001	0.033	0.075	0.108	0.167	0.205	0.229	0.182
0.000	0.000	0.000	0.000	0.000	0.045	0.076	0.114	0.134	0.182	0.218	0.230

8.3   The Henderson Filter in the Reproducing Hilbert Space (RKHS)

The study of the properties and limitations of the Henderson filter has been done in different contexts and attracted the attention of a large number of authors, among them, Cholette [6], Kenny and Durbin [23], Castles [4], Dagum and Laniel [13], Cleveland et al. [7], Dagum [9], Gray and Thomson [21], Loader [25], Dalton and Keogh [17], Ladiray and Quenneville [24], Quenneville et al. [29], Findley and Martin [19], and Dagum and Luati [15, 16]. However, none of these studies have approached the Henderson smoother from a RKHS perspective. In this regard, in 2008 Dagum and Bianconcini [11] introduced a Reproducing Kernel Hilbert Space (RKHS) representation of the Henderson filter. The Henderson kernel representation enabled the construction of a hierarchy of kernels with varying smoothing properties. Furthermore, for each kernel order, the asymmetric filters could be derived coherently with the corresponding symmetric weights or from a lower or higher order kernel within the hierarchy, if more appropriate. In the particular case of the currently applied asymmetric Henderson filters, those obtained by means of the RKHS, coherent to the symmetric smoother, were shown to have superior properties from the viewpoint of signal passing, noise suppression, and revisions. A comparison was made with real life series.

8.3.1   Linear Filters in Reproducing Kernel Hilbert Spaces

Let $\{y_t, t = 1, 2, \ldots, n\}$ denote the input series. In this study, we work under the following (basic) specification for the input time series.

Assumption 1 *The input time series* $\{y_t, t = 1, 2, \ldots, n\}$, *that is seasonally adjusted or without seasonality, can be decomposed into the sum of a systematic component, called the signal (or nonstationary mean or trend-cycle)* TC_t, *plus an irregular component* I_t, *called the noise, such that*

$$y_t = TC_t + I_t. \tag{8.19}$$

The noise I_t *is assumed to be either a white noise,* $WN(0, \sigma_I^2)$, *or, more generally, to follow a stationary and invertible AutoRegressive Moving Average (ARMA) process.*

The trend-cycle can be deterministic or stochastic, and has a global or a local representation. It can be represented *locally* by a polynomial of degree p of a variable j which measures the distance between y_t and the neighboring observations y_{t+j}.

Assumption 2 *Given* ε_t *for some time point t, it is possible to find a local polynomial trend estimator*

$$TC_{t+j} = a_0 + a_1 j + \cdots + a_p j^p + \varepsilon_{t+j}, \quad j = -m, \ldots, m, \tag{8.20}$$

where $a_0, a_1, \ldots, a_p \in \mathbb{R}$ *and* ε_t *is assumed to be purely random and mutually uncorrelated with* I_t.

The coefficients $a_0, a_1, \ldots, a_p$ can be estimated by ordinary or weighted least squares or by summation formulae. The solution for $\hat{a}_0$ provides the trend-cycle estimate $\widehat{TC}_t(0)$, which equivalently consists in a weighted average applied in a moving manner, such that

$$\widehat{TC}_t(0) = \widehat{TC}_t = \sum_{j=-m}^{m} w_j y_{t+j}, \tag{8.21}$$

where $w_j, j < n$, denotes the weights to be applied to the observations y_{t+j} to get the estimate $\widehat{TC}_t$ for each point in time $t = 1, 2, \ldots, n$.

The weights depend on: (1) the degree of the fitted polynomial, (2) the amplitude of the neighborhood, and (3) the shape of the function used to average the observations in each neighborhood.

Once a (symmetric) span $2m + 1$ of the neighborhood has been selected, the w_j's for the observations corresponding to points falling out of the neighborhood of any target point are null or approximately null, such that the estimates of the $n - 2m$ central observations are obtained by applying $2m + 1$ symmetric weights to the observations neighboring the target point. The missing estimates for the first and last m observations can be obtained by applying asymmetric moving averages of variable length to the first and last m observations, respectively. The length of the moving average or time invariant symmetric linear filter is $2m + 1$, whereas the asymmetric linear filters length is time-varying.

Using the backshift operator B, such that $B^d y_t = y_{t-d}$, Eq. (8.21) can be written as

$$\widehat{TC}_t = \sum_{j=-m}^{m} w_j B^j y_t = W(B) y_t, \qquad t = 1, 2, \dots, n, \tag{8.22}$$

where $W(B)$ is a linear nonparametric estimator.

Definition 8.1 Given $p \geq 2$, $W(B)$ is a p-th order kernel if

$$\sum_{j=-m}^{m} w_j = 1, \tag{8.23}$$

and

$$\sum_{j=-m}^{m} j^i w_j = 0, \tag{8.24}$$

for some $i = 1, 2, \dots, p$. In other words, it will reproduce a polynomial trend of degree $(p - 1)$ without distortion.

A different characterization of a p-th order nonparametric estimator can be provided by means of the RKHS methodology.

A Hilbert space is a complete linear space with a norm given by an inner product. The space of square integrable functions, denoted by L^2, and the finite p-dimensional space $\mathbb{R}^p$ are those used in the study.

Assumption 3 *The time series $\{y_t, t = 1, 2, \dots, n\}$ is a finite realization of a family of square Lebesgue integrable random variables, i.e., $\int_{\mathbb{T}} | Y_t |^2 \, dt < \infty$. Hence, the random process $\{Y_t\}_{t \in \mathbb{T}}$ belongs to the space $L^2(\mathbb{T})$.*

The space $L^2(\mathbb{T})$ is a Hilbert space endowed with the inner product defined by

$$< U(t), V(t) >= E(U(t)V(t)) = \int_{\mathbb{T}} U(t)V(t)f_0(t)dt, \tag{8.25}$$

where $U(t)V(t) \in L^2(\mathbb{T})$, and f_0 is a probability density function, weighting each observation to take into account its position in time. In the following, $L^2(\mathbb{T})$ will be indicated as $L^2(f_0)$.

Under the Assumption 2, the local trend $TC_t(\cdot)$ belongs to the $\mathbb{P}_p$ space of polynomials of degree at most p, being p a nonnegative integer.

$\mathbb{P}_p$ is a Hilbert subspace of $L^2(f_0)$, hence it inherits its inner product property given by

$$< P(t), Q(t) >= \int_{\mathbb{T}} P(t)Q(t)f_0(t)dt, \tag{8.26}$$

where $P(t), Q(t) \in \mathbb{P}_p$.

Suppose that $\{y_t, t = 1, 2, \ldots, n\} \in L^2(f_0)$ can be decomposed as in Eq. (8.19), the estimate $\widehat{TC}_t$ of Eq. (8.20) can be obtained by minimizing the distance between y_{t+j} and $TC_t(j)$, that is,

$$\min_{TC \in \mathbb{P}_p} \|y - TC\|^2 = \min_{TC \in \mathbb{P}_p} \int_{\mathbb{T}} (y(t+s) - TC_t(s))^2 f_0(s)ds, \tag{8.27}$$

where the weighting function f_0 depends on the distance between the target point t and each observation within a symmetric neighborhood of points around t.

Theorem 8.1 *Under the Assumptions 1–3, the minimization problem (8.27) has a unique and explicit solution.*

Proof By the projection theorem (see, e.g., Priestley [28]), each element y_{t+j} of the Hilbert space $L^2(f_0)$ can be decomposed into the sum of its projection in a Hilbert subspace of $L^2(f_0)$, such as the space $\mathbb{P}_p$, plus its orthogonal complement as follows

$$y_{t+j} = \Pi_{\mathbb{P}_p}[y_{t+j}] + \{y_{t+j} - \Pi_{\mathbb{P}_p}[y_{t+j}]\} \tag{8.28}$$

where $\Pi_{\mathbb{P}_p}[y_{t+j}]$ denotes the projection of the observations $y_{t+j}, j = -m, \ldots, m$, on $\mathbb{P}_p$. By orthogonality, for every $j \in T$

$$< y_{t+j}, R_p(j, 0) >=< \Pi_{\mathbb{P}_p}[y_{t+j}], R_p(j, 0) >= \Pi_{\mathbb{P}_p}[y_t] = \widehat{TC}_t(0) = \widehat{TC}_t. \tag{8.29}$$

Thus, $\widehat{TC}_t(0)$ is given by

$$\widehat{TC}_t(0) = \int_{\mathbb{T}} y(t+s)R_p(s, 0)f_0(s)ds \tag{8.30}$$

$$= \int_{\mathbb{T}} \Pi_{\mathbb{P}_p}[y(t+s)]R_p(s, 0)f_0(s)ds, \tag{8.31}$$

where R_p is the reproducing kernel of the space $\mathbb{P}_p$.

Hence, the estimate $\widehat{TC}_t$ can be equivalently seen as the projection of y_t on $\mathbb{P}_p$ and as a local weighted average of the observations for the discrete version of the filter given in Eq. (8.22), where the weights w_j are derived by a kernel function K of order $(p + 1)$,

$$K_{p+1}(t) = R_p(t, 0)f_0(t), \qquad (8.32)$$

where p is the degree of the fitted polynomial.

The following result, proved by Berlinet [1], is fundamental.

Corollary 8.1 *Kernels of order* $(p + 1)$, $p \geq 1$, *can be written as products of the reproducing kernel* $R_p(t, .)$ *of the space* $\mathbb{P}_p \subseteq L^2(f_0)$ *and a density function* f_0 *with finite moments up to order* $2p$. *That is,*

$$K_{p+1}(t) = R_p(t, 0)f_0(t).$$

Remark 8.1 (Christoffel–Darboux Formula) For any sequence $(P_i)_{0 \leq i \leq p}$ of $(p + 1)$ orthonormal polynomials in $L^2(f_0)$,

$$R_p(t, 0) = \sum_{i=0}^{p} P_i(t)P_i(0). \qquad (8.33)$$

Therefore, Eq. (8.32) becomes

$$K_{p+1}(t) = \sum_{i=0}^{p} P_i(t)P_i(0)f_0(t). \qquad (8.34)$$

An important outcome of the RKHS theory is that linear filters can be grouped into hierarchies $\{K_p, p = 2, 3, 4, \ldots\}$ with the following property: each hierarchy is identified by a density f_0 and contains kernels of order 2, 3, 4, ... which are products of orthonormal polynomials by f_0.

The weight system of a hierarchy is completely determined by specifying: (a) the bandwidth or smoothing parameter, (b) the maximum order of the estimator in the family, and (c) the density f_0.

There are several procedures to determine the bandwidth or smoothing parameter (for a detailed discussion see, e.g., Berlinet and Devroye [2]). In this study, however, the smoothing parameter is not derived by data dependent optimization criteria, but it is fixed with the aim to obtain a kernel representation of the most often applied Henderson smoothers. Kernels of any length, including infinite ones, can be obtained with the above approach. Consequently, the results discussed can be easily extended to any filter length as long as the density function and its orthonormal polynomials are specified. The identification and specification of the density is one of the most crucial task for smoothers based on local polynomial fitting by weighted least squares, as LOESS and the Henderson smoothers. The density is related to the weighting penalizing function of the minimization problem.

It should be noticed that the RKHS approach can be applied to any linear filter characterized by varying degrees of fidelity and smoothness as described by Gray and Thomson [21]. In particular, applications to local polynomial smoothers are treated in Dagum and Bianconcini [10], and to smoothing splines in Wahba [31].

8.3.2 The Symmetric Henderson Smoother and Its Kernel Representation

Recognition of the fact that the smoothness of the estimated trend-cycle curve depends directly on the smoothness of the weight diagram led Henderson to develop a formula which makes the sum of squares of the third differences of the smoothed series a minimum for any number of terms.

Henderson's starting point was the requirement that the filter should reproduce a cubic polynomial trend without distortion. Henderson proved that three alternative smoothing criteria give the same formula, as shown explicitly by Kenny and Durbin [23] and Gray and Thomson [21]: (1) minimization of the variance of the third differences of the series defined by the application of the moving average; (2) minimization of the sum of squares of the third differences of the coefficients of the moving average formula; and (3) fitting a cubic polynomial by weighted least squares, where the weights are chosen as to minimize the sum of squares of their third differences.

The problem is one of fitting a cubic trend by weighted least squares to the observations $y_{t+j}, j = -m, \ldots, m$, the value of the fitted function at $j = 0$ being taken as the smoothed observation $\widehat{TC}_t$. Representing the weight assigned to the residuals from the local polynomial regression by $W_j, j = -m, \ldots, m$, where $W_j = W_{-j}$, the problem is the minimization of

$$\sum_{j=-m}^{m} W_j[y_{t+j} - a_0 - a_1 j - a_2 j^2 - a_3 j^3]^2, \tag{8.35}$$

where the solution for the constant term $\hat{a}_0$ is the smoothed observation $\widehat{TC}_t$.

Henderson showed that $\widehat{TC}_t$ is given by

$$\widehat{TC}_t = \sum_{j=-m}^{m} \phi(j) W_j y_{t+j} = \sum_{j=-m}^{m} w_j y_{t+j}, \tag{8.36}$$

where $\phi(j)$ is a cubic polynomial whose coefficients have the property that the filter reproduces the data if they follow a cubic. Henderson also proved the converse: if the coefficients of a cubic-reproducing summation formula $\{w_j, j = -m, \ldots, m\}$ do not change their sign more than three times within the filter span, then the formula can be represented as a local cubic smoother with weights $W_j > 0$ and a cubic polynomial $\phi(j)$, such that $\phi(j) W_j = w_j$.

Henderson measured the amount of smoothing of the input series by $\sum_{t=3}^{n}((1 - B)^3 y_t)^2$ or equivalently by the sum of squares of the third differences of the weight diagram, $\sum_{j=-m+3}^{m}((1 - B)^3 w_j)^2$. The solution is that resulting from the minimization of a cubic polynomial function by weighted least squares with

$$W_j \propto \{(m + 1)^2 - j^2\}\{(m + 2)^2 - j^2\}\{(m + 3)^2 - j^2\} \qquad (8.37)$$

as the weighting penalty function of criterion (3) above. The weight diagram $\{w_j, j = -m, \ldots, m\}$ corresponding to Eq. (8.37), known as Henderson's ideal formula, is obtained, for a filter length equal to $2m' - 3$, by

$$w_j = \frac{315[(m' - 1)^2 - j^2](m'^2 - j^2)[(m' + 1)^2 - j^2](3m'^2 - 16 - 11j^2)}{8m'(m'^2 - 1)(4m'^2 - 1)(4m'^2 - 9)(4m'^2 - 25)}. \qquad (8.38)$$

This optimality result has been rediscovered several times in the modern literature, usually for asymptotic variants. Loader [25] showed that the Henderson's ideal formula (8.38) is a finite sample variant of a kernel with second order vanishing moments which minimizes the third derivative of the function given by Muller [26]. In particular, Loader showed that for large m, the weights of Henderson's ideal penalty function W_j are approximately $m^6 W(j/m)$, where $W(j/m)$ is the triweight density function. He concluded that, for very large m, the weight diagram is approximately $(315/512) * W(j/m)(3 - 11(j/m)^2)$ equivalent to the kernel given by Muller [26].

To derive the Henderson kernel hierarchy by means of the RKHS methodology, the density corresponding to W_j and its orthonormal polynomials have to be determined. The triweight density function gives very poor results when the Henderson smoother spans are of short or medium lengths, as in most application cases, ranging from 5 to 23 terms. Hence, the exact density function corresponding to W_j was derived.

Theorem 8.2 *The exact probability density corresponding to the Henderson's ideal weighting penalty function (8.37) is given by*

$$f_{0H}(t) = \frac{(m + 1)}{k} W((m + 1)t), \qquad t \in [-1, 1] \qquad (8.39)$$

where $k = \int_{-m-1}^{m+1} W(j)dj$, and $j = (m + 1)t$.

Proof The weighting function $W(j)$ in Eq. (8.37), is nonnegative in the intervals $(-m - 1, m + 1)$, $(-m - 3, -m - 2)$, $(m + 2, m + 3)$ and negative otherwise. $W(j)$ is also equal to zero if $j = \pm(m + 1), \pm(m + 2), \pm(m + 3)$, and on $[-m - 1, m + 1]$, $W(j)$ is increasing in $[-m - 1, 0)$, decreasing in $(0, m + 1]$, and reaches its maximum at $j = 0$.

Therefore, the support chosen is $[-m - 1, m + 1]$ to satisfy the positive definite condition. The integral $k = \int_{-m-1}^{m+1} W(j)dj$ is different from 1 and represents the integration constant on this support. It follows that the density corresponding to

$W(j)$ on the interval $[-m-1, m+1]$ is given by

$$f_0(j) = W(j)/k.$$

To eliminate the dependence of the support on the bandwidth parameter m, a new variable ranging on $[-1, 1]$, $t = j/(m+1)$, is considered. Applying the change of variables method,

$$f_{0H}(t) = f_0(t^{-1}(j)) \left| \frac{\partial t^{-1}(j)}{\partial t} \right|,$$

where $t(j) = \frac{j}{m+1}$, and $t^{-1}(j) = (m+1)t$. The result follows by substitution.

The density $f_{0H}(t)$ is symmetric, i.e., $f_{0H}(-t) = f_{0H}(t)$, nonnegative on $[-1, 1]$, and is equal to zero when $t = -1$ or $t = 1$. Furthermore, $f_{0H}(t)$ is increasing on $[-1, 0)$, decreasing on $(0, 1]$, and reaches its maximum at $t = 0$.

For $m = 6$, the filter is the classical 13-term Henderson and the corresponding probability function results

$$f_{0H}(t) = \frac{15}{79376}(5184 - 12289t^2 + 9506t^4 - 2401t^6), \quad t \in [-1, 1]. \tag{8.40}$$

To obtain higher order kernels the corresponding orthonormal polynomials have to be computed for the density (8.39). The polynomials can be derived by the Gram–Schmidt orthonormalization procedure or by solving the Hankel system based on the moments of the density f_{0H}. This latter is the approach followed in this study. The hierarchy corresponding to the 13-term Henderson kernel is shown in Table 8.3, where for $p = 3$ it provides a representation of the classical Henderson filter.

Since the triweight density function gives a poor approximation of the Henderson weights for small m (5–23 terms), another density function with well-known theoretical properties was searched. The main reason is that the exact density (8.39) is function of the bandwidth and need to be calculated any time that m changes together with its corresponding orthonormal polynomials. The biweight density was found to give almost equivalent results to those obtained with the exact density function without the need to be calculated any time that the Henderson smoother length changes.

Table 8.3 13-Term Henderson kernel hierarchy

Henderson kernels	Kernel orders
$\frac{15}{79376}(5184 - 12289t^2 + 9506t^4 - 2401t^6)$	$p = 2$
$\frac{15}{79376}(5184 - 12289t^2 + 9506t^4 - 2401t^6) * \left(\frac{2175}{1274} - \frac{1372}{265}t^2 \right)$	$p = 3$

Another important advantage is that the biweight density function belongs to the well-known Beta distribution family, that is,

$$f(t) = \left(\frac{r}{2B(s+1,\frac{1}{r})} \right) (1 - |t|^r)^s I_{[-1,1]}(t), \qquad (8.41)$$

where $B(a,b) = \int_0^1 t^{a-1}(1-t)^{b-1}dt$ with $a, b > 0$ is the Beta function. The orthonormal polynomials needed for the reproducing kernel associated with the biweight function are the Jacobi polynomials, for which explicit expressions for computation are available and their properties have been widely studied in the literature.

Therefore, another Henderson kernel hierarchy is obtained using the biweight density

$$f_{0B}(t) = \frac{15}{16}(1 - t^2)^2 I_{[-1,1]}(t), \qquad (8.42)$$

combined with the Jacobi orthonormal polynomials. These latter are characterized by the following explicit expression:

$$P_n^{\alpha,\beta}(t) = \frac{1}{2^n} \sum_{m=0}^{n} \binom{n+\alpha}{m} \binom{n+\beta}{n-m} (t-1)^{n-m}(t+1)^m, \qquad (8.43)$$

where $\alpha = 2$ and $\beta = 2$. The Henderson second order kernel is given by the density function f_{0B}, since the reproducing kernel $R_1(j, 0)$ of the space of polynomials of degree at most one is always equal to one.

On the other hand, the third order kernel is given by

$$\frac{15}{16}(1 - |t|^2)^2 \times \left(\frac{7}{4} - \frac{21}{4}t^2 \right). \qquad (8.44)$$

Table 8.4 shows the classical and the two kernels (exact and biweight) Henderson symmetric weights for spans of 9, 13, and 23 terms, where the central weight values are given in bold.

The small discrepancy of the two kernel functions relative to the classical Henderson smoother is due to the fact that the exact density is obtained by interpolation from a finite small number of points of the weighting penalty function W_j.

On the other hand, the biweight is already a density function which is made discrete by choosing selected points to produce the weights. The smoothing measure $\sum_{j=-m+3}^{m}((1-B)^3 w_j)^2$ for each filter span is calculated as shown in Table 8.5.

The smoothing power of the filters is very close except for the exact 9-term Henderson kernel which gives the smoothest curve. Given the equivalence for symmetric weights, the RKHS methodology is used to generate the correspondent asymmetric filters for the m first and last points.

Table 8.4 Weight systems for 9-, 13-, and 23-term Henderson smoothers

Length	Henderson filter												
9	Classical							−0.041	−0.010	0.118	0.267	**0.331**
	Exact Kernel							−0.038	−0.005	0.119	0.262	**0.325**
	Biweight Kernel							−0.039	−0.011	0.120	0.266	**0.328**
13	Classical					−0.019	−0.028	0.000	0.065	0.147	0.214	**0.240**
	Exact Kernel					−0.019	−0.027	0.001	0.066	0.147	0.213	**0.238**
	Biweight Kernel					−0.020	−0.030	0.002	0.070	0.149	0.211	**0.234**
23	Classical	−0.004	−0.011	−0.016	−0.015	−0.005	0.013	0.039	0.068	0.097	0.122	0.138	**0.144**
	Exact Kernel	−0.004	−0.011	−0.016	−0.014	−0.005	0.014	0.039	0.068	0.097	0.122	0.138	**0.144**
	Biweight Kernel	−0.005	−0.014	−0.018	−0.014	−0.001	0.019	0.045	0.072	0.098	0.118	0.132	**0.137**

Table 8.5 Sum of squares of the third differences of the weights for the classical and kernel Henderson smoother

Filters	9-term	13-term	23-term
Classical Henderson smoother	0.053	0.01	0.00
Exact Henderson kernel	0.048	0.01	0.00
Biweight Henderson kernel	0.052	0.01	0.00

8.3.3 Asymmetric Henderson Smoothers and Their Kernel Representations

The asymmetric Henderson smoothers currently in use were developed by Musgrave [27]. The asymmetric weights of the Henderson kernels are derived by adapting the third order kernel functions to the length of the last m asymmetric filters such that,

$$w_{q,j} = \frac{K(j/b)}{\sum_{j=-m}^{q} K(j/b)}, \quad j = -m, \ldots, q; q = 0, \ldots, m-1, \tag{8.45}$$

where $K(\cdot)$ denotes the third order Henderson kernel, j the distance to the target point t, b the bandwidth parameter equal to $(m+1)$, and $(m+q+1)$ is the asymmetric filter length. For example, the asymmetric weights of the 13-term Henderson kernel for the last point $(q=0)$ are given by

$$w_{0,j} = \frac{K(j/7)}{\sum_{j=-6}^{0} K(i/7)}, \quad j = -6, \ldots, 0. \tag{8.46}$$

Figure 8.6 shows the gain functions of the symmetric Henderson filter together with the classical and kernel representation for the last point superposed on the X11/X12ARIMA Seasonal Adjustment (SA) symmetric filter. This latter results from the convolution of: (1) 12-term centered moving average, (2) 3×3 m.a., (3) 3×5 m.a., and (4) 13-term Henderson filter. Its property has been widely discussed in Dagum et al. [12]. The gain function of the SA filter should be interpreted as relating the spectrum of the original series to the spectrum of the estimated seasonally adjusted series.

It is apparent that the asymmetric Henderson kernel filter does not amplify the signal as the classical asymmetric one and converges faster to the final one. Furthermore, the asymmetric kernel suppresses more noise relative to the Musgrave filter. Since the weights of the biweight and exact Henderson kernels are equal up to the third digit, no visible differences are seen in the corresponding gain and phase shift functions. Hence, we only show those corresponding to the exact Henderson kernel in Figs. 8.6 and 8.7. There is an increase of the phase shift for the low frequencies relative to that of the classical H13 but both are less than a month, as exhibited in Fig. 8.7.

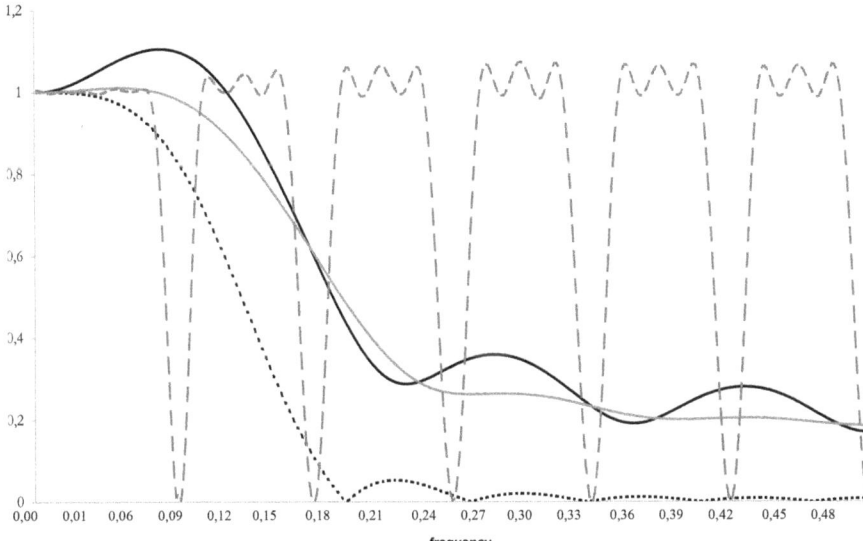

Fig. 8.6 Gain functions of the (last point) asymmetric 13-term filter (*continuous line*) and kernel (*gray continuous line*) with the corresponding symmetric (*dotted line*) and SA cascade filter (*gray dotted line*)

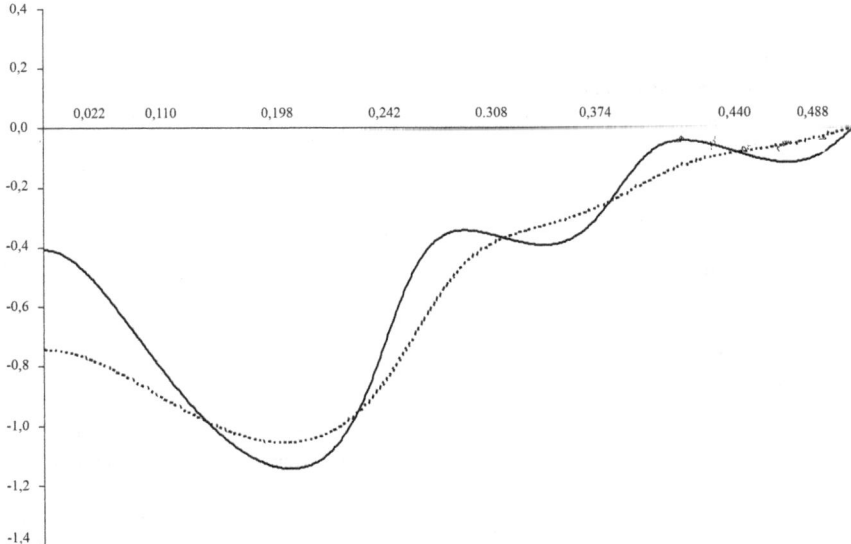

Fig. 8.7 Phase shifts of the asymmetric (end point) weights of the Henderson kernel (*dotted line*) and of the classical H13 filter (*continuous line*)

References

1. Berlinet, A. (1993). Hierarchies of higher order kernels. *Probability Theory and Related Fields, 94*, 489–504.
2. Berlinet, A., & Devroye, L. (1994). A comparison of kernel density estimates. *Publications de l'Institut de Statistique de l'Université de Paris, 38*, 3–59.
3. Box, G. E. P., & Jenkins, G. M. (1970). *Time series analysis: Forecasting and control*. San Francisco, CA, USA: Holden-Day (Second Edition 1976).
4. Castles, I. (1987). A guide to smoothing time series estimates of trend. Catalogue No. 1316. Australian Bureau.
5. Chhab, N., Morry, M., & Dagum, E. B. (1999). Results on alternative trend-cycle estimators for current economic analysis. *Estadistica, 49–51*, 231–257.
6. Cholette, P. A. (1981). A comparison of various trend-cycle estimators. In O. D. Anderson & M. R. Perryman (Eds.), *Time series analysis* (pp. 77–87). Amsterdam: North Holland.
7. Cleveland, R. B., Cleveland, W. S., McRae, J. E., & Terpenning, I. (1990). STL: A seasonal trend decomposition procedure based on LOESS. *Journal of Official Statistics, 6*(1), 3–33.
8. Dagum, E. B. (1988). *The X11ARIMA/88 seasonal adjustment method-foundations and user's manual*. Ottawa, Canada: Time Series Research and Analysis Centre, Statistics Canada.
9. Dagum, E. B. (1996). A new method to reduce unwanted ripples and revisions in trend-cycle estimates from X11ARIMA. *Survey Methodology, 22*, 77–83.
10. Dagum, E. B., & Bianconcini, S. (2006). Local polynomial trend-cycle predictors in reproducing kernel Hilbert spaces for current economic analysis. *Anales de Economia Aplicada*, 1–22.
11. Dagum, E. B., & Bianconcini, S. (2008). The Henderson smoother in reproducing kernel Hilbert space. *Journal of Business and Economic Statistics, 26*(4), 536–545.
12. Dagum, E. B., Chhab, N., & Chiu, K. (1996). Derivation and analysis of the X11ARIMA and Census X11 linear filters. *Journal of Official Statistics, 12*(4), 329–347.
13. Dagum, E. B., & Laniel, N. (1987). Revisions of trend-cycle estimators of moving average seasonal adjustment methods. *Journal of Business and Economic Statistics, 5*, 177–189.
14. Dagum, E. B., & Luati, A. (2000). Predictive performance of some nonparametric linear and nonlinear smoothers for noisy data. *Statistica, LX*(4), 635–654.
15. Dagum, E. B., & Luati, A. (2004). A linear transformation and its properties with special applications in time series filtering. *Linear Algebra and its Applications, 338*, 107–117.
16. Dagum, E. B., & Luati, A. (2009). A cascade linear filter to reduce revisions and turning points for real time trend-cycle estimation. *Econometric Reviews, 28*(1–3), 40–59.
17. Dalton, P., & Keogh, G. (2000). An experimental indicator to forecast turning points in the Irish business cycle. *Journal of the Statistical and Social Inquiry Society of Ireland, XXIX*, 117–176.
18. Darnè, O. (2002). La dèsaisonnalisation des chroniques économiques: analyse des conditions conjuncturelles (Ph.D. dissertation, Department of Economics, University of Montpellier 1, France).
19. Findley, D. F., & Martin, D. E. K. (2006). Frequency domain analysis of SEATS and X-11/X-12-ARIMA seasonal adjustment filters for short and moderate length time series. *Journal of Official Statistics, 22*, 1–34.
20. Findley, D. F., Monsell, B. C., Bell, W. R., Otto, M. C., & Chen, B. C. (1998). New capabilities and methods of the X12ARIMA seasonal adjustment program. *Journal of Business and Economic Statistics, 16*(2), 127–152.
21. Gray, A., & Thomson, P. (1996). Design of moving-average trend filters using fidelity and smoothness criteria. In P. M. Robinson & M. Rosenblatt (Eds.), *Time series analysis* (in memory of E.J. Hannan). Vol. II. Lecture notes in statistics (Vol. 115, pp. 205–219). New York: Springer.
22. Henderson, R. (1916). Note on graduation by adjusted average. *Transaction of Actuarial Society of America, 17*, 43–48.
23. Kenny, P., & Durbin, J. (1982). Local trend estimation and seasonal adjustment of economic and social time series. *Journal of the Royal Statistical Society A, 145*, 1–41.

24. Ladiray, D., & Quenneville, B. (2001). *Seasonal adjustment with the X-11 method*. Lecture notes in statistics (Vol. 158). New York: Springer.
25. Loader, C. (1999). *Local regression and likelihood*. New York: Springer.
26. Müller, H. G. (1984). Smooth optimum kernel estimators of regression curves, densities and modes. *Annals of Statistics, 12*, 766–774.
27. Musgrave, J. (1964). A set of end weights to end all end weights. Working Paper. U.S. Bureau of Census, Washington DC.
28. Priestley, M. B. (1981). *Spectral analysis and time series*. Probability and mathematical statistics. New York: Academic Press.
29. Quenneville, B., Ladiray, D., & Lefrancois, B. (2003). A note on Musgrave asymmetrical trend-cycle filters. *International Journal of Forecasting, 19*(4), 727–734.
30. Shiskin, J., Young, A. H., & Musgrave, J. C. (1967). The X-11 variant of the Census Method II seasonal adjustment program. Technical Paper 15 (revised). US Department of Commerce, Bureau of the Census, Washington, DC.
31. Wahba, G. (1990). *Spline models for observational data*. Philadelphia: SIAM.
32. Zellner, A., Hong, C., & Min, C. (1991). Forecasting turning points in international output growth rates using Bayesian exponentially weighted autoregression, time-varying parameter and pooling techniques. *Journal of Econometrics, 48*, 275–304.

Chapter 9
A Unified View of Trend-Cycle Predictors in Reproducing Kernel Hilbert Spaces (RKHS)

Abstract A common approach for trend-cycle estimation is to use nonparametric smoothing estimators which can be finite or infinite in length. The main methods presented in this chapter are based on different criteria of fitting and smoothing, namely: (1) density functions, (2) local polynomial fitting, (3) graduation theory, and (4) smoothing spline regression. A unified approach for all of these different trend-cycle nonparametric estimators is provided by means of the Reproducing Kernel Hilbert Space (RKHS) methodology. It is shown how nonparametric estimators can be transformed into kernel functions of order two, that are probability densities, and from which corresponding higher order kernels are derived. This kernel representation enables the comparison of estimators based on different smoothing criteria, and has important consequences in the derivation of the asymmetric filters which are applied to the most recent seasonally adjusted data for real time trend-cycle estimation.

In many scientific fields the observations are measured with error and in such cases the nonstationary mean of the time series is of great interest. Based on the assumption that a time series can be decomposed as a sum of a signal plus an erratic component, signal extraction deals with the problem of finding the "best" estimate of the nonstationary mean given the observations corrupted by noise.

If the time series is without seasonality or seasonally adjusted, the signal represents the trend and cyclical components, usually referred to as trend-cycle for they are estimated jointly in medium and short length series. The determination of a suitable inferential methodology will hinge on the assumptions made about the signal.

A common approach is to use nonparametric smoothing estimators which can be finite or infinite in length. They are based on different criteria of fitting and smoothing and are: (1) density functions, (2) local polynomial fitting, (3) graduation theory, and (4) smoothing spline regression.

A unified approach for all of these different nonparametric estimators is provided by means of the Reproducing Kernel Hilbert Space (RKHS) methodology.

E. Bee Dagum, S. Bianconcini, *Seasonal Adjustment Methods and Real Time Trend-Cycle Estimation*, Statistics for Social and Behavioral Sciences,
DOI 10.1007/978-3-319-31822-6_9

The main theory and systematic development of reproducing kernels and associated Hilbert spaces were laid out by Aronszajn [1], who showed that the properties of RKHS are intimately bounded up with properties of nonnegative definite functions. Parzen [26] was the first to apply these concepts to time series problems by means of a strictly parametric approach. From a nonparametric perspective, de Boor and Lynch [12] used this methodology in the context of cubic spline approximation. Later, Kimeldorf and Wahba [21, 22] proved that minimum norm interpolations and smoothing problems with quadratic constraints imply an equivalent Gaussian stochastic process. Recently, reproducing kernel methods have been prominent as a framework for penalized spline and quantile regression (see, e.g., Wahba [33]), and in the support vector machine literature, as described in [7, 14, 34], and [27].

In this chapter, we show how nonparametric estimators can be transformed into kernel functions of order two, that are probability densities, and from which corresponding higher order kernels are derived. The density function provides the initial weighting form from which the higher order kernels inherit their properties. This kernel representation enables the comparison of estimators based on different smoothing criteria, and has important consequences in the derivation of the asymmetric filters which are applied to the most recent seasonally adjusted data for real time trend-cycle estimation. In particular, those obtained by means of RKHS are shown to have superior properties from the view point of signal passing, noise suppression, and revisions relative to the currently applied that we call the classical ones.

The main methods presented here have been discussed in details by Dagum and Bianconcini [11], and we refer the readers to such study for further information.

9.1   Nonparametric Estimators in RKHS

Let $\{y_t, t = 1, 2, \ldots, n\}$ denote an n-dimensional vector of observed seasonally adjusted (or without seasonality) time series, specified as follows:

Assumption 1 *The time series* $\{y_t, t = 1, 2, \ldots, n\}$*, that is seasonally adjusted or without seasonality, can be decomposed as the sum of a systematic component, called the signal* TC_t *(or nonstationary mean or trend-cycle), plus an erratic component* I_t*, called the noise, such that*

$$y_t = TC_t + I_t, \tag{9.1}$$

where I_t *is purely random or follows an ARMA process of low order.*

Assumption 2 *The time series* $\{y_t, t = 1, 2, \ldots, n\}$ *is a finite realization of a stochastic process, whose trajectories* $\{Y(t), t \in \mathbb{T}\}$ *are square Lebesgue integrable functions with respect to a density function* f_0 *defined on* $\mathbb{T}$ *(e.g.,* $\mathbb{T} = [-1, 1], [0, +\infty),$ *or* $(-\infty, \infty)$*) that is,* $\int_{\mathbb{T}} |Y(t)|^2 f_0(t) dt < +\infty.$

In other words, $\{Y(t), t \in \mathbb{T}\}$ belong to the Hilbert space $L^2(f_0)$ endowed with the inner product defined by

$$< Y_1, Y_2 >_{L^2(f_0)} = \int_{\mathbb{T}} Y_1(t)Y_2(t)f_0(t)dt \quad \forall Y_1, Y_2 \in L^2(f_0). \tag{9.2}$$

f_0 is a density function of time, that is a nonnegative definite function integrating to unity over the time domain $\mathbb{T}$. It weights each value to take into account its temporal position. In particular, f_0 can be seen as a probabilistic density function since it allows to associate to each value in time a measure of its reliability as a function of its distance to the time point of interest.

Assumption 3 *The signal TC is globally, i.e., on the whole time span, a smooth function of time.*

Under Assumption 3, in virtue of the Taylor's expansion, TC can be locally approximated by a polynomial function, say of degree p, of the time distance j between y_t and its neighboring values y_{t+j}, such that

$$\mathrm{TC}_{t+j} = a_0 + a_1 j + \cdots + a_p j^p + \varepsilon_{t+j}, \quad j = -m, \ldots, m, \tag{9.3}$$

where $a_0, a_1, \ldots, a_p \in \mathbb{R}$, and ε_t is a random component assumed to be mutually uncorrelated with I_t. The size of the neighborhood of values around t taken into account in the estimation of the polynomial coefficients $a_0, a_1, \ldots, a_p$ is equal to $2m + 1$.

Given Eq. (9.3), the analysis of the signal can be performed in the space of polynomials of degree at most p, $\mathbb{P}_p \subseteq L^2(f_0)$, being p a nonnegative integer.

The coefficients $a_0, a_1, \ldots, a_p$ of the local polynomial trend-cycle are estimated by projecting the values in a neighborhood of y_t on the subspace $\mathbb{P}_p$, or equivalently by minimizing the weighted least square fitting criterion

$$\min_{\mathrm{TC} \in \mathbb{P}_p} \|y - \mathrm{TC}\|_{\mathbb{P}_p}^2 = \int_{\mathbb{T}} (y(t-s) - \mathrm{TC}(t-s))^2 f_0(s)ds, \tag{9.4}$$

where $\|\cdot\|_{\mathbb{P}_p}^2$ denotes the $\mathbb{P}_p$-norm. The weighting function f_0 depends on the distance between the target point t and each observation within a symmetric neighborhood of points around t.

The solution for $\hat{a}_0$ provides the estimate $\widehat{\mathrm{TC}}_t$, for which a general characterization and explicit representation can be provided by means of the RKHS methodology.

Definition 9.1 Given a Hilbert space $\mathscr{H}$, the reproducing kernel R is a function $R : \mathbb{T} \times \mathbb{T} \to \mathbb{R}$, satisfying the following properties:

1. $R(t, .) \in \mathscr{H}, \forall t \in \mathbb{T}$;
2. $< g(.), R(t, .) >= g(t), \forall t \in \mathbb{T}$ and $g \in \mathscr{H}$.

Condition 2 is called the *reproducing property*: the value of the function TC at the point t is reproduced by the inner product of TC with $R(t, .)$. $R(t, .)$ is called the *reproducing kernel* since

$$< R(t, .), R(., s) >= R(t, s). \tag{9.5}$$

Lemma 9.1 *The space* $\mathbb{P}_p$ *is a reproducing kernel Hilbert space of polynomials on some domain* $\mathbb{T}$, *that is, there exists an element* $R_p(t, .) \in \mathbb{P}_p$, *such that*

$$P(t) =< P(.), R_p(t, .) > \quad \forall t \in T \quad and \quad \forall P \in \mathbb{P}_p$$

The proof easily follows by the fact that any finite dimensional Hilbert space has a reproducing kernel [3].

Theorem 9.1 *Under the Assumptions 1–3, the minimization problem (9.4) has a unique and explicit solution given by*

$$\widehat{TC}(t) = \int_{\mathbb{T}} y(t - s)K_{p+1}(s)ds, \tag{9.6}$$

where K_{p+1} *is a kernel function of order* $p + 1$. *Note that for* $p \geq 2$, K_p *is said to be of order* p *if*

$$\int_{\mathbb{T}} K_p(s)ds = 1, \quad and \quad \int_{\mathbb{T}} s^i K_p(s)ds = 0, \tag{9.7}$$

for $i = 1, 2, \ldots, p-1$. *In other words, it will reproduce a polynomial trend of degree* $p - 1$ *without distortion.*

Proof By the projection theorem [29], each element of the Hilbert space $L^2(f_0)$ can be decomposed into the sum of its projection in a Hilbert subspace of $L^2(f_0)$, such as the space $\mathbb{P}_p$, plus its orthogonal complement as follows:

$$y(t - s) = \Pi_{\mathbb{P}_p}[y(t - s)] + \{y(t - s) - \Pi_{\mathbb{P}_p}[y(t - s)]\}, \tag{9.8}$$

where $\Pi_{\mathbb{P}_p}[y(t - s)]$ denotes the projection of the data $y(t - s)$ on $\mathbb{P}_p$. By orthogonality,

$$\widehat{TC}(t) = \Pi_{\mathbb{P}_p}[y(t)] =< \Pi_{\mathbb{P}_p}[y(t-s)], R_p(s, 0) >=< y(t-s), R_p(s, 0) > . \tag{9.9}$$

Thus, $\widehat{TC}(t)$ is given by

$$\widehat{TC}(t) = \int_{\mathbb{T}} \Pi_{\mathbb{P}_p}[y(t - s)]R_p(s, 0)f_0(s)ds \tag{9.10}$$

$$= \int_{\mathbb{T}} y(t - s)R_p(s, 0)f_0(s)ds, \tag{9.11}$$

where R_p is the reproducing kernel of the space $\mathbb{P}_p$.

Hence, the estimate $\widehat{TC}(t)$ can be seen as a local weighted average of the seasonally adjusted values, where the weights are derived by a kernel function K of order $p+1$, where p is the degree of the fitted polynomial.

Next we make use of the two following fundamental results.

First, as shown by Berlinet [2], kernels of order $(p+1)$, $p \geq 1$, can be written as products of the reproducing kernel $R_p(t, .)$ of the space $\mathbb{P}_p \subseteq L^2(f_0)$ and a density function f_0 with finite moments up to order $2p$. That is,

$$K_{p+1}(t) = R_p(t, 0)f_0(t). \tag{9.12}$$

Second, using the Christoffel–Darboux formula, for any sequence $(P_i)_{0 \leq i \leq p}$ of $(p + 1)$ orthonormal polynomials in $L^2(f_0)$, we have

$$R_p(t, 0) = \sum_{i=0}^{p} P_i(t)P_i(0). \tag{9.13}$$

Therefore, Eq. (9.12) becomes

$$K_{p+1}(t) = \sum_{i=0}^{p} P_i(t)P_i(0)f_0(t). \tag{9.14}$$

Applied to seasonally adjusted data or data without seasonality, the kernel acts as a locally weighted average or linear filter that for each target point t gives the estimate

$$\widehat{TC}_t = \sum_{i=1}^{n} w_{t,i}y_i, \quad t = 1, 2, \dots, n, \tag{9.15}$$

where $w_{t,i}$ denotes the weights to be applied to y_i to get the estimate $\widehat{TC}_t$ for each point in time t. Once a symmetric span $2m+1$ of the neighborhood has been selected, the weights for the data corresponding to points falling out the neighborhood of any target point are null or approximately null. Hence, the estimates of the $n-2m$ central values are obtained by applying $2m+1$ symmetric weights to the values neighboring the target point. That is,

$$\widehat{TC}_t = \sum_{j=-m}^{m} w_j y_{t+j} \quad t = m + 1, \dots, n - m.$$

The weights $w_j, j = -m, \dots, m$, depend on the shape of the nonparametric estimator K_{p+1} and on the value of a bandwidth parameter b fixed to ensure a neighborhood amplitude equal to $2m + 1$, such that

$$w_j = \frac{K_{p+1}\left(\frac{j}{b}\right)}{\sum_{j=-m}^{m} K_{p+1}\left(\frac{j}{b}\right)}, \qquad j = -m, \ldots, m. \qquad (9.16)$$

On the other hand, the missing estimates for the first and last m data points can be obtained by applying asymmetric linear filters of variable lengths.

Several nonparametric estimators have been developed in the literature for time series smoothing. One approach based on least squares includes: (1) kernel estimators, (2) local polynomial fitting, and (3) graduation theory. A second approach corresponds to smoothing spline regression. Smoothing splines introduce a roughness penalty term in the minimization problem (9.4), searching for an optimal solution between both fitting and smoothing of the data. This requires an adapted RKHS. Here it is shown how an equivalent kernel representation for the smoothing spline can be derived in the polynomial space $\mathbb{P}_p$.

A unified perspective is provided for all of these different nonparametric estimators according to which they are transformed into kernel functions. This enables comparisons among smoothers of different order within the same hierarchy as well as kernels of the same order, but belonging to different hierarchies. Therefore, for every estimator the density function f_0 and the corresponding reproducing kernel are identified.

9.1.1   Polynomial Kernel Regression

Local kernel regression deals with the problem of fitting a polynomial trend to $y_{t+j}, j = -m, \ldots, m$, where the value of the fitted function at $j = 0$ defines the smoothed observation $\widehat{TC}_t$. Representing the weights assigned to the residuals from the local polynomial regression by a symmetric and nonnegative function K_0, the problem is the minimization of

$$\sum_{i=1}^{n} K_0\left(\frac{t-i}{b}\right)\left[y_t - a_0 - a_1(t-i) - \cdots - a_p(t-i)^p\right]^2, \qquad (9.17)$$

where the parameter b determines the bandwidth of the weighting function.

Kernel estimators, local polynomial regression smoothers, and filters derived by the graduation theory differ in the degree of the fitted polynomial, in the shape of the weighting function, and the neighborhood of the values taken into account. We shall discuss here the properties of the most often applied trend-cycle smoothers, namely

Table 9.1 Second and third order kernel within the Gaussian, tricube (LOESS) and biweight (Henderson) hierarchies

Estimator	Density		Third order kernel
Gaussian kernel	$f_{0G}(t) = \frac{1}{\sqrt{2\pi}} \exp\left(-\frac{t^2}{2}\right)$	$t \in (-\infty, \infty)$	$K_{3G}(t) = \frac{1}{\sqrt{2\pi}} \exp\left(-\frac{t^2}{2}\right) \times \left(\frac{3-t^2}{2}\right)$
LOESS	$f_{0T}(t) = \frac{70}{81}\left(1 - |t|^3\right)^3 I_{[-1,1]}(t),$		$K_{3T}(t) = \frac{70}{81}\left(1 - |t|^3\right)^3 \left(\frac{539}{293} - \frac{3719}{638}t^2\right)$
Henderson filter	$f_{0B}(t) = \frac{15}{16}(1 - t^2)^2 I_{[-1,1]}(t).$		$K_{3B}(t) = \frac{15}{16}(1 - |t|^2)^2 \times \left(\frac{7}{4} - \frac{21}{4}t^2\right)$

the Gaussian kernel, the LOESS estimator [5], and the Henderson filter [19]. The density function and third order kernel for each estimator are shown in Table 9.1. We refer the reader to [4, 9, 10], for a detailed discussion on how these density functions and reproducing kernels have been derived.

9.1.2 Smoothing Spline Regression

Extending previous work by Whittaker [35] and Whittaker and Robinson [36], Schoenberg [30] showed that a natural smoothing spline estimator of order ℓ arises by minimizing the loss function

$$\min_{\mathrm{TC} \in W_2^\ell(\mathbb{T})} \|y - \mathrm{TC}\|_{W_2^\ell}^2 = \int_\mathbb{T} (y(t) - \mathrm{TC}(t))^2 \, dt + \lambda \int_\mathbb{T} \left(\mathrm{TC}^{(\ell)}(t)\right)^2 dt, \quad (9.18)$$

where $\mathrm{TC}^{(\ell)}$ denotes the ℓ-th derivative of the function TC, and W_2^ℓ is the Sobolev space of functions TC satisfying the following properties: (1) $\mathrm{TC}^{(i)}, i = 1, \ldots, \ell-1$, are absolutely continuous, (2) $\mathrm{TC}^{(\ell)} \in L^2(\mathbb{T})$.

The parameter λ regulates the balance between the goodness of fit and the smoothness of the curve. As $\lambda \to 0$ the solution approaches an interpolating spline, whereas as $\lambda \to \infty$, the solution tends to the least squares line.

Kimeldorf and Wahba [21, 22] showed that the minimizer of Eq. (9.18) is a polynomial spline of degree $p = 2\ell-1$ with knots at the data points. It is determined by the unique Green's function $G_\lambda(t, s)$, such that

$$\widehat{\mathrm{TC}}(t) = \int_\mathbb{T} G_\lambda(t, s) y(s) ds. \quad (9.19)$$

The derivation of $G_\lambda(t, s)$ corresponding to a smoothing spline of order ℓ requires the solution of a $2\ell \times 2\ell$ system of linear equations for each value of λ.

A simplification is provided by studying $G_\lambda(t, s)$ as the reproducing kernel $R_{\ell,\lambda}(t, s)$ of the Sobolev space $W_2^\ell(\mathbb{T})$, where $\mathbb{T}$ is an open subset of $\mathbb{R}$. When $\mathbb{T} = \mathbb{R}$, the space $W_2^\ell(\mathbb{T})$ falls into the family of Beppo-Levi spaces described in Thomas-Agnan [32]. The corresponding reproducing kernel is translation invariant, and can be expressed in terms of $R_{\ell,1}$ as follows:

$$R_{\ell,\lambda}(t,s) = R_{\ell,\lambda}(t-s) = \frac{1}{\lambda} R_{\ell,1} \left(\frac{t-s}{\lambda} \right). \tag{9.20}$$

A general formula for $R_{\ell,1}$, denoted as R_ℓ, is given by [32],

$$R_\ell(t) = \sum_{k=0}^{\ell} \frac{\exp\left(-|t|e^{i\frac{\pi}{2\ell} + k\frac{\pi}{\ell} - \frac{\pi}{2}}\right)}{2\ell e^{(2\ell-1)\left(i\frac{\pi}{2\ell} + i\frac{k\pi}{\ell}\right)}}, \quad \ell = 1,2,3\ldots \tag{9.21}$$

Equation (9.21) describes an equivalent kernel hierarchy for smoothing splines of order ℓ, $\ell = 1, 2, \ldots$. The standard Laplace density R_1 is the second order kernel when $\ell = 1$, and hence $p = 1$. Higher order kernels are derived by multiplying R_1 by a combination of trigonometric polynomials which take into account for the roughness penalty term in (9.18). The third order smoother R_2, obtained by selecting $\ell = 2$, hence $p = 3$, is familiar to the nonparametric statisticians since it is the asymptotically equivalent kernel to the cubic smoothing spline derived by Silverman [31]. It should be noticed, however, that when the neighborhood of points for the estimation is small, as in most socioeconomic cases, Eq. (9.21) poorly represents the linear approximation of the classical cubic smoothing spline, described in terms of the *influential matrix* $\mathbf{A}(\lambda)$ [18, 33]. $\mathbf{A}(\lambda)$ relates the estimated values $\widehat{TC}' = (\widehat{TC}_1, \widehat{TC}_2, \ldots, \widehat{TC}_n)$ to the values $y' = (y_1, y_2, \ldots, y_n)$ as follows:

$$\widehat{TC} = \mathbf{A}(\lambda)y. \tag{9.22}$$

Each $\widehat{TC}_t$ is a weighted linear combination of all the values y_t, with weights given by the elements of the t-th row of $\mathbf{A}(\lambda)$. In this study, we assume λ as given, and thus approximate each cubic spline predictor with time invariant linear filters. This enables us to analyze the properties of the estimators looking at the corresponding transfer functions. To obtain a reproducing kernel representation of smoothing splines coherent with that derived for local kernel regression estimators, we have to find a density function f_0 according to which higher order kernels are obtained via multiplication of f_0 with corresponding orthonormal polynomials. This density has to take into account for the smoothing term $\lambda \int_{\mathbb{T}} (g^{(\ell)}(u))^2 du$ in Eq. (9.18).

Starting from the results of Eq. (9.21), we first consider the standard Laplace density multiplied by the corresponding orthonormal polynomials. To evaluate the goodness of this approximation we calculate the Euclidean distance Δ between the linear approximation of the classical cubic smoothing spline (CSS) and the corresponding third order kernel within each hierarchy, denoted by K, both in terms of weights and gain functions. That is,

$$\Delta_{\text{weights}} = \sqrt{\sum_{j=-m}^{m} |w_{\text{CSS}j} - w_{Kj}|^2} \qquad \Delta_{\text{gain}} = \sqrt{\sum_{\omega=0}^{1/2} |G_{\text{CSS}}(\omega) - G_K(\omega)|^2},$$

Table 9.2 2-Norm distances between classical and reproducing kernel smoothing splines

Third order kernel	Filter length					
	9-term		13-term		23-term	
	Weights	Gain function	Weights	Gain function	Weights	Gain function
Sobolev space R_2	0.144	3.213	0.428	1.925	0.482	1.709
Standard Laplace	0.049	1.088	0.608	3.863	0.583	2.916
Logistic $\beta = 0.2$	0.021	0.480	0.271	0.588	0.260	0.471

where ω denotes the frequency in cycles per unit of time and $G(\omega)$ is the gain of the filter. For illustrative purposes, we compute these measures for filter spans generally applied to monthly time series, that is, 9-, 13-, and 23-term, even if filter of any length can be considered. Table 9.2 shows that the third order standard Laplace kernel presents large discrepancies for each span indicating that asymptotically it does not approach to the CSS. On the other hand, the equivalent kernel representation derived in the Sobolev space R_2 provides the worst approximation for the shortest filter length, but it has a better performance as the span increases.

Given the poor performance of the standard Laplace estimators, we looked for another density function. To be coherent with the results of Eq. (9.21), we considered the $\log F_{2m_1, 2m_2}$ distribution family introduced by Prentice [28] which includes the standard Normal ($m_1, m_2 \to 0$) and Laplace ($m_1, m_2 \to \infty$) as limiting cases. Many densities belong to this class of distributions, among others, the logistic ($m_1 = m_2 = 1$) and the exponential ($m_1 \neq 0, m_2 \to 0$). We concentrate on the former given its strong connection with the standard Laplace and other widely applied density functions, as recently shown by Lin and Hu [23]. These authors modified and extended previous work by Mudholkar and George [24], providing a characterization of the logistic density in terms of sample median and Laplace distribution, as well as in terms of the smallest order statistics and exponential density [20]. Furthermore, George and Ojo [16], and George et al. [15] studied the logistic distribution as approximation of Student's t functions. The logistic density is defined as

$$f_{0L}(t) = \left(\frac{1}{4\beta}\right)^{-1} \text{sech}^2 \left[\frac{1}{2}\left(\frac{t-\alpha}{\beta}\right)\right], \quad t \in (-\infty, \infty), \tag{9.23}$$

where α is the mean, set equal to 0, $\beta > 0$ is the dispersion parameter, and sech is the hyperbolic secant function.

Table 9.2 shows the results for the third order kernel within the hierarchy derived by f_{0L}, where the dispersion parameter has been set equal to 0.2. It is given by

$$K_{3L}(t) = \frac{5}{4}\text{sech}^2 \left(\frac{5}{2}t\right)\left(\frac{21}{16} - \frac{2085}{878}t^2\right). \tag{9.24}$$

This kernel representation closely approximates the CSS as shown in Table 9.2.

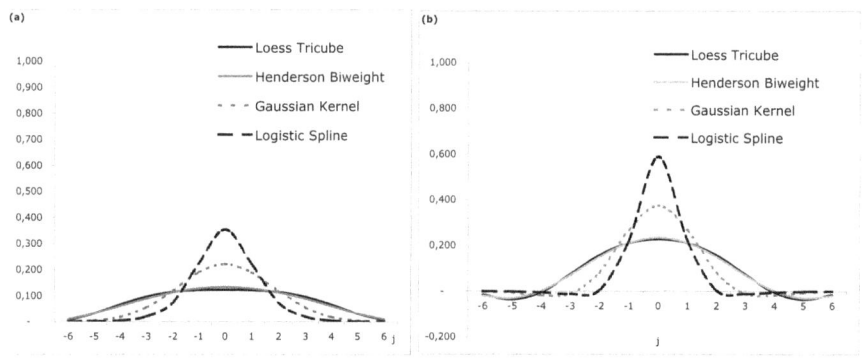

Fig. 9.1   13-Term (**a**) second and (**b**) third order kernels

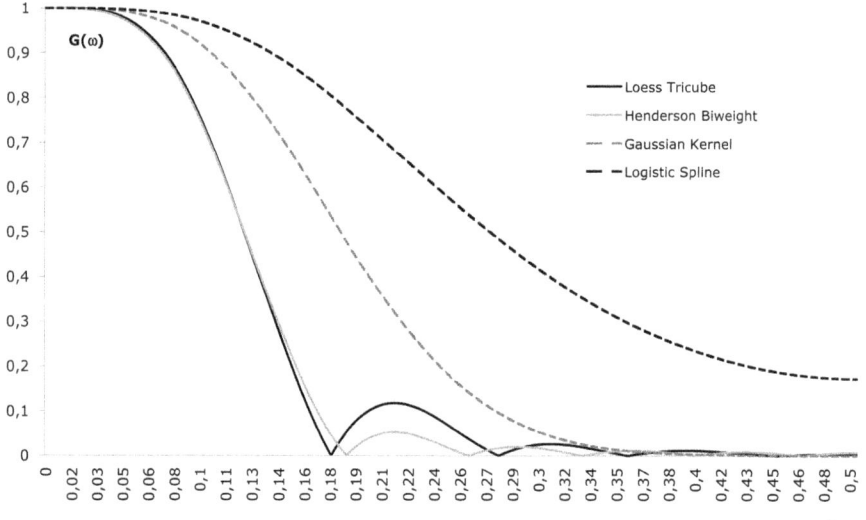

Fig. 9.2   Gain functions of symmetric 13-term third order kernels

The logistic kernel representation of smoothing splines enables to compare directly all the derived hierarchies. Figure 9.1a shows the density functions or second order kernels within each family, whereas Fig. 9.1b illustrates the third order smoothers. The 13-term filters are exhibited because the logistic kernel representation of smoothing splines enables to compare directly in the time domain all the hierarchies we have considered. But similar results are derived for 9- and 23-term smoothers. These hierarchies reproduce and describe several temporal dynamics by predicting polynomial trends of different degrees that solve several minimization problems. This is shown in Fig. 9.2, where we illustrate the gain functions of 13-term third order kernels.

LOESS and Henderson kernels present similar properties in terms of trend-cycle estimation. They eliminate a large amount of noise and pass all the power associated to the signal frequency band, here defined by $0 < \omega \leq 0.06$. However, they do not suppress power at the frequency $\omega = 0.10$, corresponding to cycles of 10 months, known as unwanted ripples. On the other hand, the third order Gaussian and spline kernels lose part of their optimality prediction properties, leaving untouched the signal but passing a lot of noise including a large number of unwanted ripples.

9.2 Boundary Behavior

The kernels derived by means of the RKHS methodology provide a unified view of presently used nonparametric trend-cycle estimators. The third order kernels in the tricube, biweight, and logistic hierarchies are equivalent kernels for the classical LOESS of degree 2 (LOESS 2), Henderson filter, and cubic smoothing spline, respectively. No comparisons can be made for the third order Gaussian estimator which is already a kernel function, and for which no counterpart exists in the literature.

The reproducing kernel representation has important consequences in the derivation of the corresponding asymmetric smoothers, which are of crucial importance in current analysis where the aim is to obtain the estimate of the trend-cycle for the most recent seasonally adjusted values. We derived the asymmetric smoothers by adapting the kernel functions to the length of the last m asymmetric filters, such that

$$w_{qj} = \frac{K_3\left(\frac{j}{b}\right)}{\sum_{j=-m}^{q} K_3\left(\frac{j}{b}\right)} \qquad j = -m, \ldots, q; \quad q = 0, \ldots, m-1,$$

where j denotes the distance to the target point t ($t = n - m + 1, \ldots, n$), b is the bandwidth parameter selected in view of ensuring a symmetric filter of length $2m + 1$, and $m + q + 1, q = 0, \ldots, m - 1$, is the asymmetric filter length.

These asymmetric filters only satisfy the condition $\sum_{j=-m}^{q} w_{qj} = 1$, hence will reproduce without distortion only a constant on the asymmetric support. Given the small number of points generally available in the boundaries, this condition can be considered sufficient in view of obtaining a better performance of the kernel in terms of mean square error.

The main goal is to analyze the goodness of the reproducing kernel representations versus the classical last point asymmetric filters.

Cleveland [5] showed that, in the middle of the series, LOESS acts as a symmetric moving average with window length $2m + 1$. At the end of the series, its window length remains $2m + 1$, rather than decreasing to $m + 1$ as in the case of the most widely applied asymmetric last point trend-cycle estimators. As discussed by Gray and Thomson [17], this implies a heavier than expected smoothing at both ends

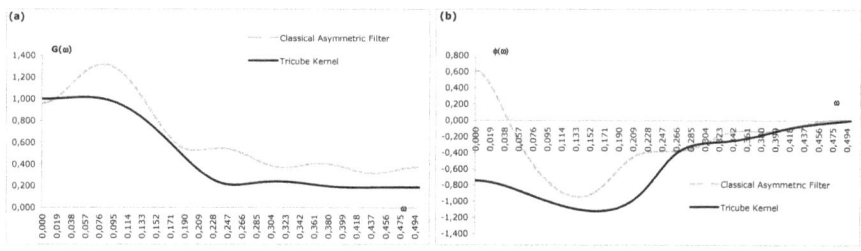

Fig. 9.3 (**a**) Gain and (**b**) phase shift functions of the asymmetric (end point) weights of the third order tricube kernel and the classical LOESS 2

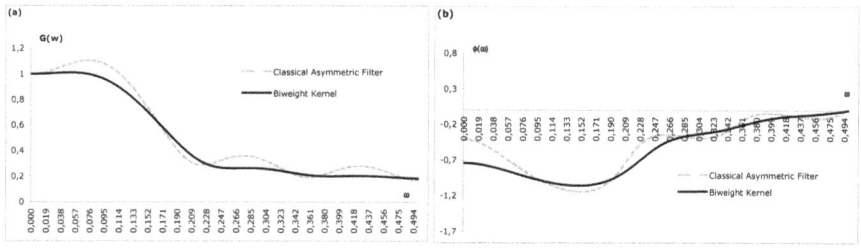

Fig. 9.4 (**a**) Gain and (**b**) phase shift functions of the asymmetric (end point) weights of the third order biweight kernel and the classical Henderson filter

of the series with respect to the middle, and represents a drawback, particularly for economic time series where turning points are important to be identified. As shown in Fig. 9.3a, the last point LOESS asymmetric kernel exhibits a gain function with better properties of signal passing and noise suppression relative to the classical one. This implies smaller filter revisions as new data are added to the series. The phase shifts for both filters, shown in Fig. 9.3b, are smaller than one month in the signal frequency band $0 < \omega \leq 0.06$, usually assumed to correspond to the trend-cycle component.

Similar conclusions can be derived for the last point asymmetric Henderson filter which were developed by Musgrave [25], as shown in Fig. 9.4. They are based on the minimization of the mean squared revisions between the final estimates (obtained by the application of the symmetric filter) and the preliminary ones (obtained by the application of an asymmetric filter) subject to the constraint that the sum of the weights is equal to one [13]. The assumption made is that at the end of the series, the seasonally adjusted values follow a linear trend-cycle plus a purely random irregular ε_t, such that $\varepsilon_t \sim NID(0, \sigma^2)$.

The asymmetric filters for the natural cubic smoothing splines are obtained by adding additional constraints, ensuring that the function is of degree one beyond the boundary knots. In this study, the asymmetric classical splines are obtained by fixing the λ parameter to have a symmetric filter of $2m + 1$ terms, and then selecting the last m rows of the influential matrix $\mathbf{A}(\lambda)$. We illustrate the results for the last point asymmetric weights corresponding to a 23-term symmetric filter because the

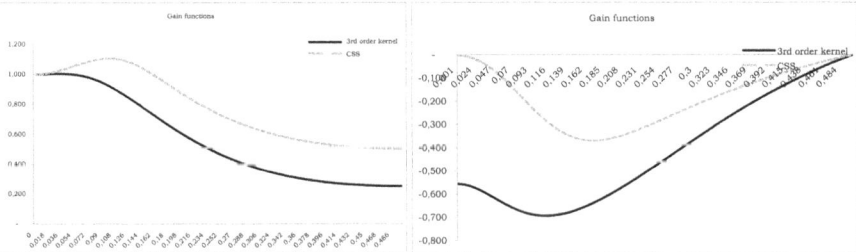

Fig. 9.5 (**a**) Gain and (**b**) phase shift functions of the asymmetric (end point) weights of the third order spline kernel and the CSS smoother

spline gives very poor results for short lengths such as 13 or less. Figure 9.5a shows that the asymmetric kernel exhibits a gain function with better properties of signal passing and noise suppression, without implying larger phase shifts. These latter are smaller than 1 month for both filters (see Fig. 9.5b).

9.2.1   Empirical Evaluation

This section examines how the kernel estimators perform on real seasonally adjusted data in comparison with the LOESS, Henderson filter, and cubic smoothing spline. The last point filters are applied to a set of fifty time series taken from the Hyd-man's time series library (http://www-personal.buseco.monash.edu.au/~hyndman/TSDL/). These series are seasonally adjusted and related to different fields (finance, labor market, production, sales, transports, meteorology, hydrology, physics, and health). The periods selected vary to sufficiently cover the various lengths published for these series. For each series, the length of the Henderson and LOESS filters is selected according to the I/C (noise to signal) ratio that ranges from 0.16 to 7.35, hence filters of length 9-, 13-, and 23-term are applied.

On the other hand, for the cubic smoothing spline the generalized cross validation (GCV) criteria [6] is used to estimate the λ parameter, and consequently the length of the filters to be applied by considering the number of non-null elements in a central row of the matrix $\mathbf{A}(\lambda)$. The kernel estimator of the same length is then calculated. The smoothing parameter λ is known as hyperparameter in the Bayesian terminology, and it has the interpretation of a noise to signal ratio: the larger the λ, the smoother the output. In our sample, λ ranges from a minimum of 0.013, at which corresponds a filter length equal to 7 terms, to a maximum of 15, which corresponds to a 43-term smoother. This enables us to analyze the pattern of the two estimators on series characterized by different degrees of variability.

The comparisons are based on the relative filter revisions between the final symmetric filter S and the last point asymmetric filter A, that is,

$$R_t = \frac{S_t - A_t}{S_t}, \quad t = 1, 2, \ldots, n \qquad (9.25)$$

For each series and for each estimator, it is calculated the ratio between the mean square error (MSE) of the revisions corresponding to the third order kernel R^K and to the corresponding classical filter R^C, that is $\frac{\text{MSE}(R^K)}{\text{MSE}(R^C)}$.

For all the estimators, the results illustrated in Table 9.3 indicate that the ratio is always smaller than one, showing that the kernel last point predictors introduce smaller revisions than the classical ones. This implies that the kernel estimates will be more reliable and efficient than the ones obtained by the application of the classical LOESS, Henderson filter, and cubic spline. In particular, for the LOESS estimator, 96 % of the sample has a ratio less than 0.7 and, in general, it is never greater than 0.806. For the Henderson filter, the ratio is less than 0.7 in 20 % of the series and for the cubic spline in 34 % of the sample. In both cases, the highest value of the ratio is close to 0.895.

For current economic analysis, it is also of great importance to assess the performance of the several kernel estimators for turning point detection. A turning point is generally defined to occur at time t if (*downturn*):

$$y_{t-k} \leq \cdots \leq y_{t-1} > y_t \geq y_{t+1} \geq \cdots \geq y_{t+m}$$

or (*upturn*)

$$y_{t-k} \geq \cdots \geq y_{t-1} < y_t \leq y_{t+1} \leq \cdots \leq y_{t+m}$$

for $k = 3$ and $m = 1$ given the smoothness of the trend-cycle data [8]. For each estimator, the time lag to detect the true turning point is obtained by calculating the number of months it takes for the revised trend series to signal a turning point in the same position as in the final trend series.

True turning points are those recognized in the economy after occurring. For instance, the Business Cycle Dating Committee of the National Bureau of Economic Research (NBER) had identified the last trough in the US economy in June 2009. To determine this date, the behavior of various monthly indicators of the economic activity has been analyzed. Among others, the committee considered a measure of monthly Gross Domestic Product (GDP) developed by the private forecasting firm Macroeconomic Advisers (MAGDP), monthly GDP and Gross Domestic Income (GDI) developed by James H. Stock and Mark W. Watson (SWGDP, SWGDI), real personal income excluding transfers (INC), the payroll (PAY) and household measures of total employment (HOU). Monthly data series for industrial production (IIP) and manufacturing trade sales (MAN) have also been taken into account. The committee designated June as the month of the trough for several of these indicators, that is, MAGDP, SWGDP, SWGDI, MAN, and IIP. On the other hand, the trough was identified in October 2009 for INC, and in December 2009 for PAY and HOU.

Table 9.3 MSE revision ratio between kernel and classical last point predictors

	Series	Ratio
Health	Number of cases of measles, New York city	0.806
	Number of cases of measles, Baltimore	0.512
	Body weight of rats	0.544
	Number of chickenpox, New York city	0.597
Hydrology	Temperature, copper mine	0.539
	Flows, Colorado river Lees ferry	0.580
	Lake Frie levels	0.568
	Flows, Chang Jiang	0.696
Labour market	Wisconsin employment series, fabricated metals	0.502
	US male (20 years and over) unemployment figures	0.540
	Number of employed persons in Australia	0.318
Macroeconomics	Consumer price index	0.346
Meteorology	Degree days per heating in Chicago	0.541
Microeconomics	Gambling expenditure in Victoria, Australia	0.584
	Logged flour price indices over the 9-years	0.558
Miscellaneous	Average daily calls to directory assistance	0.607
Physics	Zurich sunspot numbers	0.546
	Critical radio frequencies in Washington DC	0.621
	Mean thickness (Dobson units) ozone column Switzerland	0.685
Production	Basic iron production in Australia	0.696
	Production of chocolate confectionery in Australia	0.628
	Production of Portland cement	0.687
	Electricity production in Australia	0.783
	Production of blooms and slabs in Australia	0.666
	Production of blooms and slabs	0.677
Sales	Sales of Tasty Cola	0.459
	Unit sales, Winnebago Industries	0.594
	Sales of new one-family houses sold in US	0.603
	Sales of souvenir shop in Queensland, Australia	0.570
	Demand for carpet	0.502
Transport and Tourism	Portland Oregon average monthly bus ridership	0.511
	US air passenger miles	0.546
	International airline passengers	0.473
	Weekday bus ridership, Iowa city (monthly averages)	0.567
	Passenger miles flow domestic UK	0.703
Utilities	Average residential gas usage Iowa	0.633
	Total number of consumer	0.615

Table 9.4 Time lag in detecting true turning points (TP) for classical and kernel estimators

Series	True TP	LOESS			Henderson			CSS		
		Length	Classic	Kernel	Length	Classic	Kernel	Length	Classic	Kernel
MAGDP	06/09	13	6	4	13	4	3	27	6	4
SWGDP	06/09	13	6	4	13	4	3	29	6	6
SWGDI	06/09	13	6	4	13	5	3	21	4	4
MAN	06/09	9	4	3	9	4	2	33	8	8
IIP	06/09	9	4	3	9	2	2	25	6	6
INC	10/09	13	6	4	13	4	3	17	6	6
PAY	12/09	9	4	3	9	4	4	13	6	5
HOU	12/09	9	4	3	9	2	1	21	6	6

The results obtained by applying the classical and kernel estimators to all of these series are shown in Table 9.4.

It can be noticed that kernel estimators perform better than classical filters. In particular, the third order biweight kernel always detects the turning points only very few months after it has occurred, and faster than the classical Henderson filter, which is very well-known for fast turning point detection. Tricube LOESS kernel also presents good time lags, not greater than 4 months, that are smaller than those observed for the classical LOESS estimator. On the other hand, the length selected for the classical and kernel CSS estimators is generally greater than 20 terms and, as expected, they are generally poor predictors of the true turning points, with a slightly better performance for the logistic kernel than for the classical estimator.

References

1. Aronszajn, N. (1950). Theory of reproducing kernels. *Transaction of the AMS, 68*, 337–404.
2. Berlinet, A. (1993). Hierarchies of higher order kernels. *Probability Theory and Related Fields, 94*, 489–504.
3. Berlinet, A., & Thomas-Agnan, C. (2003). *Reproducing kernel Hilbert spaces in probability and statistics*. Boston, MA: Kluwer Academic Publishers.
4. Bianconcini, S. (2006). Trend-Cycle Estimation in Reproducing Kernel Hilbert Spaces (Ph.D. thesis, Department of Statistics, University of Bologna).
5. Cleveland, W. S. (1979). Robust locally regression and smoothing scatterplots. *Journal of the American Statistical Association, 74*, 829–836.
6. Craven, P., & Wahba, G. (1979). Smoothing noisy data with spline functions. *Numerical Mathematics, 31*, 377–403.
7. Cristianini, N., & Shawe-Taylor, J. (2000). *An introduction to support vector machines and other kernel-based learning methods*. New York: Cambridge University Press.
8. Dagum, E. B. (1996). A new method to reduce unwanted ripples and revisions in trend-cycle estimates from X11ARIMA. *Survey Methodology, 22*(1), 77–83.
9. Dagum, E. B., & Bianconcini, S. (2006). Local polynomial trend-cycle predictors in reproducing kernel Hilbert spaces for current economic analysis. *Anales de Economia Aplicada*, 1–22.
10. Dagum, E. B., & Bianconcini, S. (2008). The Henderson smoother in reproducing kernel Hilbert space. *Journal of Business and Economic Statistics, 26*(4), 536–545.

11. Dagum, E. B., & Bianconcini, S. (2013). A unified probabilistic view of nonparametric predictors via reproducing kernel Hilbert spaces. *Econometric Reviews, 32*(7), 848–867.
12. De Boor, C., & Lynch, R. (1966). On splines and their minimum properties. *Journal of Mathematics and Mechanics, 15*, 953–969.
13. Doherty, M. (1992). *The Surrogate Henderson Filters in X-11*. Working Paper. Wellington, New Zealand: Statistics New Zealand.
14. Evgeniou, T., Pontil, M., & Poggio, T. (2000). Regularization networks and support vector machines. *Advanced in Computational Mathematics, 13*, 1–50.
15. George, E. O., El-Saidi, M., & Singh, K. (1986). A generalized logistic approximation of the Student t distribution. *Communications in Statistics B - Simulation Computing, 15*(1), 261–277.
16. George, E. O., & Ojo, M. O. (1980). On a generalization of the logistic distribution. *Annals of the Institute of Statistical Mathematics, 32*, 161–169.
17. Gray, A., & Thomson, P. (1996). On a family of moving-average trend filters for the ends of series. In *Proceedings of the Business and Economic Statistics Section*. American Statistical Association Annual Meeting, Chicago.
18. Green, P. J., & Silverman, B. W. (1994). *Nonparametric regression and generalized linear models*. London: Chapman and Hall.
19. Henderson, R. (1916). Note on graduation by adjusted average. *Transactions of the Actuarial Society of America, 17*, 43–48.
20. Jones, M. C. (2006). The logistic and the log F distributions. In N. Balakrishan (Ed.), *Handbook of logistic distribution*. New York: Dekker.
21. Kimeldorf, G. S., & Wahba, G. (1970). A correspondence between Bayesian estimation on stochastic processes and smoothing by splines. *Annals of Mathematical Statistics, 41*, 495–502.
22. Kimeldorf, G. S., & Wahba, G. (1970). Splines functions and stochastic processes. *Sankhya, Series A, 32*, 173–180.
23. Lin, G. D., & Hu, C. Y. (2008). On the characterization of the logistic distribution. *Journal of Statistical Planning and Inference, 138*(4), 1147–1156.
24. Mudholkar, G. S., & George, E. O. (1978). A remark on the shape of the logistic distribution. *Biometrika, 65*, 667–668.
25. Musgrave, J. (1964). *A Set of End Weights to End All End Weights*. Working paper. Washington: US Bureau of the Census.
26. Parzen, E. (1959). *Statistical Inference on Time Series by Hilbert Space Methods*. Technical Report No. 53. , Stanford, CA: Statistics Department, Stanford University.
27. Pearce, N. D., & Wand, M. P. (2006). Penalized splines and reproducing kernel methods. *The American Statistician, 60*(3), 223–240.
28. Prentice, R. L. (1976). A generalization of probit and logit methods for dose response curves. *Biometrics, 32*, 761–768.
29. Priestley, M. B. (1981). *Spectral analysis and time series*. Probability and mathematical statistics. New York: Academic.
30. Schoenberg, I. (1964). Monosplines and quadrature formulae. In T. Greville (Ed.), *Theory and applications of spline functions*. Madison, WI: University of Wisconsin Press.
31. Silverman, B. (1994). Spline smoothing: The equivalent kernel method. *Annals of Statistics, 12*, 898–916.
32. Thomas-Agnan, C. (1991). Splines functions and stochastic filtering. *Annals of Statistics, 9*(3), 1512–1527.
33. Wahba, G. (1990). *Spline models for observational data*. Philadelphia: SIAM.
34. Wahba, G. (1999). Support vector machine, reproducing kernel Hilbert spaces, and randomized GACV. In B. Scholkopf, C. Burges, & A. Smola (Eds.), *Advanced in kernel methods: Support vector learning* (pp. 69–88). Cambridge, MA: MIT Press.
35. Whittaker, E. T. (1923). On a new method of graduation. In *Proceedings of the Edinburgh Mathematical Association* (Vol. 78, pp. 81–89).
36. Whittaker, E. T., & Robinson, G. (1924). *Calculus of observations: A treasure on numerical calculations*. London: Blackie and Son.

Chapter 10
Real Time Trend-Cycle Prediction

Abstract For real time trend-cycle analysis, official statistical agencies generally produce estimates derived from asymmetric moving average techniques. However, the use of the asymmetric filters introduces revisions as new observations are incorporated to the series and also delays in detecting true turning points. This chapter deals with a reproducing kernel approach to obtain time-varying asymmetric trend-cycle filters that converge fast and monotonically to the corresponding symmetric ones. It shows that the bandwidth parameter that minimizes the distance between the gain functions of the asymmetric and symmetric filters is to be preferred. Respect to the asymmetric filters incorporated in the X12ARIMA software currently applied by the majority of official statistical agencies, the new set of weights produces both smaller revisions and time delay to detect true turning points. The theoretical results are empirically corroborated with a set of leading, coincident, and lagging indicators of the US economy.

The basic approach to the analysis of current economic conditions, known as recession and recovery analysis, is that of assessing the real time trend-cycle of major socioeconomic indicators (*leading, coincident, and lagging*) using percentage changes, based on seasonally adjusted units, calculated for months and quarters in chronological sequence. The main goal is to evaluate the behavior of the economic indicators during incomplete phases by comparing current contractions or expansions with corresponding phases in the past. This is done by measuring changes of single time series (mostly seasonally adjusted) from their standing at cyclical turning points with past changes over a series of increasing spans. This differs from business-cycle studies where cyclical fluctuations are measured around a long-term trend to estimate complete business cycles. The real time trend corresponds to an incomplete business cycle and is strongly related to what is currently happening on the business-cycle stage.

In recent years, statistical agencies have shown an interest in providing trend-cycle or smoothed seasonally adjusted graphs to facilitate recession and recovery analysis. Among other reasons, this interest originated from the recent crisis and major economic and financial changes of global nature which have introduced more variability in the data. The USA entered in recession in December 2007 till June 2009, and this has produced a chain reaction all over the world, particularly,

© Springer International Publishing Switzerland 2016   243
E. Bee Dagum, S. Bianconcini, *Seasonal Adjustment Methods and Real Time Trend-Cycle Estimation*, Statistics for Social and Behavioral Sciences,
DOI 10.1007/978-3-319-31822-6_10

in Europe. There is no evidence of a fast recovery as in previous recessions: the economic growth is sluggish and with high levels of unemployment. It has become difficult to determine the direction of the short-term trend (or trend-cycle) as traditionally done by looking at month to month (quarter to quarter) changes of seasonally adjusted values, particularly to assess the upcoming of a true turning point. Failure in providing reliable real time trend-cycle estimates could give rise to dangerous drift of the adopted policies. Therefore, a consistent prediction is of fundamental importance. It can be done by means of either univariate parametric models or nonparametric techniques. The majority of the statistical agencies use nonparametric seasonally adjusted software, such as the Census II-X11 method and its variants X11/X12ARIMA and X13, and hence this chapter deals with the real time trend-cycle estimation produced by the Musgrave filters [19] available in these software.

As widely discussed in Chap. 8, the linear filter developed by Henderson [15] is the most frequently applied and has the property that fitted to exact cubic functions will reproduce their values, and fitted to stochastic cubic polynomials it will give smoother results than those estimated by ordinary least squares. The properties and limitations of the Henderson filters have been extensively discussed in Chap. 8, where we have also provided its Reproducing Kernel Hilbert Space (RKHS) representation. It consists in a kernel function obtained as the product of the biweight density function and the sum of its orthonormal polynomials that is particularly suitable when the length of the filter is rather short, say between 5 and 23 terms, which are those often applied by statistical agencies (see also, for details, Dagum and Bianconcini [6]).

At the beginning and end of the sample period, the Henderson filter of length, say $2m + 1$, cannot be applied to the first and last m data points, hence only asymmetric filters can be used. The estimates of the real time trend are then subject to revisions produced by the innovations brought by the new data entering in the estimation and the time-varying nature of the asymmetric filters, in the sense of being different for each of the m data points. The asymmetric filters applied to the first and last m observations associated with the Henderson filter were developed by Musgrave [19] on the basis of minimizing the mean squared revision between the final estimates, obtained with the symmetric Henderson weights, and preliminary estimates from the asymmetric weights, subject to the constraint that the sum of these weights is equal to one. The assumption made is that at the end of the series, the seasonally adjusted values do not follow a cubic polynomial, but a linear trend-cycle plus a purely random irregular. Several authors studied the properties and limitations of the Musgrave filters [2, 10–12, 14, 18, 20], and Dagum and Bianconcini [6, 7] were the first to introduce an RKHS representation of them.

The RKHS approach followed in this book and presented here is strictly nonparametric. As discussed in Chaps. 8 and 9, based on a fundamental result due to Berlinet [1], a kernel estimator of order p can be always decomposed into the product of a reproducing kernel R_{p-1}, belonging to the space of polynomials of degree at most $p - 1$, and a density function f_0 with finite moments up to order $2p$. Given the density function, once the length of the symmetric filter is chosen, let us say,

$2m + 1$, the statistical properties of the asymmetric filters are strongly affected by the bandwidth parameter of the kernel function from which the weights are derived. In Chap. 9, as suggested by Dagum and Bianconcini [6, 7], this bandwidth parameter has been selected to be equal to $m+1$, independently on the length of the associated asymmetric filter. The corresponding asymmetric weights approximate very well the classical Musgrave filters [19]. This chapter presents time-varying bandwidth parameters because the asymmetric filters are time-varying. Three specific criteria of bandwidth selection are chosen based on the minimization of

(1) the distance between the transfer functions of asymmetric and symmetric filters,
(2) the distance between the gain functions of asymmetric and symmetric filters, and
(3) the phase shift function over the domain of the signal.

It deals only with the reduction of revisions due to filter changes that depends on how close the asymmetric filters are respect to the symmetric one [5, 9] and do not consider those introduced by the innovations in the new data. Another important aspect dealt with is the capability of the asymmetric filters to signal the upcoming of a true turning point which depends on the time delay for its identification. This is obtained by calculating the number of months (quarters) it takes for the last trend-cycle estimate to signal a true turning point in the same position of the final trend-cycle data. An optimal asymmetric filter should have a time path that converges fast and monotonically to the final estimate as new observations are added to the series.

10.1 Asymmetric Filters and RKHS

Let $\{y_t, t = 1, 2, \ldots, n\}$ denote the input series, supposed to be seasonally adjusted where trading day variations, moving holidays, and outliers, if present, have been removed. It is assumed that it can be decomposed into the sum of a systematic component (signal) TC_t, representing the trend-cycle (usually estimated jointly) plus an erratic component I_t (noise), such that

$$y_t = TC_t + I_t. \tag{10.1}$$

The noise I_t is assumed to be white noise, $WN(0, \sigma_I^2)$, or, more generally, a stationary and invertible autoregressive moving average (ARMA) process. The signal $TC_t, t = 1, \ldots, n$, is assumed to be a smooth function of time, such that it can be represented *locally* by a polynomial of degree p in a variable j, which measures the distance between y_t and its neighboring observations $y_{t+j}, j = -m, \ldots, m$. This is equivalent to estimate the trend-cycle $\widehat{TC}_t$ as a weighted moving average as follows:

$$\widehat{TC}_t = \sum_{j=-m}^{m} w_j y_{t+j} = \mathbf{w}'\mathbf{y} \qquad t = m + 1, \ldots, n - m, \tag{10.2}$$

where $\mathbf{w}' = \begin{bmatrix} w_{-m} & \cdots & w_0 & \cdots & w_m \end{bmatrix}$ contains the weights to be applied to the input data $\mathbf{y}' = \begin{bmatrix} y_{t-m} & \cdots & y_t & \cdots & y_{t+m} \end{bmatrix}$ to get the estimate $\widehat{TC}_t$ for each point in time.

As discussed in Sect. 9.2, several nonparametric estimators, based on different sets of weights $\mathbf{w}$, have been developed in the literature. It is shown in Sect. 8.3.2 that the Henderson filter can be derived using the Reproducing Kernel Hilbert Space methodology. In this context, its equivalent kernel representation is given by

$$K_4(t) = \sum_{i=0}^{3} P_i(t)P_i(0)f_{0B}(t) \qquad t \in [-1, 1], \tag{10.3}$$

where f_{0B} is the biweight density function $f_{0B}(t) = \frac{15}{16}(1 - t^2)^2, t \in [-1, 1]$, and $P_i, i = 0, 1, 2, 3$, are the corresponding orthonormal polynomials. Equation (10.3) provides a good approximation for Henderson filters of short length, say between 5 and 23 terms, which are those used by statistical agencies [6].

Dagum and Bianconcini [8] have shown that, when applied to real data, the symmetric filter weights are given by

$$w_j = \left[\frac{\mu_4 - \mu_2 \left(\frac{j}{b}\right)^2}{S_0\mu_4 - S_2\mu_2} \right] \frac{1}{b} f_{0B}\left(\frac{j}{b}\right) \qquad j = -m, \ldots, m, \tag{10.4}$$

where b is a time invariant global bandwidth parameter (same for all $t = m + 1, \ldots, n - m$), whereas μ_2 and μ_4 are the second and fourth moments of the biweight density, respectively, being $S_r = \sum_{j=-m}^{m} \frac{1}{b}\left(\frac{j}{b}\right)^r f_0\left(\frac{j}{b}\right)$ their discrete approximations.

In this setting, once the length of the filter is selected, the choice of the bandwidth parameter b is fundamental. It has to be chosen to ensure that only, say, $2m + 1$ observations surrounding the target point will receive nonzero weights. It has also to approximate, as close as possible, the continuous density function with the discrete function as well as its moments. The bandwidth parameter selection is done to guarantee specific inferential properties of the trend-cycle estimators. In this regard, Dagum and Bianconcini [6, 7] used a time invariant global bandwidth b equal to $m + 1$, which gave good results.

The derivation of the symmetric Henderson filter has assumed the availability of $2m + 1$ input values centered at t. However, at the end of the sample period, that is $t = n - (m + 1), \ldots, n$, only $2m, \ldots, m + 1$ observations are available, and asymmetric filters of the same length have to be considered. Hence, at the boundary, the effective domain of the kernel function K_4 is $[-1, q^*]$, with $q^* \ll 1$, instead of $[-1, 1]$ as for any interior point. This implies that the symmetry of the kernel is lost, and it does not integrate to unity on the asymmetric support ($\int_{-1}^{q^*} K_4(t)dt \neq 1$). Furthermore, the moment conditions are not longer satisfied, that is, $\int_{-1}^{q^*} t^i K_4(t)dt \neq$

0 for $i = 1, 2, 3$. To overcome these limitations, several boundary kernels have been proposed in the literature.

In the context of real time trend-cycle estimation, the condition that the kernel function integrates to unity is essential, whereas the unbiasedness property can only be satisfied with a great increase in the variance of the estimates. This is a consequence of the well-known trade-off between bias and variance. This latter becomes very large because most of the contribution to the real time trend-cycle estimates comes from the current observation which gets the largest weight. Based on these considerations, Dagum and Bianconcini [6, 7] have suggested to follow the so-called cut and normalize method [13, 16], according to which the boundary kernels $K_4^{q^*}$ are obtained by cutting the symmetric kernel K_4 to omit that part of the function lying between q^* and 1, and by normalizing it on $[-1, q^*]$. That is,

$$K_4^{q^*}(t) = \frac{K_4(t)}{\int_{-1}^{q^*} K_4(t)dt} = \frac{\det(\mathbf{H}_4^0[1, \mathbf{t}])f_{0B}(t)}{\det(\mathbf{H}_4^0[1, \boldsymbol{\mu}^{q^*}])} \qquad t \in [-1, q^*] \qquad (10.5)$$

where $\boldsymbol{\mu}^{q*} = \begin{bmatrix} \mu_0^{q*} & \mu_1^{q*} & \mu_2^{q*} & \mu_3^{q*} \end{bmatrix}$ with $\mu_r^{q*} = \int_{-1}^{q^*} t^r f_{0B}(t)dt$ being proportional to the moments of the truncated biweight density f_{0B} on the support $[-1, q^*]$, which from now on we simply refer to as truncated moments. $\mathbf{H}_4^0[1, \mathbf{t}]$ is the Hankel matrix whose elements are the moments of f_{0B}, and where the first column has been substituted by the vector $\mathbf{t}' = \begin{bmatrix} 1 & t & t^2 & t^3 \end{bmatrix}$, whereas $\mathbf{H}_4^0[1, \boldsymbol{\mu}^{q*}]$) is the Hankel matrix where the first column is substituted by $\boldsymbol{\mu}^{q*}$.

Applied to real data, the "cut and normalize" method gives the following formula for the asymmetric weights:

$$w_{q,j} = \frac{K_4^{q^*}(j/b_q)}{\sum_{j=-m}^q K_4^{q^*}(j/b_q)} = \frac{\det(\mathbf{H}_4^0[1, \mathbf{j}/\mathbf{b_q}])(1/b_q)f_{0B}(j/b_q)}{\det(\mathbf{H}_a)}, \qquad (10.6)$$

$$j = -m, \ldots, q; q = 0, \ldots, m - 1,$$

where $\mathbf{H}_4^0[1, \mathbf{j}/\mathbf{b_q}]$ is the Hankel matrix whose elements are the moments of f_{0B}, and where the first column has been substituted by the vector $\mathbf{j}/\mathbf{b_q}' = \begin{bmatrix} 1 & (j/b_q) & (j/b_q)^2 & (j/b_q)^3 \end{bmatrix}$. On the other hand, $\mathbf{H}_a = \mathbf{H}_4^0[1, \mathbf{S}^q]$ with $\mathbf{S}^q = \begin{bmatrix} S_0^q & S_1^q & S_2^q & S_3^q \end{bmatrix}'$, being $S_r^q = \sum_{j=-m}^q \frac{1}{b_q} \left(\frac{j}{b_q} \right)^r f_{0B} \left(\frac{j}{b_q} \right)$ the discrete approximation of μ_r^{q*}. Finally, $b_q, q = 0, \ldots, m - 1$, is the local bandwidth, specific for each asymmetric filter. It allows to relate the discrete domain of the filter, that is, $\{-m, \ldots, q\}$, for each $q = 0, \ldots, m - 1$, to the continuous domain of the kernel function, that is $[-1, q^*]$.

Proposition 1 *Each asymmetric filter* $\mathbf{w}_q = [w_{q,-m} \cdots w_{q,q}]'$ *of length* $(m + q + 1), q = 0, \ldots, m - 1$, *admits the following matrix representation:*

$$\mathbf{w}_q' = \mathbf{e}_1' \mathbf{H}_a^{-1} \mathbf{X}_q' \mathbf{F}_q \qquad q = 0, \ldots, m - 1, \qquad (10.7)$$

where $\mathbf{X}_q$ *is a matrix of dimensions* $(m + q + 1) \times 4$, *whose generic row is given by* $\mathbf{j}/\mathbf{b_q}$, $j = -m, \ldots, q$, *and* $\mathbf{F}_q = diag\left(\frac{1}{b_q} f_{0B}\left(-\frac{m}{b_q}\right), \ldots, \frac{1}{b_q} f_{0B}\left(\frac{q}{b_q}\right)\right)$. *It can be easily shown that the generic element of* $\mathbf{w}_q$ *is*

$$w_{qj} = \left[\frac{\mu_4 - \mu_2 \left(\frac{j}{b_q}\right)^2}{S_0^q \mu_4 - S_2^q \mu_2}\right] \frac{1}{b_q} f_{0B}\left(\frac{j}{b_q}\right) \tag{10.8}$$

$$j = -m, \ldots, q; q = 0, \ldots, m - 1.$$

Proof Based on Eq. (10.6), we can write

$$w_{qj} = \frac{\det(\mathbf{H}_4^0[1, \mathbf{j}/\mathbf{b_q}]) \frac{1}{b_q} f_{0B}\left(\frac{j}{b_q}\right)}{\sum_{j=-m}^{q} \det(\mathbf{H}_4^0[1, \mathbf{j}/\mathbf{b_q}]) \frac{1}{b_q} f_{0B}\left(\frac{j}{b_q}\right)} = \frac{\det(\mathbf{H}_4^0[1, \mathbf{j}/\mathbf{b_q}]) \frac{1}{b_q} f_{0B}\left(\frac{j}{b_q}\right)}{\det\left(\mathbf{H}_4^0\left[1, \sum_{j=-m}^{q} \mathbf{j}/\mathbf{b_q} \frac{1}{b_q} f_{0B}\left(\frac{j}{b_q}\right)\right]\right)}$$

$$= \frac{\det(\mathbf{H}_4^0[1, \mathbf{j}/\mathbf{b_q}]) \frac{1}{b_q} f_{0B}\left(\frac{j}{b_q}\right)}{\det(\mathbf{H}_4^0[1, \mathbf{S}^q])} = \frac{\det(\mathbf{H}_4^0[1, \mathbf{j}/\mathbf{b_q}]) \frac{1}{b_q} f_{0B}\left(\frac{j}{b_q}\right)}{\det(\mathbf{H}_a)}. \tag{10.9}$$

The expression (10.9) is exactly the same we would obtain by solving for $\hat{\beta}_0 = \widehat{TC}_t$ the system of linear equations

$$\mathbf{H}_a \boldsymbol{\beta} = \mathbf{X}_q' \mathbf{F}_q \mathbf{y}.$$

Indeed, setting $\mathbf{c} = \mathbf{X}_q' \mathbf{F}_q \mathbf{y}$, the first coordinate of the solution vector is

$$\hat{\beta}_0 = \frac{\det(\tilde{\mathbf{H}}_4^0[1, \mathbf{c}])}{\det(\mathbf{H}_a)} = \frac{\det(\mathbf{H}_4^0[1, \mathbf{c}])}{\det(\mathbf{H}_a)}.$$

Given that $\mathbf{c} = \sum_{j=-m}^{q} \left(\frac{\mathbf{j}}{\mathbf{b_q}}\right) \frac{1}{b_q} f_{0B}\left(\frac{j}{b_q}\right) y_{t+j}$, it follows that

$$\det(\mathbf{H}_4^0[1, \mathbf{b_q}]) = \sum_{j=-m}^{q} \det(\mathbf{H}_4^0[1, \frac{\mathbf{j}}{\mathbf{b_q}}]) \frac{1}{b_q} f_{0B}\left(\frac{j}{b_q}\right) y_{t+j}$$

and therefore

$$\widehat{TC}_t = \sum_{j=-m}^{q} \frac{\det(\mathbf{H}_4^0[1, \mathbf{j}/\mathbf{b_q}]) \frac{1}{b_q} f_{0B}\left(\frac{j}{b_q}\right)}{\det(\mathbf{H}_a)} y_{t+j}.$$

Hence,

$$\hat{\beta}_0 = \mathbf{e}_1' \mathbf{H}_a^{-1} \mathbf{X}_q' \mathbf{F}_q \mathbf{y},$$

and it follows that

$$\mathbf{w}' = \mathbf{e}_1' \mathbf{H}_a^{-1} \mathbf{X}_q' \mathbf{F}_q.$$

10.1.1   Properties of the Asymmetric Filters

Since the trend-cycle estimates for the last m data points do not use $2m + 1$ observations as for any interior point, but $2m, 2m - 1, \ldots, m + 1$ data, they are subject to revisions due to new observations entering in the estimation and filters change. As said before, we will concentrate on the reduction of revisions due to filters change. The reduction of these revisions is an important property that the asymmetric filters should possess together with a fast detection of true turning points. In the specific case of the RKHS filters, Eq. (10.8) shows how the asymmetric filter weights are related to the symmetric ones given in Eq. (10.4). It is clear that the convergence depends on the relationship between the two discretized biweight density functions, truncated and non-truncated, jointly with the relationship between their respective truncated S_r^q and untruncated S_r discrete moments. The latter provide an approximation of the continuous moments μ_r, which improves as the asymmetric filter length increases. Similarly, the convergence of $S_r^q, q = 0, \ldots, m$, to the corresponding non-truncated moment S_r depends on the length of the asymmetric filter given by q, and on the local bandwidth b_q. It should be noticed that b_q plays a very important role in the convergence property. For the last trend-cycle point weight, $q = 0$, Eq. (10.8) reduces to

$$w_{0,0} = \frac{\mu_4}{S_0^0 \mu_4 - S_2^0 \mu_2} \frac{15}{16 b_0}.$$

It is apparent that the largest b_0, the smaller is the weight given to the last trend-cycle point. Since the sum of all the weights of the last point asymmetric filter, $w_{0,-m}, \ldots, w_{0,0}$, must be equal to one, this implies that the weights for the remaining points are very close to one another. This can be seen in Fig. 10.1 (right side) that shows, for $m = 6$, the truncated continuous biweight density function and its discretized version when b_0 is equal to 12. The opposite is observed when b_0 is smaller, as shown in the same figure (left side) for b_0 equal to 7. Since a larger weight is given to the last point, much smaller weights have to be assigned to the remaining ones for all of them to add to one. Next, we introduce time-varying local bandwidths to improve the properties of the asymmetric filters in terms of size of revisions and time delay to signal the upcoming of true turning points.

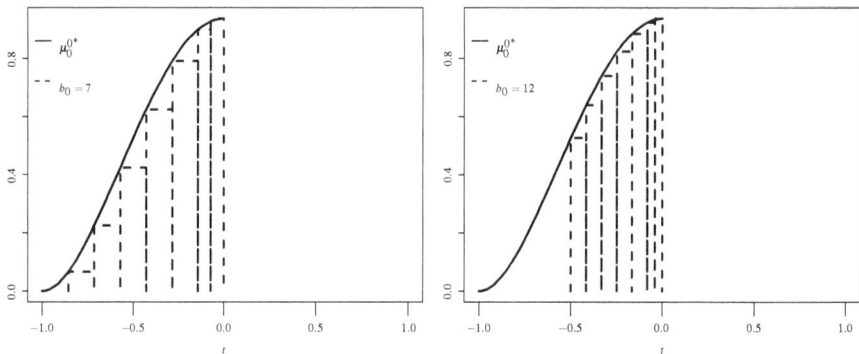

Fig. 10.1 Behavior of S_0^0 with $m = 6$, $b_0 = 7$ (*left*) and $b_0 = 12$ (*right*)

10.2   Optimal Bandwidth Selection

The main effects introduced by a linear filter on a given input can be fully described in the frequency domain by its transfer function

$$\Gamma(\omega) = \sum_{j=-m}^{m} w_j \exp(-i2\pi\omega j) \qquad \omega \in [-1/2, 1/2],$$

where, for better interpretation, the frequencies ω are given in cycles for unit of time instead of radians. $\Gamma(\omega)$ represents the Fourier transform of the filter weights, $w_j, j = -m, \ldots, m$, and it relates the spectral density $h_y(\omega)$ and $h_g(\omega)$ of the input and of the output, respectively, by

$$h_g(\omega) = \Gamma(\omega)h_y(\omega).$$

Thus, the transfer function $\Gamma(\omega)$ measures the effect of the filter on the total variance of the input at different frequencies. It is generally expressed in polar coordinates

$$\Gamma(\omega) = G(\omega) \exp(-i2\pi\phi(\omega)), \tag{10.10}$$

such that the impact of the filter on a particular (complex-valued) series $y_t = \exp(i2\pi\omega t)$, $\omega \in [-1/2, 1/2]$, is given by

$$\begin{aligned}
\widehat{TC}_t &= \Gamma(\omega) \exp(i2\pi\omega t) \\
&= G(\omega) \exp(-i2\pi\phi(\omega)) \exp(i2\pi\omega t) \\
&= G(\omega) \exp[i2\pi(\omega t - \phi(\omega))].
\end{aligned}$$

$G(\omega) = |\Gamma(\omega)|$ is called the gain of the filter and measures the amplitude of the output for a sinusoidal input of unit amplitude, whereas $\phi(\omega)$ is called the phase function and shows the shift in phase of the output compared with the input. Hence, the transfer function plays a fundamental role to measure that part of the total revisions due to filters change.

A measure of total revisions introduced by Musgrave [19] is given by

$$E\left[\sum_{j=-m}^{q} w_{q,j} y_{t-j} - \sum_{j=-m}^{m} w_j y_{t-j}\right]^2, \qquad q = 0, \ldots, m-1, \qquad (10.11)$$

where, in our specific case, $w_{q,j}$ and w_j are given by Eqs. (10.8) and (10.4), respectively. This criterion can be expressed in the frequency domain as follows:

$$E\left[\sum_{j=-m}^{q} w_{q,j} e^{i2\pi\omega(t-j)} - \sum_{j=-m}^{m} w_j e^{i2\pi\omega(t-j)}\right]^2 = E\left[(\Gamma_q(\omega) - \Gamma(\omega))e^{i2\pi\omega t}\right]^2$$

$$= \int_{-1/2}^{1/2} |\Gamma_q(\omega) - \Gamma(\omega)|^2 e^{i4\pi\omega t} h_y(\omega) d\omega, \qquad (10.12)$$

where $h_y(\omega)$ is the unknown spectral density of $y_t, t = 1, \ldots, n$, whereas $\Gamma_q(\omega)$ and $\Gamma(\omega)$ are the transfer functions corresponding to the asymmetric and symmetric filters, respectively. Similarly to (10.11), expression (10.12) shows that, as new observations become available, revisions are produced by the new innovations entering the input series and the change of the asymmetric filters. In order to improve the current trend-cycle prediction based on the asymmetric Henderson filters we study that part of the revisions due to asymmetric filters change. Because the estimation of the real time trend-cycle is done concurrently, that is using all of the data up to and including the most recent value, knowledge of the speed of convergence of the last point trend-cycle filter to the central one gives valuable information on how often the real time trend estimate should be revised.

The quantity $|\Gamma_q(\omega) - \Gamma(\omega)|^2$ accounts for the revisions due to filters change [3, 4], and it can be decomposed using the law of cosines as follows:

$$|\Gamma_q(\omega) - \Gamma(\omega)|^2 = |G_q(\omega) - G(\omega)|^2 + 2G_q(\omega)G(\omega)\left[1 - \cos(\phi_q(\omega))\right] \qquad (10.13)$$

$$= |G_q(\omega) - G(\omega)|^2 + 4G_q(\omega)G(\omega)\sin\left(\phi_q\left(\frac{\omega}{2}\right)\right)^2,$$

where the phase shift for the symmetric filter is equal to 0 or $\pm\pi$, and where $1 - \cos(\phi_q(\omega)) = 2\sin\left(\phi_q\left(\frac{\omega}{2}\right)\right)^2$. Based on Eq. (10.13), the mean square filter revision error can be expressed as follows:

$$2 \int_0^{1/2} |\Gamma_q(\omega) - \Gamma(\omega)|^2 d\omega \tag{10.14}$$

$$= 2 \int_0^{1/2} |G_q(\omega) - G(\omega)|^2 d\omega + 8 \int_0^{1/2} G_q(\omega)G(\omega) \sin\left(\phi\left(\frac{\omega}{2}\right)\right)^2 d\omega.$$

The first component reflects the part of the total mean square filter error which is attributed to the amplitude function of the asymmetric filter. On the other hand, the second term measures the distinctive contribution of the phase shift. The term $G_q(\omega)G(\omega)$ is a scaling factor which accounts for the fact that the phase function is dimensionless, i.e., it does not convey level information [21].

As previously discussed, once the length of the filter is fixed, the properties of the asymmetric filters derived in RKHS are strongly affected by the choice of the time-varying local bandwidths $b_q, q = 0, \ldots, m - 1$. A filter is said to be optimal if it minimizes both revisions and time delay to detect a true turning point.

The LHS of Eq. (10.14) is a measure of total filter revision that provides the best compromise between the amplitude function of the asymmetric filter (gain) and its phase function (time displacement) [3, 4, 9]. Optimal asymmetric filters in this sense can be derived using local bandwidth parameters selected according to the following criterion:

$$b_{q,\Gamma} = \min_{b_q} \sqrt{2 \int_0^{1/2} |\Gamma_q(\omega) - \Gamma(\omega)|^2 d\omega}. \tag{10.15}$$

Based on the decomposition of the total filter revision error provided in Eq. (10.14), further bandwidth selection criteria can be defined by emphasizing more the gain or phase shift effects, and/or by attaching varying importance to the different frequency components, depending on whether they appear in the spectrum of the input time series or not. In the context of smoothing a monthly input, the frequency domain $\Omega = \{0 \leq \omega \leq 0.50\}$ can be partitioned into two main intervals: (1) $\Omega_S = \{0 \leq \omega \leq 0.06\}$ associated with cycles of 16 months or longer attributed to the signal (trend-cycle) of the series and (2) $\bar{\Omega}_S = \{0.06 < \omega \leq 0.50\}$ corresponding to short cyclical fluctuations attributed to the noise.

A class of optimal asymmetric filters based on bandwidth parameters $b_q, q = 0, \ldots, m - 1$, is selected as follows:

$$b_{q,G} = \min_{b_q} \sqrt{2 \int_0^{1/2} |G_q(\omega) - G(\omega)|^2 d\omega}, \tag{10.16}$$

and

$$b_{q,\phi} = \min_{b_q} \sqrt{2 \int_{\Omega_S} G_q(\omega)G(\omega) \left[1 - \cos(\phi_q(\omega))\right]}. \tag{10.17}$$

Table 10.1 Optimal bandwidth values selected for each of the biweight asymmetric filters corresponding to the 9-, 13-, and 23-term Henderson symmetric filters

q	0	1	2	3							
$b_{q,\Gamma}$	6.47	5.21	4.90	4.92							
$b_{q,G}$	8.00	5.67	4.87	4.90							
$b_{q,\phi}$	4.01	4.45	5.97	6.93							
q	0	1	2	3	4	5					
$b_{q,\Gamma}$	9.54	7.88	7.07	6.88	6.87	6.94					
$b_{q,G}$	11.78	9.24	7.34	6.85	6.84	6.95					
$b_{q,\phi}$	6.01	6.01	7.12	8.44	9.46	10.39					
q	0	1	2	3	4	5	6	7	8	9	10
$b_{q,\Gamma}$	17.32	15.35	13.53	12.47	12.05	11.86	11.77	11.77	11.82	11.91	11.98
$b_{q,G}$	21.18	18.40	16.07	13.89	12.44	11.90	11.72	11.73	11.83	11.92	11.98
$b_{q,\phi}$	11.01	11.01	11.01	11.01	11.41	13.85	15.13	16.21	17.21	18.15	19.05

It has to be noticed that the minimization of the phase error in Eq. (10.17) is very close to minimizing the average phase shift in month for the signal, that is

$$b_{q,\phi} = \min_{b_q} \left[\frac{1}{0.06} \int_{\Omega_S} \frac{\phi(\omega)}{2\pi\omega} d\omega \right]. \tag{10.18}$$

Table 10.1 illustrates the bandwidth parameters $b_{q,\Gamma}, b_{q,G}, b_{q,\phi}, q = 0, \ldots, m-1$, derived as minimizers of Eqs. (10.15), (10.16), and (10.18), respectively, corresponding to the 9-, 13-, and 23-term symmetric filters. It can be noticed that, as q approaches m, the bandwidth parameters selected to optimize the criteria (10.15) and (10.16) get closer to $m + 1$, which is the global bandwidth considered for the symmetric Henderson filter. Hence, based on the relationships between truncated and untruncated discrete biweight density functions and respective discrete moments previously discussed, the asymmetric filters based on $b_{q,\Gamma}$ and $b_{q,G}$, $q = 0, \ldots, m-1$, should be characterized by a fast convergence to the symmetric filter. This is confirmed by Fig. 10.2 that illustrates, as an example, the time path of these filters corresponding to the 13-term symmetric one. Similar conclusions can be drawn for different filter lengths.

The asymmetric filters based on $b_{q,\Gamma}$ and $b_{q,G}$, $q = 0, \ldots, m-1$, converge very fast to the symmetric filter, particularly after the previous to the last point, with the main differences observed for the last point filters. For these latter, the different behavior is analyzed in the frequency domain in Fig. 10.4, that shows the corresponding gain and phase shift functions. It can be noticed that, as expected, the filter whose bandwidth $b_{0,G}$ is derived as minimizer of Eq. (10.16) shows a gain function closer to that of the symmetric Henderson filter than the one based on $b_{0,\Gamma}$, suppressing more noise at the highest frequencies, and it reproduces very well the signal in the lower frequency band.

In terms of phase shift or time delay, the filters that behave better are the ones based on the bandwidth parameters selected to minimize the average phase shift

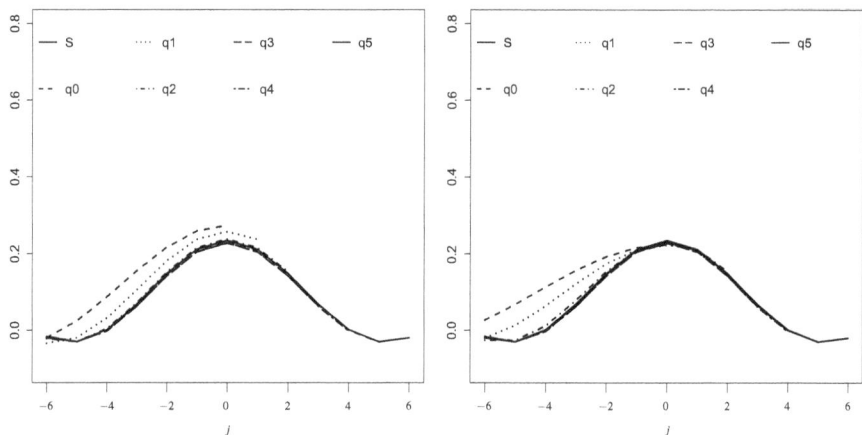

Fig. 10.2 Time path of the asymmetric filters based on $b_{q,\Gamma}$ (*left*), $b_{q,G}$ (*right*) corresponding to the 13-term symmetric filter

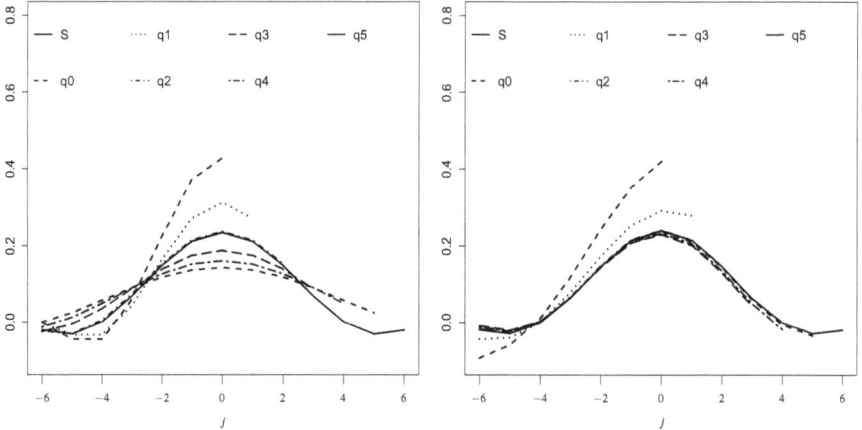

Fig. 10.3 Time path of the asymmetric filters based on $b_{q,\phi}$ (*left*) and of the Musgrave asymmetric filters (*right*) corresponding to the 13-term symmetric filter

in months over the signal domain. However, as shown in Fig. 10.3, their time path is only very close to that of the filters derived by Musgrave [19] up to $q = 2$ but there is no monotonic convergence of these asymmetric filters to their final ones. As already said, the Musgrave filters are based on the minimization of the mean squared revision between the final estimates, obtained by the application of the symmetric filter, and the preliminary estimates, obtained by the application of an asymmetric filter, subject to the constraint that the sum of the weights is equal to one [12, 18]. These filters have the good property of fast detection of turning points. This property is reflected in their phase shift function that, for the last point filter, is illustrated in Fig. 10.4. As can be seen, both the last point Musgrave filter and

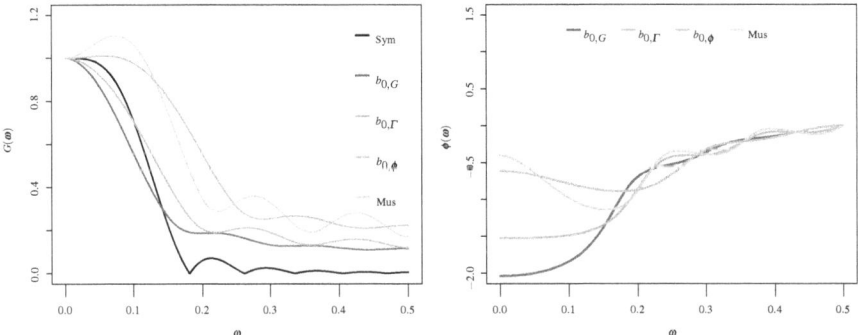

Fig. 10.4 Gain (*left*) and phase shift (*right*) functions for the last point asymmetric filters based on $b_{0,\Gamma}$, $b_{0,G}$, and $b_{0,\phi}$ compared with the last point Musgrave filter

the one based on $b_{0,\phi}$ produce almost one half of the phase shift introduced by the filter based on $b_{0,\Gamma}$ and a quarter of the one introduced by the filter based on $b_{0,G}$ at the signal frequency band. However, the reduced phase shift produced by these two filters is compensated by larger revisions introduced in the final estimates. Indeed, as shown by the corresponding gain functions, the last point Musgrave filter and the one based on $b_{0,\phi}$ suppress much less noise than the filters obtained through minimization of Eqs. (10.15) and (10.16). Furthermore, the Musgrave filter has the worst performance since it introduces a large amplification of the power attributed to the trend and suppresses less noise.

10.3 Empirical Application

The asymmetric filters previously derived can be applied in many fields, such as, macroeconomic, finance, health, hydrology, meteorology, criminology, physics, labor markets, and utilities. In fact, in any time series where the impact of the trend-cycle is of relevance. A set of leading, coincident, and lagging indicators of the US economy is chosen to corroborate the theoretical conclusions discussed before. The leading indicators are time series that have a turning point before the economy as a whole changes, whereas the coincident indicators change direction approximately at the same time as the whole economy, thereby providing information about the current state of the economy. On the other hand, the lagging indicators are those that usually change direction after the economy as a whole does. The composite indexes are typically reported in the financial and trade press, and the data analyzed in this study are from the St. Louis Federal Reserve Bank database, the Bureau of Labor

Statistics, and the National Bureau of Economic Research (NBER). In particular, the analyzed leading indicators are

- Average weekly overtime hours, manufacturing
- New orders for durable goods
- New orders for nondefense capital goods
- New private housing units authorized by building permits
- Stock prices, S&P common stocks
- Money supply, M2
- Interest rate spread, 10-year treasury bonds less federal funds
- Index of consumer expectation (University of Michigan).

We also consider the following coincident indicators:

- Employees on nonagricultural payrolls
- Personal income less transfer payments
- Industrial production index
- Manufacturing and trade sales.

Finally, the lagging indicators treated are

- Average duration of unemployment, weeks
- Ratio, manufacturing, and trade inventory to sale
- Change in labor cost per unit of output, manufacturing
- Commercial and industrial loans outstanding.

The asymmetric filters derived following the RKHS methodology versus the Musgrave filters, applied in conjunction with the symmetric Henderson filter, are evaluated as follows.

10.3.1   Reduction of Revision Size in Real Time Trend-Cycle Estimates

The reduction of revisions in real time trend-cycle estimates is very important because the estimates are preliminary and used to assess the current stage of the economy. Statistical agencies and major users of these indicators are reluctant to large revisions because these can lead to erroneous statement concerning the current economic situation. The series considered are all seasonally adjusted, where also trading day variations, moving holidays, and extreme values have been removed, if present. The socioeconomic indicators are series of different length but the periods selected sufficiently cover the various lengths published for these series.

Here, we study how the filters derived in RKHS and the classical Musgrave estimators respond to the variability of the data. For each series, the length of the filters is selected according to the I/C (noise to signal) ratio, as classically done in

Table 10.2 Ratio of the mean square percentage revision errors of the last point asymmetric filters based on $b_{0,G}$, $b_{0,\Gamma}$, and $b_{0,\phi}$, and the last point Musgrave filter

Macro-area	Series	$\frac{b_{0,G}}{Mus}$	$\frac{b_{0,\Gamma}}{Mus}$	$\frac{b_{0,\phi}}{Mus}$
Leading	Average weekly overtime hours: manufacturing	0.492	0.630	0.922
	New orders for durable goods	0.493	0.633	0.931
	New orders for nondefense capital goods	0.493	0.633	0.931
	New private housing units authorized by building permits	0.475	0.651	0.927
	S&P 500 stock price index	0.454	0.591	0.856
	M2 money stock	0.508	0.655	0.932
	10-year treasury constant maturity rate	0.446	0.582	0.849
	University of Michigan: consumer sentiment	0.480	0.621	0.912
Coincident	All employees: total nonfarm	0.517	0.666	0.951
	Real personal income excluding current transfer receipts	0.484	0.627	0.903
	Industrial production index	0.477	0.616	0.884
	Manufacturing and trade sales	0.471	0.606	0.869
Lagging	Average (mean) duration of unemployment	0.509	0.649	0.937
	Inventory to sales ratio	0.483	0.618	0.894
	Index of total labor cost per unit of output	0.515	0.663	0.983
	Commercial and industrial loans at all commercial banks	0.473	0.610	0.871

the X11/X12ARIMA procedure [17]. In the sample, the ratio ranges from 0.20 to 1.98, hence filters of length 9 and 13 terms are applied.

The comparisons are based on the relative filter revisions between the final symmetric filter S and the last point asymmetric filter A, that is,

$$R_t = \frac{S_t - A_t}{S_t}, \quad t = 1, 2, \ldots, n. \qquad (10.19)$$

For each series and for each estimator, we calculate the ratio between the mean square percentage error (MSPE) of the revisions corresponding to the filters derived following the RKHS methodology and those corresponding to the last point Musgrave filter. For all the estimators, the results illustrated in Table 10.2 indicate that the ratio is always smaller than one, indicating that the kernel last point predictors, based on time-varying bandwidth parameters, introduce smaller revisions than the Musgrave filter. This implies that the estimates obtained by the former will be more accurate than those derived by the application of the latter. In particular, as expected, the best performance is shown by the filter based on the optimal bandwidth $b_{0,G}$ obtained by minimizing the criterion (10.16). In almost all the series its ratio with the last point Musgrave filter is less than one half and, on average, around 0.480. This implies that when applied to real data, the filter based on $b_{0,G}$ produces a reduction of almost 50% of the revisions introduced in the real time trend-cycle estimates given by the Musgrave filter. The filter based on $b_{0,\Gamma}$, derived to minimize the size of total filter revisions as defined by Eq. (10.15), also performs very well with more than the 30 % of revision reduction with respect to the Musgrave filter. In this case, the ratio is greater than the one corresponding to the filter based on $b_{0,G}$, but always less than 0.7 for all the series, being, on average, around 0.636. The filter whose bandwidth parameter is selected to minimize the average phase shift over the signal domain performs more similar to the last point Musgrave filter but still shows revisions reduction, on average, around 10 %.

10.3.2   Turning Point Detection

It is important that the reduction of revisions in real time trend-cycle estimates is not achieved at the expense of increasing the time lag to detect the upcoming of a true turning point. A turning point is generally defined to occur at time t if (*downturn*):

$$y_{t-k} \leq \cdots \leq y_{t-1} > y_t \geq y_{t+1} \geq \cdots \geq y_{t+m},$$

or (*upturn*)

$$y_{t-k} \geq \cdots \geq y_{t-1} < y_t \leq y_{t+1} \leq \cdots \leq y_{t+m}.$$

Following Zellner et al. [22], it is selected $k = 3$ and $m = 1$ given the smoothness of the trend-cycle data. For each estimator, the time lag to detect the true turning point is affected by the convergence path of its asymmetric filters $\mathbf{w}_q, q = 0, \ldots, m-1$, to the symmetric one $\mathbf{w}$.

To determine the time lag needed by an indicator to detect a true turning point it is calculated the number of months it takes for the real time trend-cycle estimate to signal a turning point in the same position as in the final trend-cycle series. For the series analyzed in this chapter, the time delays for each estimator are shown in Table 10.3. It can be noticed that the filters based on the bandwidth $b_{q,\phi}$ take almost 2 months (on average), similar to the Musgrave filters, to detect the turning point. This is due to the fact that, even if $b_{q,\phi}$ filters are designed to be optimal in timeliness, their convergence path to the symmetric filter is slower and moreover not monotonic.

On the other hand, the filters based on $b_{q,\Gamma}, q = 0, \ldots, m - 1$, and $b_{q,G}, q = 0, \ldots, m - 1$, perform strongly better. In particular, whereas the former detect the turning point with an average time delay of 1.67 months, the latter takes 1.27 months.

Table 10.3 Time lag in detecting true turning points for the asymmetric filters based on $b_{q,G}$, $b_{q,\Gamma}$, and $b_{q,\phi}$, and the Musgrave filters

Macro-area	Series	$b_{q,G}$	$b_{q,\Gamma}$	$b_{q,\phi}$	Musgrave
Leading	Average weekly overtime hours: manufacturing	1	1	1	1
	New orders for durable goods	1	2	3	2
	New orders for nondefense capital goods	1	2	2	3
	New private housing units authorized by building permits	2	2	3	3
	S&P 500 stock price index	1	2	2	2
	10-Year treasury constant maturity rate	1	1	1	2
	University of Michigan: consumer sentiment	1	1	1	1
Coincident	All employees: total nonfarm	1	1	1	2
	Real personal income excluding current transfer receipts	1	1	1	1
	Industrial production index	1	1	1	1
	Manufacturing and trade sales	1	2	3	3
Lagging	Average (mean) duration of unemployment	3	3	4	3
	Inventory to sales ratio	1	1	1	2
	Index of total labor cost per unit of output	2	2	3	2
	Commercial and industrial loans at all commercial banks	1	1	1	1
	Average time lag in months	1.27	1.67	1.93	2.00

The fastest the upcoming of a turning point is detected, the fastest new policies can be applied to counteract the impact of the business-cycle stage. Failure to recognize the downturn in the cycle or taking a long time delay to detect it may lead to the adoption of policies to curb expansion when in fact, a recession is already underway.

The filters based on local bandwidth parameters selected to minimize criterion (10.16) are optimal, since they drastically reduce the total revisions by one half with respect to the Musgrave filters and, similarly, almost by one third the number of months needed to pick up the June 2009 turning point. The real time trend-cycle filter calculated with the bandwidth parameter that minimizes the distance between the asymmetric and symmetric filters gain functions is to be preferred. This last point trend-cycle filter reduced by one half the size of the revisions and by one third the time delay to detect the June 2009 turning point relative to the Musgrave filters. For illustrative purposes, Tables 10.4 and 10.5 give the weight systems of these filters for 9- and 13-term symmetric filters.

Table 10.4 Weights corresponding to the 9-term symmetric filter

0.31218	0.22278	0.13330	0.044036	0.00000	0.00000	0.00000	0.00000
0.27101	0.27101	0.15289	0.02630	−0.03965	0.00000	0.00000	0.00000
0.10931	0.32009	0.25652	0.10931	−0.01705	−0.03470	0.00000	0.00000
−0.01544	0.25957	0.32281	0.25957	0.11250	−0.01544	−0.03605	0.00000
−0.03907	0.12023	0.26574	0.32767	0.26574	0.12023	−0.01074	−0.03907
0.00000	−0.01544	0.11250	0.25957	0.32281	0.25957	0.11250	−0.01544
0.00000	−0.03470	−0.01705	0.10931	0.25652	0.32009	0.25652	0.10931
0.00000	0.00000	−0.03965	0.02630	0.15289	0.27101	0.31845	0.27101
0.00000	0.00000	0.00000	0.04404	0.13330	0.22278	0.28804	0.31184

Table 10.5 Weights corresponding to the 13-term symmetric filter

0.22362	0.21564	0.19266	0.157478	0.11444	0.06902	0.02714	0.00000	0.00000	0.00000	0.00000	0.00000	0.00000
0.21065	0.22352	0.21065	0.174523	0.12230	0.06460	0.01357	−0.01982	0.00000	0.00000	0.00000	0.00000	0.00000
0.15391	0.21013	0.23100	0.21013	0.15391	0.08000	0.01250	−0.02600	−0.025570	0.00000	0.00000	0.00000	0.00000
0.06338	0.14212	0.20452	0.22808	0.20452	0.14212	0.06338	−0.00217	−0.02978	−0.01617	0.00000	0.00000	0.00000
−0.00258	0.06319	0.14245	0.20533	0.22909	0.20533	0.14245	0.06319	−0.00258	−0.02996	−0.01593	0.00000	0.00000
−0.02983	0.00060	0.06762	0.14651	0.20848	0.23179	0.20848	0.14651	0.06762	0.00060	−0.02983	−0.01855	0.00000
−0.01986	−0.02982	0.00217	0.07010	0.14921	0.21106	0.23429	0.21106	0.149208	0.07010	0.00217	−0.02982	−0.01986
0.00000	−0.01855	−0.02983	0.00066	0.06762	0.14651	0.20848	0.23179	0.20848	0.14651	0.06762	0.00060	−0.02983
0.00000	0.00000	−0.01593	−0.02996	−0.00258	0.06319	0.14245	0.20533	0.22909	0.20533	0.14245	0.06319	−0.00258
0.00000	0.00000	0.00000	−0.01617	−0.02978	−0.00217	0.06338	0.14212	0.20452	0.22808	0.20452	0.14212	0.06338
0.00000	0.00000	0.00000	0.00000	−0.02557	−0.02600	0.01250	0.08000	0.15391	0.21013	0.23100	0.21013	0.15391
0.00000	0.00000	0.00000	0.00000	0.00000	−0.01982	0.01357	0.06460	0.12230	0.17452	0.21065	0.22352	0.21065
0.00000	0.00000	0.00000	0.00000	0.00000	0.00000	0.02714	0.06902	0.11444	0.15748	0.19267	0.21564	0.22362

References

1. Berlinet, A. (1993). Hierarchies of higher order kernels. *Probability Theory and Related Fields, 94*, 489–504.
2. Bianconcini, S., & Quenneville, B. (2010). Real time analysis based on reproducing kernel Henderson filters. *Estudios de Economia Aplicada, 28*(3), 553–574.
3. Dagum, E. B. (1982). Revisions of time varying seasonal filters. *Journal of Forecasting, 1,* 173–187.
4. Dagum, E. B. (1982). The effects of asymmetric filters on seasonal factor revisions. *Journal of American Statistical Association, 77*, 732–738.
5. Dagum, E. B. (1996). A new method to reduce unwanted ripples and revisions in trend-cycle estimates from X11ARIMA. *Survey Methodology, 22*, 77–83.
6. Dagum, E. B., & Bianconcini, S. (2008). The Henderson smoother in reproducing kernel Hilbert space. *Journal of Business and Economic Statistics, 26*(4), 536–545.
7. Dagum, E. B., & Bianconcini, S. (2013). A unified probabilistic view of nonparametric predictors via reproducing kernel Hilbert spaces. *Econometric Reviews, 32*(7), 848–867.
8. Dagum, E. B., & Bianconcini, S. (2015). A new set of asymmetric filters for tracking the short-term trend in real time. *The Annals of Applied Statistics, 9*, 1433–1458.
9. Dagum, E. B., & Laniel, N. (1987). Revisions of trend-cycle estimators of moving average seasonal adjustment methods. *Journal of Business and Economic Statistics, 5*, 177–189.
10. Dagum, E. B., & Luati, A. (2009). A cascade linear filter to reduce revisions and turning points for real time trend-cycle estimation. *Econometric Reviews, 28*(1–3), 40–59.
11. Dagum, E. B., & Luati, A. (2012). Asymmetric filters for trend-cycle estimation. In W. R. Bell, S. H. Holan, & T. S. McElroy (Eds.), *Economic time series: Modeling and seasonality* (pp. 213–230). Boca Raton, FL: Chapman&Hall.
12. Doherty, M. (2001). The surrogate Henderson filters in X-11. *Australian and New Zealand Journal of Statistics, 43*(4), 385–392.
13. Gasser, T., & Muller, H. G. (1979). Kernel estimation of regression functions. In *Lecture notes in mathematics* (Vol. 757, pp. 23–68). New York: Springer.
14. Gray, A. G., & Thomson, P. J. (2002). On a family of finite moving-average trend filters for the ends of series. *Journal of Forecasting, 21*, 125–149.
15. Henderson, R. (1916). Note on graduation by adjusted average. *Transaction of Actuarial Society of America, 17*, 43–48.
16. Kyung-Joon, C., & Schucany, W. R. (1998). Nonparametric Kernel regression estimation near endpoints. *Journal of Statistical Planning and Inference, 66*(2), 289–304.
17. Ladiray, D., & Quenneville, B. (2001). *Seasonal adjustment with the X-11 method.* New York: Springer-Verlag.
18. Laniel, N. (1985). Design Criteria for the 13-term Henderson End Weights. Working paper. Ottawa: Methodology Branch. Statistics Canada.
19. Musgrave, J. (1964). A Set of End Weights to End All End Weights. Working paper. Washington, DC: U.S. Bureau of Census.
20. Quenneville, B., Ladiray, D., & Lefrancois, B. (2003). A note on Musgrave asymmetrical trend-cycle filters. *International Journal of Forecasting, 19*(4), 727–734.
21. Wildi, M. (2008). *Real-time signal extraction: Beyond maximum likelihood principles.* Berlin: Springer.
22. Zellner, A., Hong, C., & Min, C. (1991). Forecasting turning points in international output growth rates using Bayesian exponentially weighted autoregression, time-varying parameter, and pooling techniques. *Journal of Econometrics, 49*(1–2), 275–304.

Chapter 11
The Effect of Seasonal Adjustment on Real-Time Trend-Cycle Prediction

Abstract Asymmetric nonparametric trend-cycle filters obtained with local time-varying bandwidth parameters via the Reproducing Kernel Hilbert Space (RKHS) methodology reduce significantly revisions and turning point detection respect to the currently used by statistical agencies. The best choice of local time-varying bandwidth is the one obtained by minimizing the distance between the gain functions of the RKHS asymmetric and the symmetric filter to which it must converge. Since the input to these kernel filters is seasonally adjusted series, it is important to evaluate the impact that the seasonal adjustment method can have. The purpose of this chapter is to assess the effects of the seasonal adjustment methods when the real time trend is predicted with such nonparametric kernel filters. The seasonal adjustments compared are the two officially adopted by statistical agencies: X12ARIMA and TRAMO-SEATS applied to a sample of US leading, coincident, and lagging indicators.

The RKHS trend-cycle filters discussed in Chap. 10 are applied to seasonally adjusted data, hence it is important to evaluate how their statistical properties are affected by the seasonal adjustment method used to produce the input data. In this regard, we analyze the behavior of the Dagum and Bianconcini [9] asymmetric filters in terms of size of revisions in the estimates as new observations are added to the series, and time lag to detect true turning points. We look at seasonally adjusted data from X12ARIMA and TRAMO-SEATS which are officially adopted by statistical agencies, with outliers replaced if present.

Seasonal adjustment means the removal of seasonal variations in the original series jointly with trading day variations and moving holiday effects. The main reason for seasonal adjustment is the need of standardizing socioeconomic series because seasonality affects them with different timing and intensity. Hence, the seasonally adjusted data reflect variations due only to the remaining components, namely trend-cycle and irregulars. The information given by seasonally adjusted series has always played a crucial role in the analysis of current economic conditions and provides the basis for decision making and forecasting, being of major importance around cyclical turning points.

As discussed in Chap. 3, seasonal adjustment methods can be classified as deterministic or stochastic depending on the assumptions made concerning how

© Springer International Publishing Switzerland 2016   263
E. Bee Dagum, S. Bianconcini, *Seasonal Adjustment Methods and Real Time Trend-Cycle Estimation*, Statistics for Social and Behavioral Sciences,
DOI 10.1007/978-3-319-31822-6_11

seasonality evolves through time. Deterministic methods assume that the seasonal pattern can be predicted with no error or with variance of the prediction error null. On the contrary, stochastic methods assume that seasonality can be represented by a stochastic process, that is, a process governed by a probability law and, consequently, the variance of the prediction error is not null. The best known seasonal adjustment methods belong to the following types:

(i) regression methods which assume global or local simple functions of time,
(ii) stochastic model-based methods which assume simple ARIMA models, and
(iii) moving average methods which are based on linear filtering and hence do not assume explicit parametric models.

Only methods (i), which assume global simple functions for each component, are deterministic; the others are considered stochastic. Moving averages and ARIMA model-based (AMB) methods are those mainly applied by statistical agencies to produce officially seasonally adjusted series. A brief description is given in the following and we refer the readers to Chaps. 4 and 5, where each method is discussed in detail.

11.1   Seasonal Adjustment Methods

11.1.1   X12ARIMA

The X12ARIMA is today the most often applied seasonal adjustment method by statistical agencies. It was introduced by Findley et al. [10] and is an enhanced version of the X11ARIMA method developed at Statistics Canada by Estela Bee Dagum [5].

The major modifications concern: (1) extending the automatic identification and estimation of ARIMA models for the extrapolation option to many more than the three models available in the X11ARIMA, and (2) estimating trading day variations, moving holidays, and outliers in what is called regARIMA model. The latter consists of regression models with ARIMA errors. More precisely, they are models in which the mean function of the time series (or its logs) is described by a linear combination of regressors, and the covariance structure of the series is that of an ARIMA process. If no regressors are used, indicating that the mean is assumed to be zero, the regARIMA model reduces to an ARIMA model.

Whether or not special problems requiring the use of regressors are present in the series to be adjusted, an important use of regARIMA models is to extend the series with forecasts (and backcasts) in order to improve the seasonal adjustments of the most recent (and the earliest) data. Doing this reduces problems inherent in the trend estimation and asymmetric seasonal averaging processes of the type used by the X11 method near the ends of the series. The provision of this extension was the most important improvement offered by the X11ARIMA program. Its theoretical and empirical benefits have been documented in many publications, such as Dagum [6], Bobbit and Otto [3], and the references therein.

The X12ARIMA method has all the seasonal adjustment capabilities of the X11ARIMA variant. The same seasonal and trend moving averages are available, and the program still offers the X11 calendar and holiday adjustment routines incorporated in X11ARIMA. But several important new options have been included.

The modeling module is designed for regARIMA model building with seasonal socioeconomic time series. To this end, several categories of predetermined regression variables are available, including trend constants or overall means, fixed seasonal effects, trading day effects, holiday effects, pulse effects (additive outliers), level shifts, temporary change outliers, and ramp effects. User-specified regression variables can also be included in the models. The specification of a regARIMA model requires specification of both the regression variables to be included in the model and the type of ARIMA model for the regression errors (i.e., the order $(p, d, q)(P, D, Q)_s$).

Specification of the regression variables depends on user knowledge about the series being modeled. Identification of the ARIMA model for the regression errors follows well-established procedures based on examination of various sample autocorrelation and partial autocorrelation functions produced by the X12ARIMA program.

Once a regARIMA model has been specified, X12ARIMA estimates its parameters by maximum likelihood using an Iterative Generalized Least Squares (IGLS) algorithm. Diagnostic checking involves examination of residuals from the fitted model for signs of model inadequacy. X12ARIMA produces several standard residual diagnostics for model checking, as well as provides sophisticated methods for detecting additive outliers and level shifts. Finally, X12ARIMA can produce point forecasts, forecast standard errors, and prediction intervals from the fitted regARIMA model.

Trading day effects occur when a series is affected by the different day-of-the-week compositions of the same calendar month in different years. Trading day effects can be modeled with seven variables that represent (*no. of Mondays*), ..., (*no. of Sundays*) in month t. Bell and Hillmer [1] proposed a better parameterization of the same effects using six variables defined as (*no. of Mondays*) − (*no. of Sundays*), ..., (*no. of Saturdays*) − (*no. of Sundays*), along with a seventh variable for Length of Month (LOM) or its deseasonalized version, the leap-year regressor (*lpyear*). In X12ARIMA the six variables are called the *tdnolpyear* variables. Instead of using a seventh regressor, a simpler and often better way to handle multiplicative leap-year effects is to re-scale the February values of the original time series before transformation to $\bar{m}_{Feb} y_t / m_t$, where y_t is the original time series before transformation, m_t is the length of month t (28 or 29), and $\bar{m}_{Feb} = 28.25$ is the average length of February. If the regARIMA model includes seasonal effects, these can account for the length of month effect except in Februaries, so the trading day model only has to deal with the leap-year effect. When this is done, only the *tdnolpyear* variables need be included in the model. X12ARIMA allows explicit choice of either approach, as well as an option (*td*) that makes a default choice of how to handle length of month effects. When the time series being modeled represents the aggregation of some daily series (typically unobserved) over calendar months they are called *monthly flow series*. If the series instead represents the value of some daily series at the end of the month, called a *monthly stock series*, then different regression variables are appropriate.

Holiday effects in a monthly flow series arise from holidays whose dates vary over time if: (1) the activity measured by the series regularly increases or decreases around the date of the holiday, and (2) this affects two (or more) months depending on the date the holiday occurs each year. Effects of holidays with a fixed date, such as Christmas, are indistinguishable from fixed seasonal effects. *Easter effects* are the most frequently found holiday effects in American and European economic time series, since the date of Easter Sunday varies between March 22 and April 25.

X12ARIMA provides four other types of regression variables to deal with abrupt changes in the level of a series of a temporary or permanent nature: additive outliers (AO), level shifts (LS), temporary changes (TC), and ramps. Identifying the location and nature of potential outliers is the object of the outlier detection methodology implemented. This methodology can be used to detect AOs, TCs, and LSs (not ramps); any outlier that is detected is automatically added to the model as regression variable. Prespecified AOs, LSs, TCs, and ramps are actually simple forms of interventions as discussed by Box and Tiao [4].

The regARIMA modeling routine fulfills the role of prior adjustment and forecast extension of the time series. That is, it first estimates some deterministic effects, such as calendar effects and outliers using predefined built-in options or user-defined regressors, and it removes them from the observed series. An appropriate ARIMA model is identified for the pre-adjusted series in order to extend it with forecasts. Then, this extended pre-adjusted series is decomposed into the unobserved components of the series using moving averages, also accounting for the presence of extreme values, as follows:

1. a preliminary estimate of the trend-cycle is obtained using a centered thirteen (five) term weighted moving average of the monthly (quarterly) original series;
2. this trend-cycle series is removed from the original one to give an initial estimate of the seasonal and irregular components, often called SI ratios;
3. initial preliminary seasonal factors are estimated from these initial SI ratios by applying weighted seasonal moving averages for each month over several years, separately;
4. an initial estimate of the irregular component is obtained by removing the initial preliminary seasonal factors from the initial seasonal irregular (SI) ratios;
5. extreme values are identified and (temporarily) adjusted to replace the initial SI ratios with the modified ones;
6. steps 3. and 4. are repeated on the modified SI ratios, initial seasonal factors are derived by normalizing the initial preliminary seasonal factors;

7. an initial estimate of the seasonally adjusted series is obtained by removing the initial seasonal factors from the original series;
8. a revised trend-cycle estimate is obtained by applying Henderson moving averages to the initial seasonally adjusted series;
9. steps 2.–8. are again repeated twice to produce the final estimates of the trend-cycle, seasonal, and irregular components.

Various diagnostics and quality control statistics are computed, tabulated, and graphed in order to assess the seasonal adjustment results. If they are acceptable based upon the diagnostics and quality measures, then the process is terminated. If this is not the case, the above steps have to be repeated to search for a more satisfactory seasonal adjustment of the series. The procedure is summarized in the LHS of Fig. 11.1.

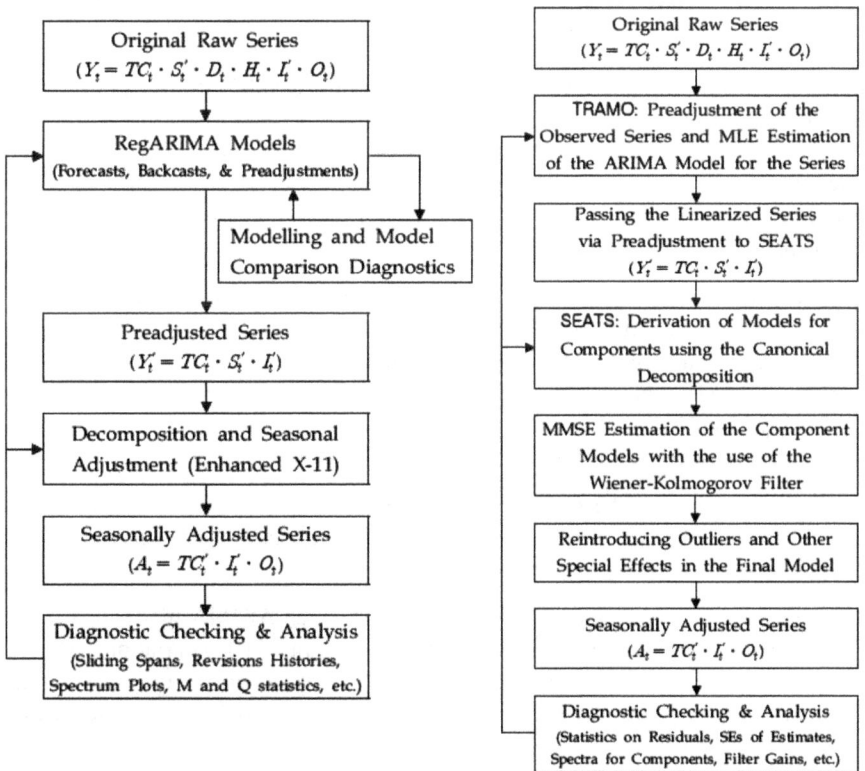

Fig. 11.1 X12ARIMA (*left*) and TRAMO-SEATS (*right*) procedures

11.1.2   *TRAMO-SEATS*

Gomez and Maravall [12] developed at the Bank of Spain, a seasonal adjustment software called TRAMO-SEATS which is currently applied mainly by European statistical agencies. TRAMO stands for "Time series Regression with ARIMA noise, Missing observations and Outliers," and SEATS for "Signal Extraction in ARIMA Time Series." First, TRAMO estimates via regression the deterministic components, which are after removed from the input data. In a second round, SEATS estimates the seasonal and trend-cycle components from the ARIMA model fitted to the data where the deterministic components are removed. SEATS uses the filters derived from the linearized ARIMA model that describes the stochastic behavior of the time series.

It should be mentioned that Eurostat, in collaboration with the National Bank of Belgium, the US Bureau of the Census, the Bank of Spain, and the European Central Bank, has developed an interface of TRAMO-SEATS and X12ARIMA called Demetra+. In the Bank of Spain and Eurostat websites it is also possible to find a considerable number of papers relevant to TRAMO-SEATS as well as in the European Statistical System (ESS) Guideline.

TRAMO is a regression method that performs the estimation, forecasting, and interpolation of missing observations with ARIMA errors, in the presence of possibly several types of outliers. The ARIMA model can be identified automatically or by the user. It consists of the following steps:

1. automatic detection of several types of outliers and, if necessary, estimation of other regression variables, such as calendar effects;
2. automatic identification of an ARIMA model;
3. then, TRAMO removes all the estimated components (trading days, moving holidays, and outliers) from the original series and passes the linearized series to SEATS, where the actual decomposition is performed.

SEATS belongs to the class of procedures based on ARIMA models for the decomposition of time series into unobserved components and consists of the following steps:

1. *ARIMA model estimation.* SEATS starts by fitting an ARIMA model to a series not affected by deterministic components, such as trading day variations, moving holidays, and outliers. Let y_t denote this linearized series, and consider an additive decomposition model (multiplicative if applied to the log transformation of y_t), such that

$$z_t = (1 - B)^d (1 - B^s)^D y_t \qquad (11.1)$$

represent the "differenced" series. The corresponding model for the differenced linearized series z_t can be written as

$$\phi_p(B)\Phi_P(B^s)(z_t - \bar{z}) = \theta_q(B)\Theta_Q(B^s)a_t, \qquad (11.2)$$

where $\bar{z}$ is the mean of z_t, a_t is a series of innovations, normally distributed with zero mean and variance σ_a^2, $\phi_p(B)\Phi_P(B^s)$, and $\theta_q(B)\Theta_Q(B^s)$ are autoregressive and moving average polynomials in B, respectively, which are expressed in multiplicative form as the product of a regular polynomial in B and a seasonal polynomial in B^s. The complete model can be written in detailed form as

$$\phi_p(B)\Phi_P(B^s)(1-B)^d(1-B^s)^D y_t = \theta_q(B)\Theta_Q(B^s)a_t + c, \qquad (11.3)$$

and, in concise form, as

$$\Psi(B)y_t = \pi(B)a_t + c, \qquad (11.4)$$

where c is equal to $\Psi(B)\bar{y}$, being $\bar{y}$ the mean of the linearized series y_t. In words, the model that SEATS assumes is a linear time series with Gaussian innovations. When used with TRAMO, estimation of the ARIMA model is made by the exact maximum likelihood method described in Gomez and Maravall [11].

2. *Derivation of the ARIMA models for each component.* The program proceeds by decomposing the series that follows the ARIMA model (11.4) into several components. The decomposition can be multiplicative or additive. Next we shall discuss the additive model, since the multiplicative relation can be taken care with the log transformation of the data. That is,

$$y_t = T_t + C_t + S_t + I_t, \qquad (11.5)$$

where T_t denotes the trend component, C_t the cycle, S_t represents the seasonal component, and I_t the irregulars. The decomposition is done in the frequency domain. The spectrum (or pseudospectrum) of y_t is partitioned into additive spectra, associated with the different components which are determined, mostly, from the AR roots of the model. The trend component represents the long-term evolution of the series and displays a spectral peak at frequency 0, whereas the seasonal component captures the spectral peaks at seasonal frequencies (e.g., for monthly data these are 0.524, 1.047, 1.571, 2.094, 2.618, and 3.142). The cyclical component captures periodic fluctuations with period longer than a year, associated with a spectral peak for a frequency between 0 and $(2\pi/s)$, and short-term variation associated with low order MA components and AR roots with small moduli. Finally, the irregular component captures white noise behavior, and hence has a flat spectrum. The components are determined and fully derived from the structure of the (aggregate) ARIMA model (11.4) for the linearized series directly identified from the data. The program is aimed at monthly or quarterly frequency data and the maximum number of observations that can be processed is 600. One important assumption is that of orthogonality among the components, and each one will have in turn an ARIMA model. In order to identify the components, the *canonical decomposition* is used which implies that the variance of the irregulars is maximized, whereas the trend, seasonal, and cycle are as stable as possible (compatible with the stochastic nature of model (11.2)). The canonical condition on the trend, seasonal, and cyclical components identifies a unique decomposition, from which the ARIMA models for the components are obtained (including the component innovation variances). The trend-cycle and seasonal components are estimated based on minimum mean squared error (MMSE) criterion and using Wiener–Kolmogorov (WK) filters; the detected outliers and some special effects are finally reintroduced into the components.

As in the case of X12ARIMA, diagnostic checking and analyses of the decomposition accuracy and adequacy are performed using graphical, descriptive, nonparametric, and parametric measures included in the output of the program. The procedure is summarized in the RHS of Fig. 11.1.

11.2 Trend-Cycle Prediction in Reproducing Kernel Hilbert Space (RKHS)

Due to major economic and financial changes of global nature, seasonally adjusted data are not smooth enough to be able to provide a clear signal of the short-term trend. Hence, further smoothing is required, but the main objections for trend-cycle estimation are

1. the size of the revisions of the most recent values (generally much larger than for the corresponding seasonally adjusted estimates), and
2. the time lag in detecting true turning points.

With the aim to overcome such limitations, Dagum and Bianconcini [7–9] have recently developed nonparametric trend-cycle filters that reduce significantly revisions and turning point detection with respect to the Musgrave [13] filters which are applied by X12ARIMA. These authors rely on the assumption that the input series $\{y_t, t = 1, 2, \ldots, n\}$ is seasonally adjusted (where trading day variations and outliers, if present, have been also removed), such that it can be decomposed into the sum of a systematic component (*signal*) TC_t, that represents the trend and cycle usually estimated jointly, plus an irregular component I_t, called the *noise*, as follows:

$$y_t = TC_t + I_t, \qquad t = 1, \ldots, n. \tag{11.6}$$

The noise I_t is assumed to be either white noise, $WN(0, \sigma_I^2)$ or, more generally, a stationary and invertible ARMA process.

On the other hand, the signal $TC_t, t = 1, \ldots, n$, is assumed to be a smooth function of time that can be estimated as a weighted moving average as follows:

$$\widehat{TC}_t = \sum_{j=-m}^{m} w_j y_{t+j}, \qquad t = m + 1, \ldots, n - m, \tag{11.7}$$

where $\{w_j, j = -m, \ldots, m\}$ are the weights applied to the $(2m + 1)$ observations surrounding the target observation y_t. Using the reproducing kernel Hilbert space methodology [2], the weights can be derived from the following kernel function:

$$K_4(s) = \sum_{i=0}^{3} P_i(s)P_i(0)f_0(s), \qquad s \in [-1, 1], \qquad (11.8)$$

where $f_0(s) = \frac{15}{16}(1 - s^2)^2, s \in [-1, 1]$, is the biweight density function, and $P_i, i = 0, 1, 2, 3$, are the corresponding orthonormal polynomials. In particular, the generic weight $w_j, j = -m, \ldots, m$, is given by

$$w_j = \left[\frac{\mu_4 - \mu_2 \left(\frac{j}{b}\right)^2}{S_0\mu_4 - S_2\mu_2} \right] \frac{1}{b} f_0 \left(\frac{j}{b}\right), \qquad (11.9)$$

where $\mu_r = \int_{-1}^{1} s^r f_0(s)ds$ are the moments of f_0, and $S_r = \sum_{j=-m}^{m} \frac{1}{b} \left(\frac{j}{b}\right)^r f_0 \left(\frac{j}{b}\right)$ their discrete approximation that depends on the bandwidth parameter b whose choice is of fundamental importance to guarantee specific inferential properties to the trend-cycle estimator. It has to be selected to ensure that only $(2m + 1)$ observations surrounding the target point will receive nonzero weight and to approximate, as close as possible, the continuous density function f_0, as well as the moments μ_r with S_r. Dagum and Bianconcini ([7] and [8]) have suggested a time invariant global bandwidth b equal to $m + 1$, which gave excellent results.

At the end of the sample period, that is, $t = n - (m + 1), \ldots, n$, only $2m, \ldots, m + 1$ observations are available, and time-varying asymmetric filters have to be considered. At the boundary, the effective domain of the kernel function K_4 is $[-1, q^*]$, where $q^* << 1$, instead of $[-1, 1]$ as for any interior point. This implies that the symmetry of the kernel is lost, and it does not integrate to unity on the asymmetric support $(\int_{-1}^{q^*} K_4(s)ds \neq 1)$. Based on these considerations, Dagum and Bianconcini [8] have suggested to derive the asymmetric weights by "cutting and normalizing" the symmetric kernel K_4, that means by omitting that part of the function lying between q^* and 1, and by normalizing it on $[-1, q^*]$. Hence, the corresponding asymmetric weights result

$$w_{qj} = \left[\frac{\mu_4 - \mu_2 \left(\frac{j}{b_q}\right)^2}{S_0^q\mu_4 - S_2^q\mu_2} \right] \frac{1}{b_q} f_0 \left(\frac{j}{b_q}\right), \qquad (11.10)$$

$$j = -m, \ldots, q; q = 0, \ldots, m - 1.$$

where $S_r^q = \sum_{j=-m}^{q} \frac{1}{b_q} \left(\frac{j}{b_q}\right)^r f_0 \left(\frac{j}{b_q}\right)$ is the discrete approximation of $\mu_r^{q*} = \int_{-1}^{q*} s^r f_0(s)ds$ that are proportional to the moments of the truncated biweight density f_0 on the support $[-1, q^*]$.

Dagum and Bianconcini [9] have analyzed in detail the statistical properties of these asymmetric filters as function of the bandwidth parameters $b_q, q = 0, \ldots, m - 1$. They have shown, both theoretically and empirically, that filters based on local time-varying bandwidth parameters $b_q, q = 0, \ldots, m - 1$, selected to minimize the distance between the gain functions of each asymmetric filter $\{w_{q,j}, j = 0, \ldots, q, q = 0, \ldots, m\}$ and the symmetric one $\{w_j, j = -m, \ldots, m\}$ have excellent properties. In particular, the last point trend-cycle filter reduces around a half the size of the total revisions as well as the time delay to detect a true turning point with respect to the Musgrave [13] filter. For a detailed discussion of the properties of these filters and of the corresponding weight systems, we refer the reader to Chap. 10.

11.3 Empirical Application

Since the RKHS trend-cycle filters discussed in the previous section are applied to seasonally adjusted data, it is important to evaluate how their statistical properties are affected by the seasonal adjustment method used to produce the input data. In this regard, in this section, it is analyzed the behavior of the Dagum and Bianconcini [9] asymmetric filters in terms of size of revisions in the estimates as new observations are added to the series, and time lag to detect true turning points. We looked at seasonally adjusted data with outliers replaced if present.

The filters are applied to a sample of series that cover various sectors, such as labor, imports, exports, housing, industrial production, and inventories. The series have been observed on different time periods. Specifically, we chose a set of leading, coincident, and lagging indicators of the US economy. The leading indicators are time series that have a turning point before the economy as a whole changes, whereas the coincident indicators change direction approximately at the same time as the whole economy, thereby providing information about the current state of the economy. On the other hand, the lagging indicators are those that usually change direction after the economy as a whole does. The composite indexes are typically reported in the financial and trade press, and the data analyzed in this study are from the St. Louis Federal Reserve Bank database, the Bureau of Labor Statistics, and the National Bureau of Economic Research (NBER). In particular, the analyzed leading indicators are

- Average weekly overtime hours, manufacturing
- New orders for durable goods
- New orders for nondefense capital goods
- New private housing units authorized by building permits

- Stock prices, S&P common stocks
- Money supply, M2
- Interest rate spread, 10-year treasury bonds less federal funds
- Index of consumer expectation (University of Michigan).

We also consider the following coincident indicators:

- Employees on nonagricultural payrolls
- Personal income less transfer payments
- Industrial production index
- Manufacturing and trade sales.

Finally, the lagging indicators treated are

- Average duration of unemployment, weeks
- Ratio, manufacturing, and trade inventory to sale
- Change in labor cost per unit of output, manufacturing
- Commercial and industrial loans outstanding.

As an example, Figs. 11.2 and 11.3 illustrate the original and seasonally adjusted (SA) monthly series of US Unemployment Rate for Males (16 years and over) for the period January 1992–December 2013. This series is provided monthly by the Bureau of Labor Statistics of the US Department of Labor. The unemployment rate is a measure of unemployment and it is calculated as a percentage by dividing the

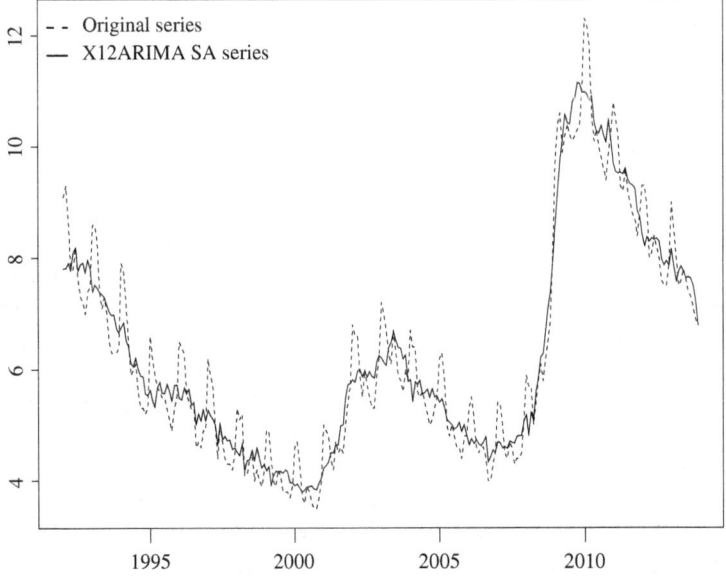

Fig. 11.2 Original and X12ARIMA (default option) seasonally adjusted US Unemployment Rate for Males (16 years and over) from January 1992 to December 2013

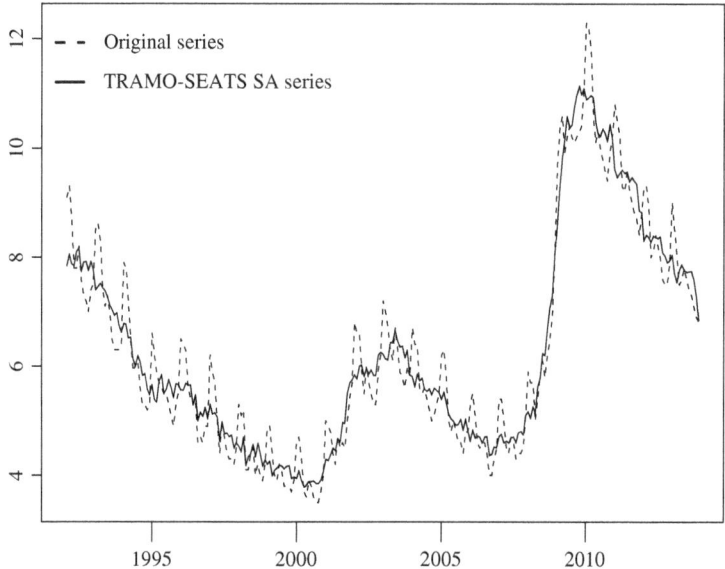

Fig. 11.3 Original and TRAMO-SEATS (default option) seasonally adjusted US Unemployment Rate for Males (16 years and over) from January 1992 to December 2013

number of unemployed individuals by all individuals currently in the labor force, employed plus unemployed. The unemployment rate for males is a series of the Index of Coincident Economic Indicators, produced by the Conference Board to provide an indication of the future direction of the US Economy. Indeed, if the economy is just beginning to grow out of a recession, businesses will tend to hold off on hiring new workers until they are more confident that economic growth is improving. As the economy continues to improve, eventually businesses will be forced to add more workers. This increase in employment will reinforce the positive trend in economic growth and the unemployment rate will decrease. In contrast, if the economy is just beginning to slow down, employers wishing to maintain employee loyalty will try to keep their workers by reducing hours worked, rather than immediately laying off workers. If the slowdown deepens into a recession, then eventually businesses are forced to lay off workers, which reinforces the negative trend in economic growth and the unemployment rate will increase. Thus unemployment rate is a coincident indicator changing direction approximately at the same time as the whole economy, thereby providing information about the current state of the economy.

The seasonal adjustment is done using the default options of the officially adopted X12ARIMA and TRAMO-SEATS. The outliers in the seasonally adjusted data are replaced, if present. Figures 11.2 and 11.3 exhibit both original and seasonally adjusted data. It is apparent that both seasonally adjusted series are very close to one another. The total male unemployment rate shows a peak at the middle of 2009, and underwent thenceforth a fast decline in the subsequent years.

11.3.1 Reduction of Revisions in Real Time Trend-Cycle Estimates

The reduction of revisions in real time trend-cycle estimates is very important because the estimates are preliminary and used to assess the current stage of the economy. For the seasonally adjusted series by X12ARIMA, the length of the filters is selected according to the I/C (noise to signal) ratio, as classically done in the X12ARIMA software. The I/C ranges from 0.19 to 3.22, hence symmetric filters of 9- and 13-term and corresponding asymmetric filters have been applied. The comparisons are based on the relative filter revisions between the final symmetric filter S and the last point asymmetric filter A, that is,

$$R_t = \frac{S_t - A_t}{S_t}, \quad t = 1, 2, \dots, n. \tag{11.11}$$

For each series, it is calculated the root mean square percentage error (RMSPE), $\sqrt{\frac{100}{N} \sum_{t=1}^{N} R_t^2}$, and the mean absolute percentage error (MAPE) of the revisions, $\frac{100}{N} \sum_{t=1}^{N} |R_t|$, corresponding to the filters derived following the RKHS methodology. Table 11.1 gives the RMSPE and MAPE of the revisions introduced by the RKHS filters when applied to the SA data produced by X12ARIMA and TRAMO-SEATS. The size of the revisions is small and very close to each other. An indication is that the seasonal adjustment method chosen has no impact on the final results from the revisions viewpoint.

11.3.2 Turning Point Detection

It is important that the reduction of revisions in real time trend-cycle estimates is not achieved at the expense of increasing the time lag to detect true turning points. A turning point is generally defined to occur at time t if (*downturn*):

$$y_{t-k} \leq \cdots \leq y_{t-1} > y_t \geq y_{t+1} \geq \cdots \geq y_{t+m},$$

or (*upturn*)

$$y_{t-k} \geq \cdots \geq y_{t-1} < y_t \leq y_{t+1} \leq \cdots \leq y_{t+m}.$$

Following Zellner et al. [14], $k = 3$ and $m = 1$ are chosen given the smoothness of the trend-cycle data. For the values given by the RKHS method, the time lag to detect the true turning point is affected by the convergence path of its asymmetric filters $\{w_{q,j}, j = -m, \dots, q; q = 0, \dots, m-1\}$ to the symmetric one $\{w_j, j = -m, \dots, m\}$.

Table 11.1 Values of RMSPE and MAPE of the revisions induced by RKHS filters applied on different SA series

Macro-area	Series	RMSPE		MAPE	
		X12ARIMA	TRAMO-SEATS	X12ARIMA	TRAMO-SEATS
Leading	Average weekly overtime hours: manufacturing	0.489	0.501	3.135	3.160
	New orders for durable goods	0.438	0.450	3.075	3.210
	New orders for nondefense capital goods	0.497	0.508	3.844	3.979
	New private housing units authorized by building permits	0.577	0.586	4.160	4.294
	S&P 500 stock price index	0.936	0.945	6.169	6.674
	10-Year treasury constant maturity rate	0.986	1.046	7.685	8.055
	University of Michigan: consumer sentiment	0.544	0.583	4.297	4.431
Coincident	All employees: total nonfarm	0.049	0.049	0.419	0.420
	Unemployment rate for males (16 years and over)	0.604	0.635	2.241	2.420
	Real personal income excluding current transfer receipts	0.340	0.356	2.866	3.000
	Industrial production index	0.128	0.125	0.962	0.959
	Manufacturing and trade sales	0.319	0.281	2.497	2.354
Lagging	Average (mean) duration of unemployment	0.364	0.390	2.778	3.021
	Index of total labor cost per unit of output	0.150	0.183	1.202	1.336
	Commercial and industrial loans at all commercial banks	0.242	0.240	2.127	2.102
	Mean	0.395	0.406	2.858	3.019

Table 11.2 Ratio in the time delay in identifying true turning points for RKHS filters applied on seasonally adjusted series

Macro-area	Series	Time lag X12ARIMA	TRAMO-SEATS
Leading	Average weekly overtime hours: manufacturing	2.60	3.00
	New orders for durable goods	2.33	2.50
	New orders for nondefense capital goods	2.33	2.43
	New private housing units authorized by building permits	2.67	3.00
	S&P 500 stock price index	2.50	2.75
	10-Year treasury constant maturity rate	1.00	2.00
	University of Michigan: consumer sentiment	1.00	1.00
Coincident	All employees: total nonfarm	1.00	1.00
	Unemployment rate for males (16 years and over)	1.61	1.94
	Industrial production index	1.80	2.33
	Manufacturing and trade sales	1.75	2.33
Lagging	Average (mean) duration of unemployment	1.00	1.00
	Index of total labor cost per unit of output	2.67	3.29
	Commercial and industrial loans at all commercial banks	1.20	1.50
	Mean	1.759	2.177

To determine the time lag to detect a true turning point, we calculate the number of months it takes for the real time trend-cycle estimate to signal a turning point in the same position as in the final trend-cycle series. We have used a very large sample of series and, but we illustrate here only the results for the leading, coincident, and lagging indicators shown in Table 11.2. For the three sets in total 40 turning points were detected and the average delay in month is equal to 1.759 when the input is seasonally adjusted with X12ARIMA and 2.177 when done with TRAMO-SEATS. This latter takes longer to identify a true turning point. In fact, two series of the leading indicator set and one of the lagging indicator set show time delays of 3 and 3.29 months.

We can conclude that the seasonally adjusted series with X12ARIMA is to be preferred relative to that from TRAMO-SEATS concerning the time delay to detect a true turning point.

References

1. Bell, W. H., & Hillmer, S. C. (1984). Issues involved with the seasonal adjustment of economic time series. *Journal of Business and Economic Statistics, 2*, 291–320.
2. Berlinet, A. (1993). Hierarchies of higher order kernels. *Probability Theory and Related Fields, 94*, 489–504.

3. Bobbit, L., & Otto, M. C. (1990). Effects of forecasts on the revisions of seasonally adjusted values using the X11 seasonal adjustment procedure. In *Proceedings of the American Statistical Association Business and Economic Statistics Session* (pp. 449–453).
4. Box, G. E. P., & Tiao, G. C. (1975). Intervention analysis with applications to economics and environmental problems. *Journal of the American Statistical Association, 70*, 70–79.
5. Dagum, E. B. (1978). Modelling, forecasting and seasonally adjusting economic time series with the X-11 ARIMA method. *The Statistician, 27*(3), 203–216.
6. Dagum, E. B. (1988). *The X-11 ARIMA/88 seasonal adjustment method-foundations and user's manual*. Ottawa, Canada: Time Series Research and Analysis Centre, Statistics Canada.
7. Dagum, E. B., & Bianconcini, S. (2008). The Henderson smoother in reproducing kernel Hilbert space. *Journal of Business and Economic Statistics, 26*(4), 536–545.
8. Dagum, E. B., & Bianconcini, S. (2013). A unified probabilistic view of nonparametric predictors via reproducing kernel Hilbert spaces. *Econometric Reviews, 32*(7), 848–867.
9. Dagum, E. B., & Bianconcini, S. (2015). A new set of asymmetric filters for tracking the short-term trend in real time. *The Annals of Applied Statistics, 9*, 1433–1458.
10. Findley, D. F., Monsell, B. C., Bell, W. R., Otto, M. C., & Chen, B. C. (1998). New capabilities and methods of the X12ARIMA seasonal adjustment program. *Journal of Business and Economic Statistics, 16*(2), 127–152.
11. Gomez, V., & Maravall, A. (1994). Estimation, prediction and interpolation for nonstationary series with the Kalman filter. *Journal of the American Statistical Association, 89*, 611–624.
12. Gomez, V., & Maravall, A. (1996). Program TRAMO and SEATS: Instructions for users. Working Paper 9628. Service de Estudios, Banco de Espana.
13. Musgrave, J. (1964). A set of end weights to end all end weights. Working Paper. U.S. Bureau of Census, Washington, DC.
14. Zellner, A., Hong, C., & Min, C. (1991). Forecasting turning points in international output growth rates using Bayesian exponentially weighted autoregression, time-varying parameter, and pooling techniques. *Journal of Econometrics, 49*(1–2), 275–304.

Glossary

ARIMA Model A model that belongs to the ARIMA (Autoregressive integrated moving average) process which is characterized by being homogeneous stationary after a finite difference operator has being applied.

Backshift Operator It is generally denoted by B or L (as lag operator) and when applied to an element of a time series will produce the previous element, that is, $By_t = y_{t-1}$.

Business Cycle It is a quasi-periodic oscillation characterized by period of expansion and contraction of the economy, lasting on average from 3 to 5 years.

Census II-X11 Variant (X11) Method Seasonal adjustment software based on linear filters or moving averages applied in an iterative manner.

Deterministic Process It is a system in which no randomness is involved in the development of future states. A deterministic model will thus always produce the same output from a given starting condition or initial state.

Difference Operator The first difference operator $(1 - B)$ is a special case of lag polynomial, such that when applied to an element of a time series it produces the difference between the element and the previous one, $(1 - B)y_t = y_t - y_{t-1}$.

Extreme Value The smallest (minimum) or largest (maximum) value of a function, either in an arbitrarily small neighborhood of a point in the function domain or on a given set contained in the domain. In a time series, it identifies a value the probability of which falls in either tail of the distribution. It follows the same probability distribution of all the other elements of the time series.

Flow Series A series where the observations are measured via the accumulation of daily activities. Specifically, a flow variable is measured over an interval of time, for example, a month or a quarter.

Gain Function A function that relates the spectrum of the original series to the spectrum of the output obtained with a linear time invariant filter. It measures the amplitude of the output for a sinusoidal input of unit amplitude.

© Springer International Publishing Switzerland 2016
E. Bee Dagum, S. Bianconcini, *Seasonal Adjustment Methods and Real Time
Trend-Cycle Estimation*, Statistics for Social and Behavioral Sciences,
DOI 10.1007/978-3-319-31822-6

Irregular Component It represents variations related to unpredictable events of all kinds. Most irregular values have a stable pattern in the sense of a constant mean and finite variance but some extreme values may be present. The irregulars are due to unforeseeable events of all kind and often arise from sampling error, nonsampling error, unseasonable weather, natural disasters, strikes, etc.

Kalman Filter It is an algorithm that uses a series of measurements observed over time, containing noise and other inaccuracies, and produces estimates of unknown variables that tend to be more precise than those based on a single measurement alone. The algorithm works in a two-step process. In the prediction step, the Kalman filter produces estimates of the current state variables, along with their uncertainties. Once the outcome of the next measurement (necessarily corrupted with some amount of error, including random noise) is observed, these estimates are updated using a weighted average, with more weight being given to estimates with higher certainty. The algorithm is recursive. It can run in real time, using only the present input measurements and the previously calculated state and its uncertainty matrix; no additional past information is required.

Linear Filter see *Moving average*

Moving Average It consists of weighted or unweighted averages usually with a fixed number of terms. It is called moving because the average is applied sequentially by adding and subtracting one term at a time.

Moving Holiday Effects The moving holiday or moving festival component is attributed to calendar variations, namely due to the fact that some holidays change date in successive years. Example of moving holidays are Easter that moves between months of March or April, the Chinese New Year date that depends on the lunar calendar, and Ramadan that falls 11 days earlier from year to year.

Moving Seasonality Seasonality characterized by changes in the seasonal amplitude and/or phase.

Noise to Signal Ratio The ratio of the average absolute month-to-month (quarter-to-quarter) change in the noise to that in the signal.

Outlier An observation that is well outside the expected range of values in a study or experiment. They are similar to extreme values but the value treated as outlier is characterized by a different distribution function from the rest of the data. Outliers can often be traced to identifiable causes, for example, strikes, droughts, floods, and data processing errors. Some outliers are the result of displacement of activity from one month to the other.

Phase Shift Function that shows the shift in phase of the output compared with the input.

Recession and Recovery Analysis Technique to assess the real time trend-cycle of major socioeconomic indicators using percentage changes based on seasonally adjusted units calculated for months and quarters in chronological sequence.

RegARIMA Models Models that combine a regression model for deterministic variables such as moving holidays, trading day variations, and/or outliers with a seasonal ARIMA model for the error term.

Reproducing Kernel Hilbert Space It is a Hilbert space characterized by a kernel function that reproduces, via inner product, every function of the space or, equivalently, a Hilbert space of real-valued functions with the property that every point evaluation functional is bounded and linear.

Seasonal Adjustment The process of identifying, estimating, and removing the seasonal variations (and other calendar-related effects, such as trading day and moving holidays) from a time series. The basic goal of seasonal adjustment is to standardize socioeconomic time series to assess the stage of the cycle at which the economy stands.

Seasonal Effects Variations due to seasonal causes that are reasonably stable in terms of timing, direction, and magnitude from year to year. Seasonality originates from climate seasons and conventional events of religious, social, and civic nature, which repeat regularly from year to year.

SEATS It stands for Signal Extraction in ARIMA Time Series and is a seasonal adjustment software that uses ARIMA models to estimate the trend-cycle and seasonality of a time series.

Second Order (or Weak) Stationary Process Stochastic process whose first two order moments are not time dependent, that is, the mean and the variance are constant and the autocovariance function depends only on the time lag and not on the time origin.

Spectrum of a Time Series It is the distribution of variance of the series as a function of frequencies. The object of *spectral analysis* is to estimate and study the spectrum. The spectrum contains no new information beyond that in the autocovariance function (ACVF), and in fact the spectrum can be computed mathematically by transformation of the ACVF. The ACVF summarizes the information in the time domain and the spectrum in the frequency domain.

Stable Seasonality Seasonality that can be represented by a strictly periodic function of time.

State Space A state space representation is a mathematical model of a system as a set of input, output, and state variables related by first order differential equations. State space refers to the space whose axes are the state variables. The state of the system can be represented as a vector within that space. To abstract from the number of inputs, outputs, and states, these variables are expressed as vectors. Additionally, if the dynamical system is linear, time invariant and finite-dimensional, then the differential and algebraic equations may be written in matrix form.

Stochastic Process In probability theory, a stochastic process, or often random process, is a collection of random variables, representing the evolution of some system of random values over time. This is the probabilistic counterpart to a deterministic process (or deterministic system). It is a set of random variables indexed in time.

Stock Series A stock variable is measured at one specific time, and represents a quantity existing at that point in time (e.g., December 31, 2015).

Structural Time Series Model Model that consists of a number of linear stochastic processes that stand for the trend, cycle, seasonality, and remaining stationary dynamic features in an observed times series.

Time Series Set of observations on a given phenomenon that are ordered in time. Formally, it is defined as a sample realization of a stochastic process, that is, a process that evolves according to a probability law.

Trading Day Effects Variations associated with the composition of the calendar which are due to the fact that the activities of some days of the week are more important than others. Trading day variations imply the existence of a daily pattern analogous to the seasonal pattern. However, these daily factors are usually referred to as daily coefficients. This occurs because only non-leap year Februaries have four of each day—four Mondays, four Tuesdays, etc. All other months have an excess of some types of days. If an activity is higher on some days compared to others, then the series can have a trading day effect.

TRAMO-SEATS It is a seasonal adjustment software composed of two main parts: (1) TRAMO that automatically selects a regARIMA model to estimate deterministic components such as trading day variations, moving holidays, and outliers that are removed from the time series and (2) SEATS that uses the time series where the deterministic components have been removed in order to estimate the trend-cycle and seasonality and produces a seasonally adjust series.

Transfer Function It measures the effect of a linear filter on the total variance of the input at different frequencies, and it is generally expressed in polar coordinates.

Trend The concept of trend is used in economics and other sciences to represent long-term smooth variations. The causes of these variations are often associated with structural phenomena such as population growth, technological progress, capital accumulation, new practices of business, and economic organization. For most economic time series, the trends evolve smoothly and gradually, whether in a deterministic or stochastic manner.

Trend-Cycle The trend-cycle is a short-term trend that includes the impact of the long-term trend plus that of the business cycle. This is due to the fact that most series are relatively short and then it is impossible to estimate the long-term trend by itself. It is important to assess the current stage of the business cycle, particularly to forecast the coming of a turning point.

Unwanted Ripple A 10-month cycle identified by the presence of high power at the frequency 0.10 which, due to its periodicity, often leads to the wrong identification of a true turning point.

White Noise Process Stochastic process characterized by random variables that have a constant expected value (usually zero), a finite variance, and are mutually uncorrelated. If the variance is assumed constant (*homoscedastic condition*), the process is referred as white noise in the strict sense. If the variance is finite but not constant (*heteroscedastic condition*), the process is called white noise in the weak sense.

X11ARIMA The X11ARIMA is an enhanced version of the Census Method II–X11 variant that was developed mainly to produce a better current seasonal adjustment of series with seasonality that changes rapidly in a stochastic manner. The latter characteristic is often found in main socioeconomic indicators, e.g., retail trade, imports and exports, unemployment, and so on. It basically consists of modeling the original series with an ARIMA model and extending the series with forecasts up to 3 years ahead. Then each component of the extended series is estimated using moving averages that are symmetric for middle observations and asymmetric for both end years. The latter are obtained via the convolution of Census II-X11 variant weights and the ARIMA model extrapolations.

X12ARIMA The X12ARIMA is an enhanced version of the X11ARIMA seasonal adjustment software. The major modifications concern: (1) extending the automatic identification and estimation of ARIMA models for the extrapolation option to many more than the three models available in X11ARIMA, and (2) estimating trading day variations, moving holidays, and outliers in what is called the regARIMA model. The X12ARIMA method has all the seasonal adjustment capabilities of the X11ARIMA variant. The same seasonal and trend moving averages are available, and the programs till offer the Census II-X11 calendar and holiday adjustment routines incorporated in X11ARIMA. Several other new options have been included, such as: sliding spans diagnostic procedures, capability to produce the revision history of a given seasonal adjustment, new options for seasonal and trend-cycle filters, several new outlier detection options for the irregular component, and a pseudo-additive seasonal adjustment mode.

X13ARIMA-SEATS The newest in the X11 family of seasonal adjustment software, it integrates an enhanced version of X12ARIMA with an enhanced version of SEATS to provide both nonparametric X11-type seasonal adjustments and ARIMA model-based SEATS-type adjustments, combined with the diagnostics available in X12ARIMA.

CPSIA information can be obtained
at www.ICGtesting.com
Printed in the USA
LVHW011415020820
662183LV00003B/141

9 783319 318202